Springer Oceanography

Series Editor

Andrés H. Arias, Argentinean Institute of Oceanography, National South University, Bahia Blanca, Argentina

The Springer Oceanography series seeks to publish a broad portfolio of scientific books, aiming at researchers, students, and everyone interested in marine sciences. The series includes peer-reviewed monographs, edited volumes, textbooks, and conference proceedings. It covers the entire area of oceanography including, but not limited to, Coastal Sciences, Biological/Chemical/Geological/Physical Oceanography, Paleo-ceanography, and related subjects.

Michio Aoyama · Chikako Cheong ·
Akihiko Murata
Editors

Chemical Reference Materials for Oceanography

History, Production, and Certification

 Springer

Editors
Michio Aoyama
Physical and Chemical Oceanography
Research Group
Global Ocean Observation Research Center
Japan Agency for Marine-Earth Science
and Technology
Yokosuka, Japan

Chikako Cheong
National Institute of Advanced Industrial
Science and Technology
Tsukuba, Japan

Akihiko Murata
Physical and Chemical Oceanography
Research Group
Global Ocean Observation Research Center
Japan Agency for Marine-Earth Science
and Technology
Yokosuka, Japan

ISSN 2365-7677   ISSN 2365-7685 (electronic)
Springer Oceanography
ISBN 978-981-96-2519-2   ISBN 978-981-96-2520-8 (eBook)
https://doi.org/10.1007/978-981-96-2520-8

This work was supported by Japan Agency for Marine-Earth Science and Technology and National Institute of Advanced Industrial Science and Technology.

Preface

The oceans cover 70% of the earth's surface and hold about 97% of the water on the planet. Because of its vastness, we have yet to fully understand how the ocean's processes affect and are affected by global and regional climates. Despite this, we have casually disposed of innumerable substances into the environment, many of which reach the ocean. It would be greatly beneficial to the planet and its inhabitants to improve our understanding of human activities' full impact on the ocean system.

This book is intended for those who require chemical Reference Materials (RMs), available in natural seawater, for analyzing oceanic variables. It includes the history and production of RMs, including Certified Reference Materials (CRMs) used as reference standards for aligning and comparing data sets generated worldwide. By employing CRMs, we can confidently compare data collected by different groups and institutions at various times and places, thus improving our ability to observe and interpret the ocean's spatial and temporal variations. As such, CRMs are indispensable for gaining a true understanding of the ocean that can be reliably shared with society. In an era of global environmental change, the need for CRMs is even greater; this is especially true as researchers exchange data more freely (Pendleton et al. 2020; *Proceedings of the National Academy of Sciences, 117*, 9652–9655. https://doi.org/10.1073/pnas.2005485117). Consequently, a significant investment has been made internationally in producing and distributing CRMs, particularly for properties that reflect ocean climate change, such as salinity, CO_2, and nutrients. With this in mind, we hope this book serves as a significant contribution to and motivation for the ongoing development and use of CRMs.

Finally, we must inform you of the unfortunate passing of Dr. Michio Aoyama, one of the editors of this book and the individual responsible for its initial inspiration and vision. He passed away on September 5, 2022. Given his indispensable role as our chief editor, we were on the verge of abandoning the publication. However, we recalled his assertion that the results of research should always be documented in writing. Consequently, we elected to continue writing and editing the book, which we dedicate to his memory. Chapter 1, which Michio was scheduled to write, was completed by referencing his existing body of published and oral works. In response to the unfortunate news of his passing, we received many messages from his friends and colleagues:

> We are very honoured that the late Prof. Dr. Michio Aoyama asked us to contribute to this book. We are pleased to recommend using Certified Reference Materials (CRMs) in which Michio Aoyama personally invested a lot of time and effort, not only to have them produced, but also towards their international acceptance. Michio's creation of nutrient CRMs was a major step towards routinely producing comparable measurements and data between different laboratories worldwide
>
> —The authors of the chapters

We hope this book is useful to oceanographers now and in the future, as Michio would have wished.

Yokosuka, Japan   Michio Aoyama
Tsukuba, Japan   Chikako Cheong
Yokosuka, Japan   Akihiko Murata

About This Book

Concentrations of inorganic macro-nutrients, carbonate system parameters (total alkalinity, pH, the concentration of dissolved inorganic carbon, and the partial pressure of CO_2), salinity, and the concentration of dissolved oxygen are important characteristics of seawater that can be used to define biogeographic provinces, monitor ocean health, and detect decadal-scale changes of oceanic climate. Without reference materials for these parameters, it is difficult to produce reliable datasets or carry out long-term baseline studies that are essential to quantify the extent to which changes in the marine environment may have occurred. This book will provide readers with the latest information about reference materials (RMs) and certified RMs (CRMs) and the knowledge of how to use the RMs/CRMs in their studies.

We also describe efforts to develop new reference materials for density, dissolved oxygen, dissolved organic carbon (DOC), dissolved organic matter (DOM), and trace metals, which are also important in oceanographic studies. The lead authors of each of these chapters are top scientists in the respective fields. This book will therefore be a comprehensive characterization of chemical reference materials for ocean science written by top experts in each field.

This book does not provide a standard definition of a "CRM". Metrologists may argue that a CRM can only be a product that is manufactured in accordance with the International Organization for Standardization (ISO) guidelines. However, there may be an RM that is distributed widely within a discipline and is used sufficiently to maintain comparability of measurements. In that case, the term CRM is used to refer to a conventionally designated RM. In fact, some RMs are used as CRMs in oceanography. Although the term "CRM" is clearly defined in each chapter, it should be noted that there are some differences in the meaning of CRM between disciplines.

In Chap. 1, Aoyama et al. present a history of the development of RM/CRMs for nutrients along with a history of improvements in nutrient measurements. Because CRMs are essential for accurate measurements of nutrient concentrations, they are indispensable for assessing changes in the marine environment associated with global warming and ocean health. CRMs developed in Japan are now becoming an international standard for high-precision nutrient measurements. However, it is clear from this chapter that despite the long history of nutrient measurements, RMs/CRMs have only recently been introduced to the ocean community. This chapter is translated from the Japanese reports of the late Dr. M. Aoyama: Murata et al. (2020) Current situation and future perspective for environmental standards of seawater: commencing with Certified Reference Materials (CRMs) for nutrient, *Oceanography in Japan, 20*(5), 153–187 and Aoyama (2020) Comparability of nutrients data in the world ocean: past, present and future, *Gekkan Kaiyo, 52*(8), 390–401.

Chapter 2 describes the production and certification of reference materials for nutrients in seawater (RMNS) by a private Japanese company, KANSO TECHNOS Co., Ltd. (KANSO). KANSO began developing RMNS in 1993, and it began their production and distribution in 2000. The RMNS, which are made from 100% natural seawater, contain all nutrients in one bottle and can be used directly without dilution. Multiple concentrations of seawater are mixed, autoclaved twice, cooled, and then mixed for at least two weeks to ensure homogeneity and stability. Quality control is conducted through culture tests and measurements of nutrient concentrations. Stability is confirmed for seven years. The reliability of the RMNS is assured through a co-certification system with the Japan Agency for Marine-Earth Science and Technology.

In Chap. 3, Murata et al. describe how CRMs developed in Japan are distributed worldwide, especially to countries around the Atlantic Ocean, under the framework of the Scientific Committee on Ocean Research. The Japanese CRMs are made with seawater from the Pacific Ocean, and the nutrient concentrations differ from the concentrations in seawater from the Atlantic Ocean. Researchers and technicians usually hope to have CRMs made with seawater collected near their sampling area. The authors therefore produced and distributed CRMs made with seawater from the Atlantic Ocean. In doing so, the authors discuss some issues that must be resolved for the sustainable production and distribution of CRMs. They also describe the techniques used for temporary sterilization of seawater.

In Chap. 4, Bakker et al. discuss what needs to be done to maintain inter-comparability of nutrient measurements. They state that each laboratory must first confirm that nutrient measurements are consistent within a set of analyses. The authors discuss how to maintain comparability between different sets of analyses and state that CRMs are required to assess interlaboratory comparability. This chapter emphasizes that, in addition to the use of CRMs, analyses must be carried out carefully in each laboratory to ensure the comparability of nutrient measurements.

Chapter 5 describes the development and certification of certified reference materials for nutrient analysis by the National Metrology Institute of Japan (NMIJ). The CRMs, which are designated 7601-a, 7602-a, and 7603-a, were the first CRMs in the world to be certified by a national metrology institute for a wide range of nutrient concentrations from nearly zero to high concentrations found in regions such as the deep layer of the Pacific Ocean. The characteristic values for the four nutrient components, nitrate, nitrite, phosphate, and dissolved silica, were determined based on the International System of Units (SI)-traceable national primary standards using multiple analytical methods that were developed and validated for this certification. Homogeneity and long-term stability were evaluated by analysis of variance and regression analysis based on the International Organization for Standard (ISO) guide 35.

Chapter 6 describes the development of a silicon standard solution as a national standard by the NMIJ. A silicon standard solution is in high demand for industrial and environmental analyses as well as for analysis of silicon concentrations in the ocean. The NMIJ CRM 3645-a silicon standard solution was prepared by alkali fusion of high-purity silicon dioxide, which had a purity of 99.648% based on gravimetric analysis, and dissolution in water. ICP-OES and continuous flow analysis indicated that the stability and homogeneity of the CRM were 0.13% and 0.017%, respectively, and the expanded uncertainty was 0.28%. The CRM was validated using the National Institute of Standards and Technology (NIST) SRM 3150 silicon standard solution. This CRM enables accurate and consistent measurements of silicon concentrations to be made in a variety of applications, including oceanographic observations.

Chapter 7 describes the development of a mass supply system for silicon standard solutions at KANSO. To improve the comparability of silicate analyses in seawater, an SI-traceable silicon standard solution (KANSO-Si standard solution) was developed. High-purity silicon dioxide was fused with sodium carbonate and dissolved in water. The procedure was extended to a production scale of 100 mL $\times$ 1000 bottles per lot, with a mass fraction of 1000 mg kg^{-1} as silicon. The certified value of the KANSO-Si standard solution was determined by continuous flow analysis calibrated with NMIJ CRM 3645-a with a relative expanded uncertainty of 0.4 %. The KANSO-Si standard solution is currently used for the certification of silicate in the RMNS distributed by KANSO and as a standard solution for silicate analyses in seawater.

Salinity is one of the fundamental parameters that defines the physical and chemical characteristics of seawater. Salinity is regarded as a measure of the mass of dissolved salts in a sample of seawater and is essential for determining various physical and chemical properties of seawater. For this reason, salinity has been measured since the early days of scientific oceanographic observations, and the RM necessary for consistent determination of salinity has also been examined. In Chap. 8, Jenkins and Williams provide a comprehensive review of the historical development of a salinity RM, which is now widely recognized and used as Standard Seawater. This chapter is particularly valuable not only to oceanographers but also to physical chemists engaged in research on samples of water.

In Chap. 9, Uchida et al. review the preparation of the International Association for the Physical Sciences of the Oceans (IAPSO) Standard Seawater, which is a commonly used standard in the oceanographic community for the calibration of conductivity–temperature–depth sensors. Given the long history of the production of Standard Seawater, the various batches are expected to be consistent. Nevertheless, some reports have indicated that there are differences between batches of Standard Seawater. Such discrepancies are of particular significance when examining temporal changes in the deep-sea layer. The authors discuss the application of batch offset values for the correction of batch-to-batch differences.

The Absolute Salinity scale, which reflects the quantity of dissolved substances in a sample of seawater, is now recommended for use in place of the Practical Salinity Scale 1978 (PSS-78). The reason is that the Absolute Salinity can be determined based on density measurements, which are traceable to the SI. Because Absolute Salinity is based on the Reference Composition Seawater linked to the IAPSO Standard Seawater (SSW), it is necessary to assess any changes in the composition of SSW. In Chap. 10, Uchida et al. survey and quantify the changes in the composition of SSW in terms of various dissolved substances, including dissolved inorganic carbon, total alkalinity, nutrients, dissolved organic matter, and the mass fractions of major constituents. They also examine the causes of the changes from the perspective of inorganic chemistry while focusing on processes that occur within a closed system.

IAPSO SSW, which is used worldwide in the oceanographic community, is the internationally recognized standard for salinity measurements. However, it is possible that provision of the standard may be unexpectedly interrupted. World War II, for example, limited the distribution of Standard Seawater in some countries. As a result, some countries had to produce their own standard seawater. Japan was one of the countries that produced its own standard seawater, which was used until the 1980s. In Chap. 11, Uchida provides a comprehensive review of the history of Japanese standard seawater.

Chapter 12 describes the development of Multiparametric Standard Seawater (MSSW) for measuring practical salinity, density, pH, total alkalinity, and concentrations of dissolved inorganic carbon, dissolved oxygen, and organic matter. The preparation and storage of MSSW are intended to address the limitations of existing standard seawaters and provide a more accurate and comparable reference for oceanographic measurements. The MSSW is stored without the use of toxic preservatives such as mercuric chloride in an aluminum bottle with a plastic inner cap that is highly impermeable to gas and water vapor. The long-term stability of MSSW has been good for most parameters, although dissolved oxygen and practical salinity have decreased slightly with time. It will be necessary to clarify the cause of the decreasing trend urgently.

How much anthropogenic CO_2 has been taken up by the ocean? An accurate answer to this question is required to predict future global warming. To achieve this objective, several research groups specializing in oceanic CO_2 have used an internationally accepted RM provided by a single institution to measure oceanic CO_2. However, the interruption of the supply of RM due to the COVID-19 pandemic has necessitated a re-evaluation of the current framework for RM production and certification. This incident prompted RM users to recognize the necessity of a more resilient framework for production and certification to ensure a stable supply. In Chap. 13, García-Ibáñez and Easley present a new strategy of framework based on metrology to achieve this objective.

The production and distribution of reference materials by a single organization can cause major problems in maintaining the accuracy of observations in the event of an unusual situation, such as the COVID-19 pandemic. If each institution could produce secondary RMs, the impact could be minimized. In Chap. 14, Murata et al. present the production of in-house and commercially available RMs in Japan and their actual use in a research cruise. It is introduced that the consistency of the measured values of total dissolved inorganic carbon and total alkalinity obtained from observations during the COVID-19 pandemic was ensured by using the commercially available RM. In addition, the points that become constraints in the supply of RMs are discussed.

DOM plays a pivotal role in the marine biogeochemical cycle and in the cycle of dissolved inorganic matter. However, it is more challenging to ensure the reliability of DOM measurements because of the greater number of variables that must be considered in the analysis of DOM. In Chap. 15, Hansell and Yoshimura discuss the potential for RMs to enhance the reliability of DOM measurements. The authors review existing RMs and RMs under development for organic nutrients and examine the potential for RMs for inorganic nutrients to be applied to RMs for DOM. The suitability of multiple containers for the storage of RMs is also evaluated.

Chapter 16 describes a CRM for the determination of trace elements in seawater developed by the NMIJ: NMIJ CRM 7204-a. This CRM is expected to facilitate quality control in analysis for environmental conservation and to complement other CRMs such as the Nutrients in Seawater Reference Material (NASS) and the CRM for Trace Metals in Seawater (CASS) series and GEOTRACES standards. The CRM was designed to have elevated concentrations of trace elements similar to regulatory standards. These elevated concentrations make the CRM suitable for direct determination by ICP-MS after dilution. The CRM includes ten regulated trace elements (Cr, Mn, Fe, Ni, Cu, Zn, As, Se, Cd, and Pb), and it provides technical information on major elements and density values at various temperatures. The validity of the CRM was confirmed by a joint analysis with NMIJ, the Korea Research Institute of Standards and Science (KRISS), and the National Institute of Metrology, China (NIM).

Contents

Abbreviations

AAS	Atomic Absorption Spectrometry
AIST	National Institute of Advanced Industrial Science and Technology (Japan)
ANOVA	Analysis of Variance
AS	Atlantic Seawater
ASSW	American Standard Seawater
ASW	Artificial Seawater
BATS	Bermuda Atlantic Time-series Study
BEAGLE	Blue Earth Global Expedition
BIPM	Bureau International des Poids et Mesures (in English: International Bureau of Weights and Measures)
CARINA	Carbon Dioxide in the Atlantic Ocean
CCs	Consultative Committees
CCQM	Consultative Committee for Amount of Substance: Metrology in Chemistry and Biology
CDR	Carbon Dioxide Removal
CFA	Continuous Flow Analysis
CIPM	Comité International des Poids et Mesures (in English: International Committee for Weights and Measures)
CMC	Calibration and Measurement Capability
CMO	Central Meteorological Observatory (Japan)
CO_2	Carbon Dioxide
CP SSW	Chinese Primary Standard Seawater
CRIEPI	Central Research Institute of Electric Power Industry (Japan)
CRM	Certified/Consensus Reference Material
CSK	Co-operative Study of the Kuroshio and Adjacent Regions
CTD	Conductivity–Temperature–Depth
DIC	(Total) Dissolved Inorganic Carbon
DIs	Designated Institutes
DL	Detection Limit
DO	Dissolved Oxygen

DOC	Dissolved Organic Carbon
DOM	Dissolved Organic Matter
DON	Dissolved Organic Nitrogen
DOP	Dissolved Organic Phosphorus
DSR	DOM CRM from the deep ocean
DSW	Deep Seawater
EAWG	Electrochemical Analysis Working Group
ELSW	Extremely Low Nutrient Concentration Seawater
EOS-80	Equation of State of Seawater 1980
EOVs	Essential Ocean Variables
EPA	U.S. Environmental Protection Agency
EURAMET	European Association of National Metrology Institutes
FAJ	Fisheries Agency of Japan
$f\mathrm{CO_2}$	Fugacity of $\mathrm{CO_2}$
FDOM	Fluorescent Dissolved Organic Matter
FLPE	Fluorinated High-Density Polyethylene
FOM	Fluorescent Organic Matter
FTIR	Fourier-Transform Infrared Spectroscopy
GAWG	Gas Analysis Working Group
GC-FID	Gas Chromatography with Flame Ionization Detection
GEOSECS	Geochemical Ocean Sections Study
GF-AAS	Graphite Furnace Atomic Absorption Spectrometry
GLODAP	Global Ocean Data Analysis Project
GOA-ON	Global Ocean Acidification Observing Network
GO-SHIP	Global Ocean Ship-based Hydrographic Investigations Program
GUM	Guide to the Expression of Uncertainty in Measurement
HDPE	High-Density Polyethylene
HOT	Hawaii Ocean Time-series
HPLC	High-Performance Liquid Chromatography
HSW	High nutrient concentration Seawater
HTP	2-Hydroxyterephthalate
IAPO	International Association of Physical Oceanography
IAPSO	International Association for the Physical Sciences of the Oceans
IAPWS	International Association for the Properties of Water and Steam
IAWG	Inorganic Analysis Working Group
IC	Inter-comparison/Ion chromatograph
ICES	International Council for the Exploration of the Sea
ICOS	Integrated Carbon Observation System
ICP-MS	Inductively Coupled Plasma Mass Spectrometry
ICP-OES	Inductively Coupled Plasma Optical Emission Spectrometer
ID	Isotopic Dilution
IDMS	Isotopic Dilution Mass Spectrometry
IEC	International Electrotechnical Commission
IOC	Intergovernmental Oceanographic Commission
IOCCP	International Ocean Carbon Coordination Project

IOS	Institute of Oceanographic Sciences (U.K.)
IPCC	Intergovernmental Panel on Climate Change
IQR	Interquartile Range
IRWG	Isotope Ratio Working Group
ISFET	Ion-Selective Field Effect Transistor
ISO	International Organization for Standardization
IWGOA	Interagency Working Group on Ocean Acidification
JAMSTEC	Japan Agency for Marine-Earth and Technology
JCG	Japan Coast Guard
JCS	Joint SCOR/IAPWS/IAPSO Committee on the Properties of Seawater
JCSS	Japan Calibration Service System
JGOFS	Joint Global Ocean Flux Study
JIS	Japanese Industrial Standards
JMA	Japan Meteorological Agency
JPOTS	Joint Panel on Oceanographic Tables and Standards
JSPS	Japan Society for the Promotion of Science
JSSW	Japanese Standard Seawater
JSSWC	Japanese Standard Seawater Committee
KANSO	KANSO TECHNOS CO., LTD., Japan
KCs	Key Comparisons
KRISS	Korea Research Institute of Standards and Science
LCL	Lower Control Limit
LCW	Low Carbon Water
LDPE	Low-Density Polyethylene
LMIs	Legal Metrology Institutes
LNSW	Low Nutrient Seawater
LWL	Lower Warning Limit
MEXT	Ministry of Education, Culture, Sports, Science and Technology (Japan)
MRA	International Mutual Recognition Agreement
MRI	Meteorological Research Institute (Japan)
MSR	DOM CRM from mid-depths of the ocean
MSSW	Multiparametric Standard Seawater
MSW	Middle nutrient concentration Seawater
MWJ	Marine Works Japan Ltd.
NED	N-(1-naphthyl)ethylenediamine Dihydrochloride
NIM	National Institute of Metrology (China)
NIOZ	Royal Netherlands Institute for Sea Research
NIST	National Institute of Standards and Technology (U.S.)
NMIs	National Metrology Institutes
NMIJ	National Metrology Institute in Japan
NRC/NRCC	National Research Council Canada
NSF	National Science Foundation (U.S.)
NTRM	NIST Traceable Reference Material

OAWG	Organic Analysis Working Group
OSIL	Ocean Scientific International Limited
OSJ	Oceanographic Society of Japan
PAN	Polyacrylonitrile
PC	Polycarbonate
PCTFE	Polychlorotrifluoroethene
pCO_2	Partial pressure of CO_2
PET	Polyethylene Terephthalate
PFA	Perfluoroalkoxy
PGVP	EPA Protocol Gas Verification Program
pH	A measure of the acidity or basicity of an aqueous solution, expressed as the negative logarithm (base 10) of the hydrogen ion concentration
pH_T	Seawater pH on the total hydrogen ion scale, expressed as $-\log_{10}([H^+_T])$, where $[H^+_T]$ is a measure of the sum of the free hydrogen ion and hydrogen sulfate concentrations in seawater
PML	Plymouth Marine Laboratory (U.K.)
POM	Particulate Organic Matter
PP	Polypropylene
PSS-78	Practical Salinity Scale 1978
PTB	Physikalisch-Technische Bundesanstalt (Germany)
PTFE	Polytetrafluoroethylene
PVC	Polyvinyl Chloride
QCMs	Quality Control Materials
QUASIMEME	Quality Assurance of Information for Marine Environmental Monitoring in Europe
RM	Reference Material
RMNS	Reference Materials for Nutrients in Seawater
RMOs	Regional Metrological Organizations
RMPs	Reference Measurement Procedures
RSD	Relative Standard Deviation
S_A	Absolute Salinity
s_{bb}	Standard deviation of between-bottle homogeneity
S_p	Practical Salinity
S_R	Reference-Composition Salinity
SCOR	Scientific Committee on Oceanic Research
SD	Standard Deviation
SDGs	Sustainable Development Goals
SGONS	Study Group on Nutrient Standards
SI	International System of Units
SIO	Scripps Institution of Oceanography (U.S.)
SOCAT	Surface Ocean CO_2 Atlas
SOP	Standard Operating Procedure
SPE	Solid Phase Extraction
SRM	Standard Reference Material

SSR	DOM CRM from the surface ocean
SSW	IAPSO Standard Seawater
SSWS	IAPSO Standard Seawater Service
TA	Total Alkalinity
TDN	Total Dissolved Nitrogen
TDP	Total Dissolved Phosphorus
TEOS-10	Thermodynamic Equation of Seawater 2010
TMAH	Tetramethyl Ammonium Hydroxide
TN	Total Nitrogen
ToF	The Ocean Foundation (U.S.)
TP	Terephthalate
TPX	Polymethylpentene
TUMSAT	Tokyo University of Marine Science and Technology
u_{bb}	Standard uncertainty due to between-bottle homogeneity
UCL	Upper Control Limit
UNEP	United Nations Environment Program
UNESCO	United Nations Educational, Scientific and Cultural Organization
UV	Ultraviolet
UWL	Upper Warning Limit
VKI	Water Quality Institute (Denmark)
WMO	World Meteorological Organization
WOCE	World Ocean Circulation Experiment
WWII	World War II

Chapter 1
Development and Use of Certified Reference Materials for Nutrients in Seawater

Michio Aoyama, Akihiko Murata, Tomomi Sone, and Yasuhiro Arii

Abstract This work outlines the development and use of Certified Reference Materials (CRMs) for nutrients in seawater, along with the history of nutrient measurement. Despite a long history of nutrient measurement, the use of nutrient data to assess ocean climate change is rare. One reason for this is the low comparability of nutrient measurements. The ocean community has made efforts to improve comparability through intercomparison exercises. At present, the CRMs developed and marketed by a private Japanese company are the most suitable for improving comparability. How to use these CRMs for highly accurate measurement of nutrients is also described.

Keywords Nutrients · CRM · RMNS · WOCE · Intercomparison exercise

1.1 Introduction

Nutrients in the ocean have been measured extensively since the early days of scientific oceanographic observations because they are not only essential for the activities of marine organisms but also useful as indicators of ocean circulation. However, measurements vary widely from one analyst to another, and despite the long history of measurements, using nutrient data to detect and understand medium- to long-term changes in the marine environment has been difficult and challenging and, therefore,

M. Aoyama · A. Murata (✉)
Physical and Chemical Oceanography Research Group, Global Ocean Observation Research Center, Japan Agency for Marine-Earth Science and Technology, Yokosuka, Japan
e-mail: murataa@jamstec.go.jp

T. Sone · Y. Arii
Marine Chemical Analysis Section, Office of Marine and Earth Environmental Analysis, Department of Marine and Earth Sciences, Marine Works Japan LTD., Yokosuka, Japan
e-mail: sone.tomomi@mwj.co.jp

Y. Arii
e-mail: arii.yasuhiro@mwj.co.jp

© The Author(s) 2025
M. Aoyama et al. (eds.), *Chemical Reference Materials for Oceanography*, Springer Oceanography, https://doi.org/10.1007/978-981-96-2520-8_1

rarely done. It is now recognized that nutrient reference materials (RMs) are needed to address this problem, as stated in the IPCC 4th Assessment Report (Bindoff et al., 2007). Over the past 25 years, the Japanese oceanographic community has pioneered the studies of nutrient RMs, beginning with individual research by one of the authors of this review and continuing with subsequent inter-laboratory comparison studies of nutrient measurements, and measurement of nutrient standards at all stations during repeat hydrographic cruises by the R/V *Mirai* of the Japan Agency for Marine-Earth Science and Technology (JAMSTEC). As a result, the RMs currently produced by KANSO TECHNOS Co., Ltd. (KANSO) have been found to be the most suitable for improving the comparability of nutrient data. To make the certified RMs (CRMs) for nutrients produced by KANSO widely available to the international oceanographic community, JAMSTEC, in collaboration with the Scientific Committee on Oceanic Research (SCOR), began distributing the SCOR-JAMSTEC CRMs in 2015 (see Chap. 3). In parallel with this set of activities, the development of CRMs traceable to SI units for analysis of nutrients in seawater equivalent to national standards was also undertaken (see Chap. 5). In this chapter, we review the development and distribution of nutrient CRMs and introduce their use in the field of nutrient analysis.

1.2   Nutrients in Seawater

Nitrogen, phosphorus, and silicon, which are essential elements for living organisms, are present in seawater as the nutrients nitrate, phosphate, and silicate, respectively. Their spatial distribution and temporal variation determine ocean productivity. Nutrients are also involved in the carbon cycle through biological activities such as photosynthesis and respiration. Therefore, nutrients in seawater are considered important from the perspective of global warming and ocean acidification.

In addition, marine anoxia, a recent focus of research, is closely related to biological activity in the ocean. As well as nutrients, oxygen is essential for life in the ocean and is linked to the major nutrient and carbon cycles through the C:O:N:P molar ratio. It has been estimated that 2% of the total amount of oxygen in the open ocean has been lost over the past 50 years, and dissolved oxygen minima of less than 70 $\mu mol\ kg^{-1}$ have been reported to be expanding at a depth of 200 m in the open ocean (https://www.ocean-oxygen.org/de/declaration). Unfortunately, to date, no reports have fully examined the relationship between nutrient levels in such areas and oxygen levels.

Nutrient distributions also play an important role as tracers in oceanography. For example, in early studies, general ocean circulation images were derived from the global distributions of nutrients with large spatial concentration variations (Broecker, 1991). In addition, the recently published dataset for the global distribution of nutrients, GND13 (http://www.godac.jamstec.go.jp/resource/data_catalog/GND13/GND13.zip; Aoyama, 2020), which is based on nutrient standards, may provide an image of ocean circulation in the mid-Pacific Ocean, which cannot be obtained from physical oceanographic data alone. Thus, nutrients are indispensable for understanding environmental variability in the oceans. However, although nutrients have been measured since the dawn of scientific oceanographic observations at the end of the eighteenth century, and they are still being measured around the world today, environmental RMs have only recently begun to be widely used.

1.3 Scientific Background and History

In studies of global environmental variability, especially nutrient variability in the marine environment, it is possible to observe differences in nutrient concentrations. However, determining whether the differences are due to environmental variability has been a difficult task because of the delay in the establishment of RMs and internationally agreed definitions and scales. Of course, the nutrient community has not neglected efforts to establish comparability in nutrient analysis, but it has long conducted international, national, and bilateral joint experiments to improve the quality and comparability of nutrient data. Some countries and projects have also developed and distributed nutrient standards to ensure comparability. These collaborative experiments and the development of nutrient standards were naturally closely linked. For example, the technique for making nutrient standard solutions, which was developed for the 5th collaborative experiment conducted by the International Council for the Exploration of the Sea (ICES), was passed on as a VKI (Water Quality Institute)–certified standard for a salinity of 35 (VKI Reference Materials for Environmental Chemical Analyses, 2020). An international joint experiment was conducted as a collaboration between the National Oceanic and Atmospheric Administration of the United States and the National Research Council of Canada in 2000 and 2002 for certification through joint experiments (Clancy & Willie, 2004). The joint experiment used the sample MOOS-1, which was a candidate CRM for silicate, phosphate, nitrite, and nitrate + nitrite. Co-operative Study of the Kuroshio and Adjacent Regions (CSK) standard solutions were prepared by the Sagami Chemical Research Center in Japan for the CSK in which 11 countries, including Japan, participated. Although there have been some restrictions on production of CSK standard solutions, such as they should have a pure water base or artificial seawater based on sodium chloride, and each property should be packaged in a separate bottle, they helped to improve comparability in nutrient measurements. The organizers of the 3rd ICES International Joint Experiment (1969–1970) realized that raw seawater, when distributed as a joint experiment sample, was unstable; therefore, they adopted the Japanese CSK standard as the joint experiment standard sample. The fact that these nutrient standards are still used in some laboratories was confirmed during an international collaborative experiment co-organized in 2018 by the International Ocean Carbon Coordination Project (IOCCP) and JAMSTEC. Six of the 69 participating laboratories indicated that they had used the three nutrient standards (VKI, MOOS-1 and CSK) described here.

The oceanic accumulation of anthropogenic CO_2 is estimated by calculation because it cannot be measured directly. In this calculation, the effect of biological activity is often estimated by calculations based on the amount of oxygen present. However, this approach has the disadvantage that the uncertainty of the oxygen content is large because of air–sea gas exchange and solubility changes. The estimation of biological effects from nutrients would allow the anthropogenic CO_2 accumulation to be estimated with smaller uncertainties. However, the comparability of historical nutrient data is relatively low, and greater comparability is needed to estimate the amount of anthropogenic CO_2 in the interior of the ocean. According to the IPCC 4th Report, a lack of comparability is why nutrient changes in the deep ocean cannot be studied (Bindoff et al., 2007).

The development of a nutrient CRM has become a fundamental solution to the poor comparability of nutrient measurements. During the 1990s, when an international observation program, the World Ocean Circulation Experiment (WOCE), was conducted, CRMs were discussed as necessary, but no one was able to supply them. In Japan, a prototype nutrient RM, which had been prepared before the cruise, was used for the first time in JAMSTEC's WOCE cruise in 1993. Subsequently, Japanese efforts to develop nutrient standards continued, and the homogeneity of nitrate and silicate in lot K, produced in 2000, was better than 0.4%, a value comparable to that of the standards previously distributed by ICES. However, the homogeneity of phosphate was greater than 1%, so more work was needed. Development continued, and as a result, we were able to report at the Ocean Sciences Meeting in 2002 that a nutrient standard made from natural seawater with good homogeneity and preservation lasting four years had been developed, and that we could supply 300 bottles per lot. At this stage, we were unable to certify the standard, but we posed the question of who would certify it and how it would be certified. (As a side note, M. Aoyama recalled people saying, "I can't believe it," when he announced that we had created a nutrient standard derived from natural seawater). We raised the certification issue because we firmly believe that nutrient standard solutions should be traceable to the International System of Units (SI) for the purpose of global environmental research. Although certification through joint experiments is possible, as was done with the MOOS-1 and VKI nutrient standards in the past, we chose not to go that route but instead to work with the National Metrology Institute in Japan (NMIJ), because we were convinced that it is theoretically correct and legitimate to establish a traceability system by supplying RMs as certified RMs. As a result, Japanese RMs are now supplied by NMIJ as certified standards, which has established SI traceability of nutrient reference materials in accordance with the international nitrogen and phosphorus standards for phosphate and nitrate. For silicate, NMIJ certified Japanese nutrient RM in accordance with the silicate standard solution of the US National Institute of Standards and Technology (NIST), because no other suitable silicate standard solution was available. At present, JAMSTEC and KANSO, in cooperation with NMIJ, have succeeded in preparing a silicate standard solution by melting 5 N high-purity silicon dioxide with alkali at high temperature, and the silicate analysis protocol has been modified accordingly. The silicate analysis can now be performed by the full gravimetric method in a laboratory on land and by the volumetric method in onboard laboratories. The produced silicon standard solutions are valued by NMIJ and are SI traceable (see Chaps. 6 and 7). This method has been applied to the nutrient CRMs certified and used for silicate analysis during JAMSTEC's R/V *Mirai* cruises, starting with the Arctic cruise in September 2019.

We distributed the developed nutrient standards to several Japan Meteorological Agency observation vessels in 2003 and 2004, and we conducted additional experiments to ensure comparability, even if the nutrient standards were used in different laboratories after a lapse of time. In addition, the intercomparison exercise with the developed nutrient standards was first conducted in 2003. It has since been implemented in 2006, 2008, 2012, 2014/2015, 2017/2018 (Aoyama et al., 2007, 2008, 2010, 2016, 2018).

The Intergovernmental Oceanographic Commission (IOC) has established a system to promote international cooperation, and in 2009, under the IOC, the Joint ICES–IOC Study Group on Nutrient Standards (SGONS) was established to promote international cooperation. To further develop our cooperative efforts, we proposed SCOR Working Group, which became active in 2015. The SCOR Working Group was granted a one-year extension to have continued its activities until 2019. During the Blue Earth Global Expedition (BEAGLE) of the R/V *Mirai* (2003–2004) in the southern hemisphere, a dataset was created using nutrient standards at all stations to ensure inter-station comparability. Nutrient standards have also been used as calibration curves on subsequent cruises of the R/V *Mirai* to ensure comparability between stations and between voyages. Furthermore, the organized use of nutrient standards allows for the estimation of concentration-dependent uncertainties in nutrient measurements (Sect. 1.5.2).

1.4   Results of Recent Intercomparison Exercises

Nutrient measurements are required to show comparability within 1% of the relative standard uncertainty and to be traceable to moles or masses in SI units. CRMs for nutrients in seawater have been available for only about 10 years, and their worldwide use is essential to achieve this 1% goal. With the aim of improving the comparability of seawater nutrient data at the scale of the whole ocean, the IOCCP and JAMSTEC have jointly conducted two international collaborative experiments on nutrient CRMs, in 2014 and 2018 (Aoyama et al., 2016, 2018). The number of labs using CRMs has increased rapidly over the past four years, with 28 of the 69 labs participating in the 2018 Inter-laboratory Comparison Study indicating that they use CRMs. In addition, the measurement results reported by the laboratories using CRMs have shown improved comparability. S-plots of the normalized nitrate concentrations reported by the laboratories participating in the inter-laboratory comparisons (Fig. 1.1) show the improvement of comparability obtained by repeated intercomparison studies. A similar improvement was also observed for phosphate and silicate. However, both the 2014 and 2018 Inter-laboratory Comparison Study results also revealed that the handling of the nonlinearity of the calibration curves during nutrient

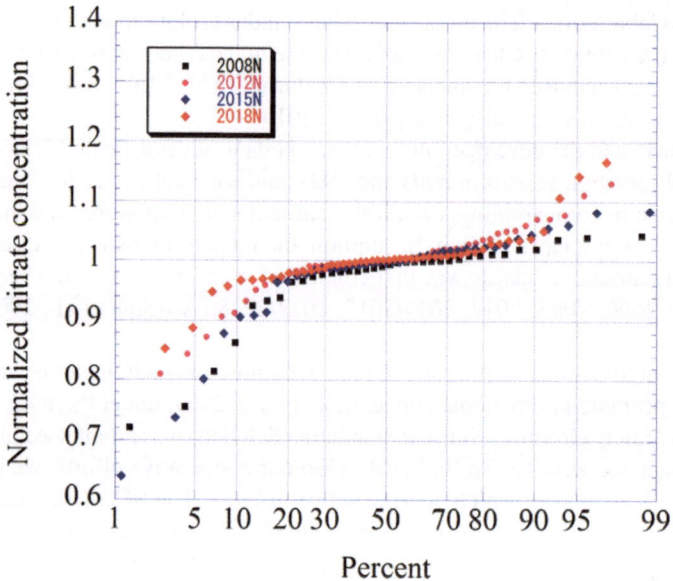

Fig. 1.1 S-plots of normalized concentrations of nitrate (μmol kg^{-1}) in inter-laboratory comparison studies

measurements in some laboratories was not appropriate. These results imply that a set of CRMs covering the full range of nutrient measurements is needed to maintain comparability across the full range of nutrient concentrations in the world's oceans. Nutrient datasets with global ocean-scale comparability are expected to enable the detection of global-scale ocean environmental variability.

1.5 High Accuracy Observations with Nutrient CRMs

To obtain highly accurate data on physical and chemical parameters of the entire ocean, the WOCE program was conducted in the 1980s and 1990s. To build on the highly accurate data obtained during WOCE, WOCE revisit, Climate Variability and Predictability-CO_2, and Global Ocean Ship-based Hydrographic Investigations Program voyages have been continuously conducted since to understand decadal-scale ocean environmental changes. Among chemical parameters, nutrients have also been measured in this series of observation programs, but, as noted by the IPCC 4th Report, a lack of comparability makes it difficult to detect ocean environmental variability from the measurements (Bindoff et al., 2007). For this reason, JAMSTEC has been measuring nutrient CRMs simultaneously with nutrient samples since 2003 during the high-accuracy observation cruises by the R/V *Mirai*, with the aim of ensuring comparability of data between stations and between cruises. For these measurements, the reference materials for nutrients in seawater (RMNSs) provided by KANSO (Chap. 2) were used. In Sect. 1.5.1, we introduce the actual use of nutrient CRMs/RMNSs and the quality evaluation of the measurements conducted during these cruises.

1.5.1 Use of RMNSs

Table 1.1 shows the RMNSs used during R/V *Mirai*'s high-accuracy observation cruises since 2003. For each observation area, five lots of RMNSs with low to high concentrations covering the nutrient range were used, and another RMNS lot was measured at all stations on each cruise to maintain comparability across cruises. As a result, a total of six lots of RMNSs were measured on each voyage. Naturally, the observations made over the 19-year period from 2003 to 2021 necessitated a change in lots. When a change was necessary, efforts were made to maintain comparability between cruises by measuring the RMNSs from before and after the change before the cruise in which the new lot was first used. Since 2005, RMNSs have been used for calibration curves; in addition to ensuring comparability, the use of RMNSs has greatly reduced the labor required to prepare in-house standard solutions. In addition, empirical equations were used to estimate uncertainties from the measurements of different RMNS lots, and an uncertainty value was assigned to each measurement (see Sect. 1.5.2), thereby making it possible to discuss temporal variations. The joint certification of RMNSs by JAMSTEC and KANSO started in 2015 (see Chap. 2); at that time, the RMNSs became CRMs, which enabled traceability to SI units. In addition, the measured values of RMNSs were compared with the uncertainties assigned to CRMs, and intermittent monitoring led to improved measurement accuracy.

1.5.2 Uncertainty Estimation by RMNS

RMNSs were used to determine the uncertainty of the measurements made during each cruise. The relative standard uncertainty (%), which was derived from the mean and standard deviation of the values measured during the MR15-05 cruise, is shown in Fig. 1.2 as a function of the silicate concentration in the RMNS (μmol/kg). By using the equation obtained by least-squares fitting of a curve to the data, it is possible to estimate the uncertainty of individual sample concentrations. It is clearly shown in the figure that for samples with silicate concentrations lower than 20 μmol/kg, the relative standard uncertainty increases rapidly. Similar plots were made to estimate the uncertainties for other measured properties, and the uncertainties thus obtained were used as a basis for judging the quality of the data.

A summary of the silicate uncertainties measured during cruises from 2005 to 2015 is shown in Fig. 1.3. These results confirm that the measurements were made with the same level of accuracy on each cruise. Even the largest uncertainties for silicate, those for R/V *Mirai* cruise MR11-08, were within 1% for silicate concentrations

Table 1.1 List of RMNSs/CRMs used during WOCE revisit cruises by R/V *Mirai*

Cruise	Year	WOCE line	RMNS lots
MR03-K04	2003/2004	P06, A10, I03/I04	T, AN, AK, AM, O, AH
MR0502	2005	P10	BA, AY, AX, AV, BC, AZ
MR05-05	2005/2006	P03	BA, AY, AX, AV, BC, AZ
MR07-04	2007	P01	BA, AY, AX, AV, BC, BF
MR07-06	2007	P01, P14	BA, AY, AX, AV, BC, BF
MR09-01	2009	P21	AS, BJ, AX, AV, BE, AZ
MR11-08	2011/2012	P10	BS, BU, BT, BD, BE, BF
MR12-05	2012/2013	S04I, P14S	BS, BU, BT, BD, BV, BF
MR14-04	2014	P01	BY, BU, CA, BW, BV, BZ
MR15-05	2015/2016	I10	BY, BU, CA, BW, BV, BZ
MR16-09	2017	P17E	BY, CD, CA, BW, CC, BZ
MR19-04	2019/2020	I08N, I07S[a]	CE, CJ, CG, CB, BZ, CF
MR21-04	2021	P01	CE, CL, CO, CG, CB, CF

[a] Line I07S was not one of the original WOCE lines, but was newly established during cruise MR19-04

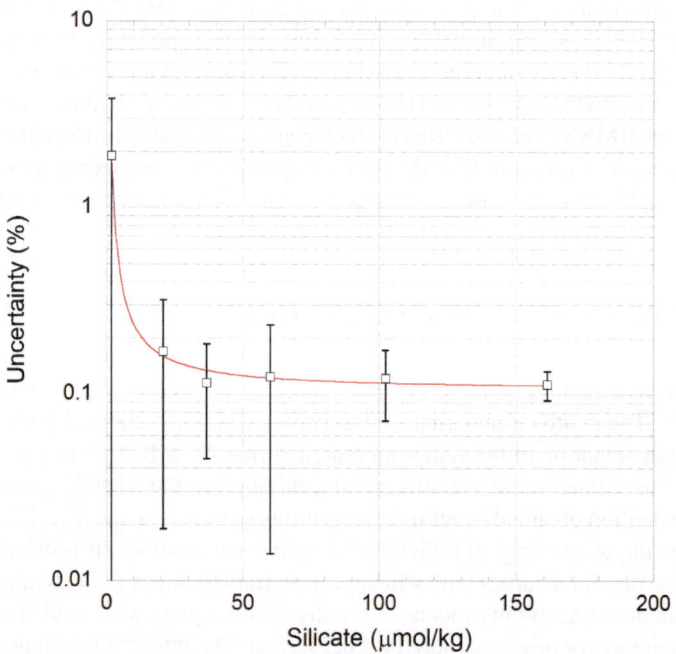

Fig. 1.2 Estimate of the uncertainty of silicate concentrations measured during cruise MR15-05. The curve was obtained by fitting $y = a + b \times (1/C_{Si}) + c \times (1/C_{Si})^2$ to the data of six RMNSs (Table 1.1), where y and C_{Si} are the uncertainty and concentration of silicate, respectively, and a, b, and c are parameters determined by the least-squares method. The error bars indicate coefficients of variance

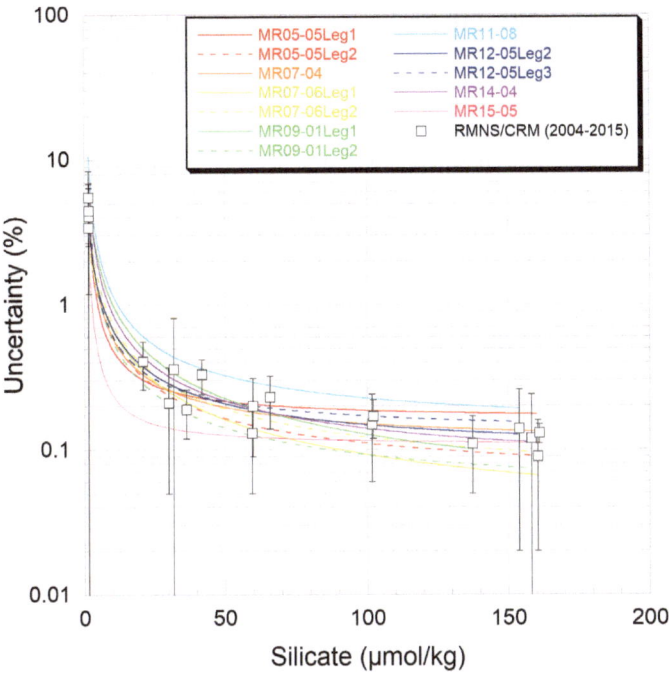

Fig. 1.3 Estimate of silicate concentration uncertainties during the 12 cruises of the R/V *Mirai*

above 11.93 μmol/kg; therefore, even low-concentration samples could be measured with small uncertainty. This level of accuracy was achieved by measuring the RMNS along with every sample measurement.

1.5.3 Future Issues

Since 2003, RMNSs have been used in conjunction with all nutrient measurements made during the high-accuracy observation cruises of the R/V *Mirai*. As a result, the comparability of data acquired by the R/V *Mirai* has been ensured, at least to the extent that nutrient data can be used to detect temporal changes in the ocean environment. As more institutions use RMNSs, it will become possible to clarify ocean environmental changes at the general ocean circulation scale. To achieve this aim, the use of RMNSs needs to be promoted more urgently.

1.6   Outlook for the Future

Nutrient CRMs, seawater reference materials that were developed in Japan, are now being widely used in marine communities around the world. The day will soon come when measurements based on nutrient CRMs will reveal long-term changes in the marine environment. This accomplishment is the result of 25 years of development, national certification, international dissemination, and efforts by many people, mainly in Japan. To obtain a more reliable picture of the marine environment in the future, however, we must not only continue these efforts at the current level but develop them further.

For global problems of the ocean environment such as global warming, ocean acidification, and ocean hypoxia, it is important to predict the future ocean environment; for this purpose, it will be necessary to distinguish effects due to biogeochemistry, which result from the activities of living organisms, from effects due to physics or physical chemistry, such as the solubilities of oxygen and CO_2 and the ionization equilibrium of carbonic acid. With accurate nutrient data, it will be possible to estimate biogeochemical effects from C:O:N:P ratios. For this purpose, high-quality nutrient concentration data verified at the whole-ocean scale by nutrient CRMs are needed.

Competing Interests The authors have no conflicts of interest to declare that are relevant to the content of this chapter.

References

Aoyama, M. (2020). Global certified-reference-material- or reference-material-scaled nutrient gridded dataset GND13. *Earth System Science Data, 12*, 487–499. https://doi.org/10.5194/essd-12-487-2020.

Aoyama, M., Becker, S., Dai, M., Daimon, H., Gordon, L. I., Kasai, H., Kerouel, R., Kress, N., Masten, D., Murata, A., Nagai, N., Ogawa, H., Ota, H., Saito, H., Saito, K., Shimizu, T., Takano, H., Tsuda, A., Yokouchi, K., & Youenou, A. (2007). Recent comparability of oceanographic nutrients data: Results of a 2003 intercomparison exercise using reference materials. *Analytical Sciences, 23*, 1151–1154. https://doi.org/10.2116/analsci.23.1151.

Aoyama, M., Barwell-Clarke, J., Becker, S., Blum M., Braga, E. S., Coverly, S. C., Czobik, E., Dahllöf, I., Dai, M. H., Donnell, G. O., Engelke, C., Gong, G. C., Hong, G.-H., Hydes, D. J., Jin, M. M., Kasai, H., Kerouel, R., Kiyomono, Y., Knockaert, M., Zhang, J.-Z. (2008). *2006 Inter-laboratory comparison study for reference material for nutrients in seawater* (No. 58, p. 104). Tsukuba, Japan: Technical Reports of the Meteorological Research Institute.

Aoyama, M., Anstey, C., Barwell-Clarke, J., Baurand, F., Becker, S., Blum, M., Coverly, S. C., Czobik, E., d'Amico, F., Dahllöf, I., Dai, M., Dobson, J., Pierre-Duplessix, O., Duval, M., Engelke, C., Gong, G.-C., Grosso, O., Hirayama, A., Inoue, H., Ishida, Y., Zhang, J.-Z. (2010). *2008 inter-laboratory comparison study for reference material for nutrients in seawater* (No. 60, p. 134pp). Tsukuba, Japan: Technical Reports of the Meteorological Research Institute.

Aoyama, M., Abad, M., Anstey, C., Ashraf P, M., Bakir, A., Becker, S. Bell, S., Berdalet, E., Blum, M., Briggs, R., Garadec, F., Cariou, T. Church, M., Coppola, L., Crump, M., Curless, S., Dai, M., Daniel, A. Davis, C., Zhang, J-Z. (2016). *IOCCP-JAMSTEC 2015 inter-laboratory calibration exercise of a certified reference material for nutrients in seawater.* IOCCP Report Number 1/ 2016, ISBN 978–4–901833–23–3.

Aoyama, M., Abda, M., Aguilar-Islas, A., Ashraf P, M., Azetsu-Scott, K., Bakir, A., Becker, S., Benoit-Cattin-Breton, A., Berdalet, E., Björkman, K., Blum, M., de Santis Braga, E., Caradec, F., Cariou, T., Chiozzini, V. G., Collin, K., Coppola, L., Crump, M., Dai, M., Zhang, J-Z. (2018). *IOCCP-JAMSTEC 2018 inter-laboratory calibration exercise of a certified reference material for nutrients in seawater.* Yokosuka, Japan, International Ocean Carbon Coordination Project/ Japan Agency for Marine-Earth Science and Technology (JAMSTEC), p. 214. (IOCCP Report Number 1/2018).

Bindoff, N. L., Willebrand, J., Artale, V., Cazenave, A., Gregory, J., Gulev, S., Hanawa, K., Le Quéré, C., Levitus, S., Nojiri, Y., Shum, C. K., Talley, L. D., & Unnikrishnan, A. (2007). Observations: Oceanic climate change and sea level. In S. Solomon, D. Qin, M. Manning, Z. Chen, M. Marquis, K. B. Averyt, M. Tignor, & H. L. Miller (Eds.), *Climate change 2007: The physical science basis. Contribution of working group I to the fourth assessment report of the intergovernmental panel on climate change* (pp. 385–433). Cambridge University Press.

Broecker, W. S. (1991). The great ocean conveyer. *Oceanography, 4*, 79–89.

Clancy, V., & Willie, S. (2004). Preparation and certification of a reference material for the determination of nutrients in seawater. *Analytical and Bioanalytical Chemistry, 378*, 1239–1242.

VKI Reference Materials for Environmental Chemical Analyses. https://www.eurofins.dk/miljoe/ vores-ydelser/vki-certificerede-referencematerialer/information-in-english/. Retrieved June 26, 2020.

Chapter 2
Production History of Certified Reference Materials of Nutrients in Seawater Developed by KANSO TECHNOS CO., LTD.

Hitoshi Mitsuda, Takeshi Fujii, Chisato Nagaya, and Naoyuki Tahara

Abstract The development of reference materials for nutrients in seawater (RMNS) started in 1993, and production and distribution of RMNS began in 2000. In 2011, KANSO TECHNOS CO., LTD. (KANSO) became accredited as a reference material producer (International Organization for Standardization, ISO 17034). Since then, RMNS has been distributed worldwide as an SI-traceable certified reference material. RMNS (1) is 100% natural seawater (no additives); (2) contains all nutrients in one bottle; (3) is used directly without any dilution step; (4) is available as several lots of low to high nutrient concentrations; and (5) has guaranteed homogeneity and stability for 7 years. With the current measurement precision, we now can obtain homogeneity results of 0.1 to 0.2% at mid- to high-nutrient-concentration ranges. In this chapter, RMNS's development history, production and certification method are described.

Keywords Certified reference material · Nutrients · Seawater · SI-traceable

2.1 Development Background

KANSO TECHNOS CO., LTD. (KANSO) has been involved in several oceanographic surveys, starting with the cruises operated by the Japan Agency for Marine-Earth Science and Technology (JAMSTEC) in the 1990s. Through experience, reference materials using natural seawater with the same matrix as that of actual samples are recognized as essential for comparing data obtained at different times, in different

The e-mail address (rminfo@kanso.co.jp) is the representative contact of CRM-RMNS production team. You can contact the corresponding author and co-authors at this address.

H. Mitsuda (✉) · T. Fujii · C. Nagaya · N. Tahara
Laboratory for Instrumentation and Analysis, KANSO TECHNOS CO., LTD., 3-1-1 Higashikuraji, Katano, Osaka 576-0061, Japan
e-mail: rminfo@kanso.co.jp

© The Author(s) 2025
M. Aoyama et al. (eds.), *Chemical Reference Materials for Oceanography*, Springer Oceanography, https://doi.org/10.1007/978-981-96-2520-8_2

places, and by different analysts. However, seawater-based reference materials for nutrient measurement had been underdeveloped until the 1990s. Thus, distinguishing the results of nutrient data as actual global-scale change or changes from analysis precision is difficult.

2.2   RMNS Production

The development of reference materials for nutrients in seawater (RMNS) began in 1993. For these reference materials, homogeneity and stability are required, and the main factors that cause changes in these properties are microbiological activity, evaporation, and/or sorption–desorption processes. A history of reaction chambers (A–H) is shown in Fig. 2.1. The glass chamber A was built in 1994 and had several problems, including silicate elution, and poor air tightness, which allowed water vapor loss during autoclaving treatment, resulting in unacceptable RMNS homogeneity. Meanwhile, the plastic chamber B also had the problem of water vapor loss. The metal chamber C was adopted to further improve air tightness. RMNS stored in the metal chamber C showed fairly good storage stability. The metal chambers D and E were developed to increase production capacity. However, nutrient concentration changes were observed after high-temperature and high-pressure treatment with an autoclave to stop microbiological activities. Thus, several containers were tested to find the best reaction chamber that can withstand autoclave treatment. Consequently, nutrient concentration changes were reduced by electropolishing the inner surface of the stainless-steel reaction chamber (Ota et al., 2006). At present, the reaction chambers F (230 L), G (320 L), and H (500 L) made of stainless steel are used.

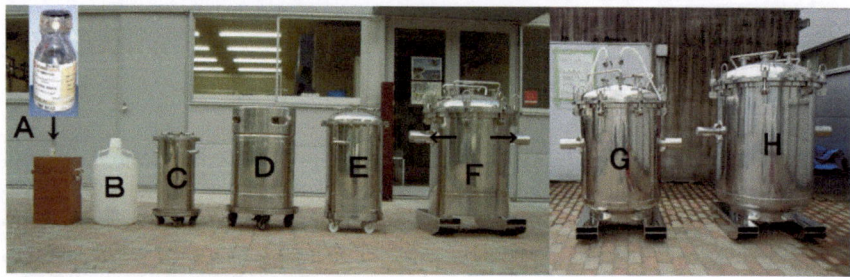

Fig. 2.1 History of reaction chambers

2.3 RMNS Production Facility

The RMNS production facility was built in a laboratory of KANSO in Katano, Osaka, Japan, in 2004 (Fig. 2.2). This facility includes a giant autoclave, as shown in Fig. 2.3 (W: 140 cm; H: 165 cm; D: 280 cm). The autoclave has two doors: one opens into a room at normal levels of laboratory cleanliness used for seawater storage, whereas the other opens into a class 10,000 clean area comprising a cooling room (current conditions: 20,000 counts m^{-3} [557 counts ft^{-3}]) and a bottling room (2000 counts m^{-3} [56 counts ft^{-3}]). Figure 2.4 shows the floor plan of the RMNS production facility. The seawater storage room is used to store source seawaters. The source seawaters are filtered, and the filtered seawater and the stainless-steel reaction chamber are weighed on an electronic balance (CAW1S4-1500LL-I; 1000 mm × 1000 mm; W × D; Combics 1; Sartorius, Goettingen, Germany) in the filtration room before the autoclaving. The structure of the RMNS production facility described in Ota et al. (2010) was not changed, but there were several new installments made since then. Three new stainless-steel reaction chambers (two 320-L chambers and one 500-L chamber) were added. The 320-L stainless-steel reaction chamber enables the production of 2500 RMNS bottles per lot. This facility is used not only for RMNS production but also for the development of reference materials, such as dissolved oxygen in seawater, and silicon standard solution, which became a certified reference material (CRM) in 2023.

Fig. 2.2 Photograph of the RMNS production facility at KANSO, Katano, Osaka, Japan

Fig. 2.3 Photograph showing the giant autoclave and reaction chamber

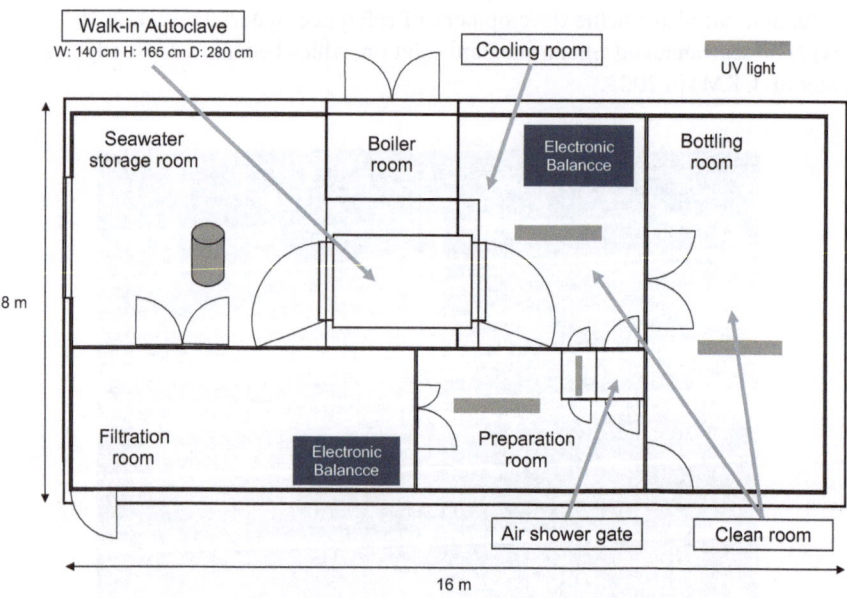

Fig. 2.4 Floor plan of the RMNS production facility

2.4 Production of RMNS of a Target Concentration

RMNS is produced in multiple concentrations. Thus, there are always challenges, and some restrictions to sampling source seawaters of multiple concentrations, both from oceanographic vessels, and pumping stations. On oceanographic vessels, the need to sample, and treat seawater from several depths considerably increases the workload. In seawater pumping stations, the water depths for seawater sampling are predetermined and fixed.

Thus, a method was developed to achieve target nutrient concentrations by mixing source seawaters of different concentrations and varying their mixing ratios to produce RMNS of several concentrations under the given restrictions. This method was patented in Japan (Ota, 2003).

A stainless-steel reaction chamber with a volume of 320 L can be used to treat 300 kg of seawater at a time. The following equations show the calculations used to produce an RMNS with a target nutrient concentration.

$$C_{RMNS} = (C_H \times W_H + C_L \times W_L)/(W_H + W_L) \tag{2.1}$$

where

C_{RMNS} = nutrient concentration of RMNS (μmol kg^{-1}).

C_H = nutrient concentration in high-nutrient-concentration seawater (μmol kg^{-1}).

C_L = nutrient concentration in low-nutrient-concentration seawater (μmol kg^{-1}).

W_H = mass of the high-nutrient-concentration seawater used (kg).

W_L = mass of the low-nutrient-concentration seawater used (kg).

Because W_L and W_H are measured using a high-accuracy electronic balance with a minimum resolution of 50 g and $W_H + W_L$ is about 300 kg, the mixing ratio can be controlled to within 0.0002.

2.5 Halting Biological Activity

In our method of RMNS preparation, there are two 2-h autoclaving treatments at 120 °C under the assumption that even if diapausing microorganisms are activated in the first round of autoclaving, their activities will be stopped in the second autoclaving. After the autoclaving process, the stainless-steel reaction chamber is left in a cooling room to gradually cool to room temperature. This minimizes the temperature gradient in the stainless-steel reaction chamber. The occurrence of a sharp temperature gradient in the stainless-steel reaction chamber generates gradients in density and nutrient concentrations. Any such gradients would diminish the RMNS homogeneity. It is important, therefore, to cool the reaction chamber as slowly as possible to enhance the homogeneity of the RMNS. During this cooling process, thorough mixing is conducted for at least two weeks to dissolve any salt precipitation that may have formed after seawater autoclaving. A polytetrafluoroethylene rotor is mounted inside the stainless-steel reaction chamber to constantly agitate the RMNS during the cooling and bottling processes to improve its homogeneity.

2.6 Seawater Transfer to Sterilized Bottles and Packaging of Bottles

Polypropylene bottles with a volume of 100 mL are used to distribute RMNS. These have been proven to be hermetic even though they are not equipped with inner lids. The bottles are cleaned with surfactant and ultrapure water, dried, and sealed in double bags in a class-100 clean room at the "As One Wakayama CIC Laboratory," Kaisougun, Japan. After being delivered to KANSO, the bottles are sterilized with UV lamps in the preparation room of the RMNS facility. The total UV exposure is set to 8800 μW min cm^{-2} or more by adjusting the length of the bottle irradiation time. The appropriate irradiation time is determined by measuring the UV intensity at each irradiation point with a digital UV meter (Sentry ST-512; MK Scientific, Yokohama, Japan). A UV exposure of 8800 μW min cm^{-2} was selected by doubling the UV germicide level of 4400 μW min cm^{-2} required to kill 99.99% of black mold (*Aspergillus niger*) (Toshiba Lighting and Technology Co. 2003).

After the seawater has gone through two autoclaving treatments in the reaction chamber and after thorough mixing, 90-mL aliquots are transferred into sterilized polypropylene bottles through a 0.22-μm pore-size filter in a class-100 clean bench (CCV-1311; Hitachi Appliances, Tokyo, Japan). The production flow diagram is shown in Fig. 2.5.

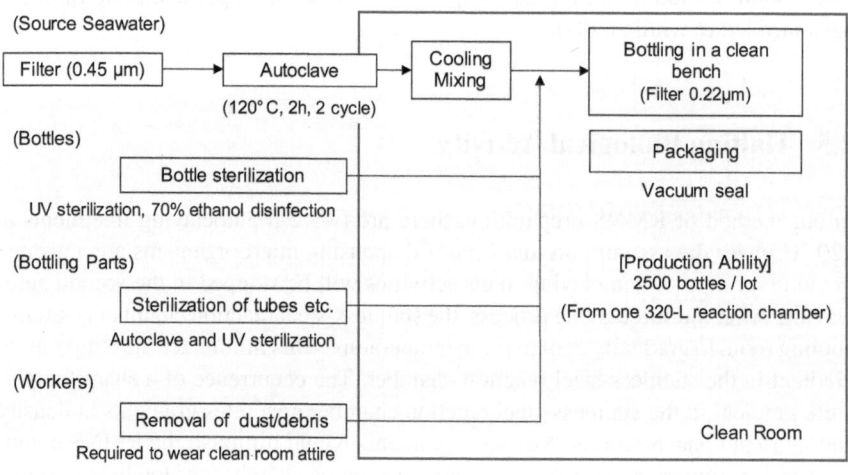

Fig. 2.5 RMNS production flow diagram

Table 2.1 Salinity change and types of outer package

Lot	Elapsed years	Initial salinity (PSU)	Salinity at elapsed year (PSU)	Change in salinity (PSU)	Percent change	Percent change per year	Type of outer package (bag)
AV	4.9	34.332	34.522	0.190	0.553	0.113	Nylon film
BG	2.3	34.336	34.334	−0.002	−0.006	−0.002	MAL
	13.5	34.336	34.340	0.005	0.013	0.001	
BT	0.4	34.413	34.422	0.009	0.026	0.065	HRS
	1.2	34.413	34.429	0.016	0.046	0.039	
BY	0.3	34.692	34.693	0.001	0.003	0.009	MAL
	7.4	34.692	34.695	0.004	0.011	0.001	
CA	7.0	34.376	34.379	0.003	0.010	0.001	
CJ	3.4	34.357	34.359	0.002	0.006	0.002	

An aluminum-film bag (MAL-1320S) manufactured by Meiwa Sanshou Co., Ltd. (http://sansho.mpx-group.jp/index.html) is used to package each RMNS bottle. The bottles are hermetically sealed in MAL-1320S bags using a vacuum heat-sealer in the clean room of the facility. This ensured homogeneity and stability by holding water evaporation down to one-tenth compared to the bottles without aluminum-film bag packaging. There was a slight difference in the evaporation rate between the types of aluminum-film bags as well. An HRS-1320S aluminum-film bag with a thicker aluminum-film layer was used from 2010 to 2011, expecting a better barrier. However, HRS-1320S showed more evaporation, thus MAL-1320S bags with a better seal are currently being used (Table 2.1).

The produced RMNS has the following characteristics: (1) 100% natural seawater is used as source (no additives); (2) it contains all nutrients in one bottle; (3) no dilution step is necessary, and it can be used directly after opening; (4) by changing the mixing ratio of source seawaters, low to high nutrient concentrations can be produced (Ota, 2003); (5) homogeneity and stability are maintained for 7 years (Aoyama et al., 2012). These are characteristics that could not be fulfilled in several of the past-developed nutrient reference materials mentioned in the previous chapter.

2.7 Quality Assurance of RMNS from a Producer's Perspective

After a new lot is produced, a series of quality checks are conducted. This section describes KANSO's standard quality check protocols for the produced RMNS. The quality control of RMNS is conducted through (1) visual inspection of RMNS bottles that have had culture medium injected and (2) measurement of nutrient concentrations. Table 2.2 lists the quality checks conducted for RMNS. During RMNS production, a culture medium (potato dextrose agar; Merck, Whitehouse Station, NJ, USA) is injected into one of every 50 RMNS bottles, and each of those bottles is hermetically sealed in a transparent package (Polyflexbag Hiryuu, type N1; Asahi Kasei Pax Co.; http://www.asahi-kasei.co.jp/pax/biz1d1.html). The bottles containing the added culture medium are visually inspected 1 week, 2 weeks, 1 month, 2 months, and 3 months after preparation to check whether any suspended solids have been generated. The bottles are opened after 3 months, and 1 mL from each bottle is transferred to a separate solid culture plate to test for microbial growth.

The first measurements for nitrate, nitrite, phosphate, silicate, and ammonium concentrations are conducted within 1 week of the production day. Then, nutrient concentrations are measured after 1, 2, 3, 6, 12, 24, 36, 48, 60, 72, and 84 months. During the 3 months after production, all bottles are visually inspected. Any bottles with signs of biocontamination are eliminated, and the nutrient concentrations in the eliminated bottle are measured so that the magnitude of the change in nutrient concentrations is known. In the past lots with the production of 2500 bottles, zero to three bottles were eliminated because of biocontamination. For 7 years from production, nutrient concentrations are measured annually as long as a stock of the RMNS bottles is available. Through this series of measurements, the long-term stability is tested following the ISO Guide 35 (ISO 2006). Once this stability test determines that a lot of RMNS is unstable and exceeds the expanded uncertainty of the certified value, KANSO will stop selling such lot.

Table 2.2 List of quality checks conducted for RMNS

CRM-RMNS quality checks		Other measurements/analyses
Homogeneity	Within 1 week from production (30 bottles)	Salinity and density (10 bottles)
↓		Culture test-bottle visual inspection (50 bottles); 1 week, 2 weeks
Stability-1	30 bottles analyzed after 1 month	Culture bottles visual check (1m)
↓		
Stability-2	30 bottles analyzed after 2 months	Culture bottles visual check (2m)
↓		
Stability-3	30 bottles analyzed after 3 months	Culture bottles visual check (3m)
↓		1 mL of each culture bottle cultured and checked for (no) growth
Inspection	Visual bottle inspection (all bottles) initiated 3 months from production	Salinity (10 bottles)
Certification: Results from JAMSTEC and KANSO evaluated. A goal is to certify within a year from production.		
Stability	10 bottles analyzed after 0.5, 1, 2, 3, 4, 5, 6, and 7 years from production	

2.8 Certification Method of KANSO-CRM-RMNS

As mentioned in the previous chapter, the National Metrology Institute of Japan (NMIJ) developed three levels of CRMs and the building of an SI-traceable system was established. At the same time, there was a demand to indicate SI traceability for RMNS. Thus, KANSO became accredited as a reference material producer (International Organization for Standardization, ISO 17034) in 2011 through the Accreditation System of National Institute of Technology and Evaluation Since then, RMNS was distributed worldwide as an SI-traceable CRM.

In 2015, a co-certification system by JAMSTEC and KANSO was constructed to certify the nutrient property values of RMNS. The current condition for certification is that the results of 30 bottles each measured in duplicate by both laboratories must be within the expanded uncertainty with a coverage factor (k) of 2. As shown in Fig. 2.6, the measurement results of three lots by both laboratories are close to one another. Standard uncertainties of the measurements are assigned on the basis of the actual test results from each laboratory or from published values combined. Furthermore, NMIJ CRMs and already certified RMNS are measured at the same time and checked to make sure they are within each CRM's expanded uncertainty to confirm the validity of the measurement values. Table 2.3 shows the nitrate results of lot CF, CG, and CJ as examples when the lots were under certification evaluation. The nutrient measurement methods used by both laboratories were based on the Manual of Oceanographic Observation (Japan Meteorological Agency 1999). The basis of the certification method followed the ISO Guide 35 (ISO 2006). With the addition of a co-certification system in place, highly reliable CRMs are now available.

The uncertainties of the certified values were calculated using Eq. (2.2). Each uncertainty component is described in Table 2.4. As an example, the calculated results for lot CG nitrate, silicate, and phosphate are shown. Major uncertainty components differed according to the type of determinants and concentration measured for certification.

$$u_{CRM}^2 = u_{JCSS}^2 + \left(u_{char1}^2 + u_{char2}^2\right)/2 + u_{lts}^2 + u_{sts}^2$$
$$u_{char}^2 = u_{standard}^2 + u_{red}^2 + u_{NO2}^2 + u_{clib}^2 + u_{repeat}^2 + u_{density}^2 \qquad (2.2)$$

2.9 Homogeneity

The homogeneity of RMNS has improved during the last several years, as shown in Fig. 2.7. This improvement in homogeneity mainly resulted from modifications to the stainless-steel reaction chamber used for autoclaving. In particular, the improvement in homogeneities to 0.2% for nitrate, phosphate, and silicate was achieved by electropolishing the inner surface of the reaction chamber. A patent application has been filed for the specifications of this reaction chamber (Ota et al., 2006). We found

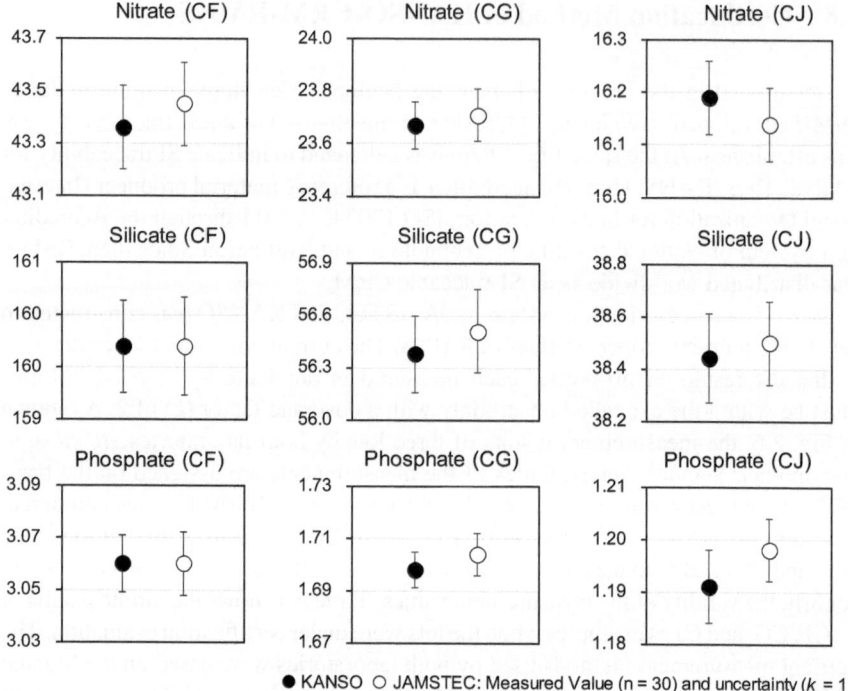

● KANSO ○ JAMSTEC: Measured Value (n = 30) and uncertainty (*k* = 1)

Fig. 2.6 Comparison of the measured concentrations (μmol kg^{-1}) of nitrate, silicate, and phosphate for the three lots of the nutrients CRM (CF, CG, and CJ) between KANSO and JAMSTEC (Murata et al., 2020)

that the electropolished inner surface of the stainless-steel reaction chamber deteriorates after the chamber has been used several times for autoclaving. However, RMNS homogeneity can be maintained by repeating the electropolishing treatment on the inner surface. Including this production improvement and developments in current measurement precision, we now can obtain homogeneity results better than 0.2% at the mid- to high-nutrient-concentration range.

2.10 Stability

The long-term stability of the RMNS was evaluated following clause 8.3 of ISO Guide 35 (ISO 2006). Table 2.5 shows the stability evaluation test results. From clause 8.5 of ISO Guide 35 (ISO 2006), uncertainty contributions due to long-term stability were estimated for 6 and 7 years using the uncertainties associated with the slope of the measured data derived from clause 8.3 with the time in months. The values were expressed in terms of the percentage of certified values. Because

Table 2.3 Nitrate concentrations (μmol kg^{-1}) of CRMs analyzed as a validation check in the same measurement run of lots CF, CG, and CJ certification by KANSO and JAMSTEC

	Series of CRMs measured for validation check Lot name							
	BY	7601-a	CD	7602-a	CC	CB	7603-a	BZ
Certified value	0.024[a]	0.021[b]	5.498	15.30	30.88	35.79	44.50	43.35
Expanded uncertainty (k = 2)	0.019	–	0.050	0.48	0.24	0.27	0.58	0.33
Validation results from lot CF certification								
KANSO	0.039	0.038	5.479	15.49	30.91	35.75	44.14	43.38
JAMSTEC	0.046	0.042	5.494	15.49	30.85	35.84	44.21	43.38
Validation results from lot CG certification								
KANSO	0.033	0.035	5.484	15.48	30.90	35.76	44.12	43.31
JAMSTEC	0.002	−0.004	5.461	15.48	30.88	35.73	44.18	43.41
Validation results from lot CJ certification								
KANSO	0.031	0.036	5.494	15.55	31.01	35.93	44.35	43.55
JAMSTEC	0.045	0.059	5.483	15.48	30.88	35.77	44.20	43.40

The value with [a] is below the quantifiable detection limit; thus, the value is not certified and shown as reference value

The value with [b] is not certified and is just shown as information

the uncertainty was derived from the measurement results, the accuracy of nutrient measurements at KANSO's laboratory was not high enough yet to evaluate the long-term stability to a standard uncertainty as low as 0.1 to 0.2%. We thus conclude that the long-term stability of RMNS has reached a level of uncertainty equal to the best available measurement techniques.

From the long-term monitoring of RMNS data, a gradual increase in nitrite and ammonium concentration was observed. For nitrite, an increasing trend (0.004 $\pm$ 0.002 μmol kg^{-1} per year) was confirmed and announced in 2019.

Meanwhile, ammonium was unstable (i.e., ensuring its stability was difficult); however, from long-term monitoring through the examination of production methods and storage tests, ammonium in RMNS was confirmed to be stable for up to 4 years under refrigerated storage. Figure 2.8 shows the ammonium concentrations of the test samples from storage tests. About 120 bottles of test samples were produced using the same method as RMNS, and 10 bottles in duplicates were measured for each storage condition each time. From 2021, the produced RMNS lots (from CQ and CR) are stored in a refrigerator to reduce ammonium concentration change. In the certificates of lots CQ and CR, an ammonium value was included as a reference value.

Table 2.4 Calculation formula for the uncertainty determination of lot CG

	Factor uncertainty	Uncertainty (%)		
		Nitrate	Silicate	Phosphate
u_{CRM}	Combined standard uncertainty of a property value	0.397	0.406	0.469
u_{JCSS}	Uncertainty of Standard Solution (concentration, 1000 mg L^{-1})	0.35	0.36	0.35
KANSO standard uncertainty				
u_{char1}	Due to characterization	0.148	0.135	0.252
$u_{standard1}$	Due to the calibration solution	0.045	0.042	0.109
u_{red1}	Due to the rate of reduction	Insignificant	–	–
u_{NO2_1}	Due to the nitrite concentration	Insignificant	–	–
u_{clib1}	Due to the calibration curve	0.082	0.060	0.176
$u_{repeat1}$	Within-bottle standard uncertainty	0.113	0.112	0.144
$u_{density1}$	Due to density conversion	0.018	0.018	0.018
JAMSTEC standard uncertainty				
u_{char2}	Due to characterization	0.210	0.211	0.349
$u_{standard2}$	Due to the calibration solution	0.107	0.149	0.283
u_{red2}	Due to the rate of reduction	0.034	–	–
u_{NO2_2}	Due to the nitrite concentration	0.005	–	–
u_{clib2}	Due to the calibration curve	0.122	0.097	0.165
$u_{repeat2}$	Within-bottle standard uncertainty	0.127	0.112	0.117
$u_{density2}$	Due to density conversion	0.018	0.018	0.018
u_{bb}	Standard uncertainty due to between-bottle homogeneity	0.044	0.060	0.064
u_{lts}	Standard uncertainty due to long-term stability	0.020	0.020	0.020
u_{sts}	Standard uncertainty due to short-term stability	Insignificant	Insignificant	Insignificant

2.11 Traceability

The current traceability for nitrate, nitrite, phosphate and ammonium is established using one of the Japan Calibration Service System (JCSS) standard solutions for each nitrate, nitrite, phosphate, and ammonium ion. For silicate, from 2015 to 2019, Merck KGaA silicon standard solution 1000 mg L^{-1} Si-traceable to the National Institute of Standards and Technology silicon standard reference material was used. However, a standard with smaller uncertainty was needed to monitor RMNS stability and traceability. In 2019, a silicon standard solution traceable to NMIJ primary standard

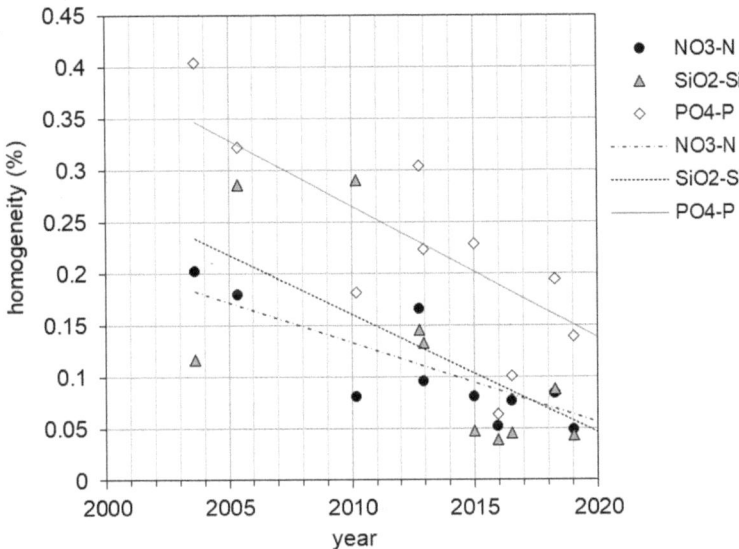

Fig. 2.7 Temporal changes of RMNS homogeneity for nitrate, silicate, and phosphate

solution was used. This silicon standard solution was co-prepared by JAMSTEC and KANSO, and the dissolved silica value was assigned by NMIJ. The development of the silicon standard solution is described in a later chapter.

2.12 Distribution of KANSO-CRM-RMNS

The number of bottles distributed worldwide is in an increasing trend as shown in Fig. 2.9. By 2022, the number of institutes provided with such bottles was 72 in total (domestic: 30; overseas: 42; 14 countries). Particularly, the number of such overseas institutes greatly increased over recent years. The overall number of bottles distributed from 2005 to 2022 reached 46,923 (Fig. 2.9).

2.13 Future of KANSO-CRM-RMNS

Over a quarter of a century has passed since the RMNS development, and over 100 lots have been produced since (Table 2.6). With the cooperation and understanding of many people involved and through repeated trial-and-error procedures, production, measurement, and certification methods were developed and improved. The current production can prepare 2500 bottles per lot. In the future, KANSO aims to maintain a sustainable global supply system while keeping in mind a way to distribute to

Table 2.5 Stability analysis results following ISO Guide 35: 2006

Lot[a]	Time	ISO Guide 35	Certified value	Expanded uncertainty	Estimated uncertainty (%)[c]	
No	(months)	Stability[b]	(μmol kg^{-1})	(%)	6 years	7 years
Nitrate						
BZ	95	Stable	43.35	0.8	–	0.4
CF	70	Instable	43.4	0.9	0.6	0.6
CN	39	Stable	43.6	0.9	0.2	0.2
CG	75	Stable	23.7	0.8	0.2	0.2
CH	47	Stable	16.94	0.9	0.4	0.5
Phosphate						
BZ	95	Stable	3.056	1.1	–	<0.1
CF	70	Stable	3.06	1.0	0.4	0.5
CN	39	Stable	2.94	1.0	0.3	0.4
CG	75	Stable	1.70	1.2	0.2	0.2
CH	47	Stable	1.172	1.3	0.5	0.6
Silicate						
BZ	95	Stable	161.0	0.6	–	0.1
CF	70	Stable	159.7	0.6	<0.1	<0.1
CN	39	Stable	152.7	0.5	0.6	0.7
CG	75	Stable	56.4	0.9	0.3	0.3
CH	47	Stable	29.84	0.5	0.1	0.1

[a]. Please refer to the supplementary information (2.14) at the end of this chapter for an explanation of the RMNS lot numbering system

[b]. (In)stability is determined by calculating the standard uncertainty associated with the slope ($s[b_1]$) and with the use of Student's t-factor at n-2 degrees of freedom, 95% level of confidence. When slope (b_1) is less than the t-test value ($|b_1| < t0.95$, n-2 $\times$ $s[b_1]$), the slope is insignificant and, therefore, stable. Slope (b_1) is determined using the measured nutrient concentration as the dependent value (Y) and time in months as the independent value (X)

[c]. The estimated uncertainty is computed using the following: standard uncertainty due to long-term (in)stability (u_{lts}) = uncertainty associated with slope (s_{b1}) $\times$ time (months); estimated uncertainty (%) = u_{lts}/certified value $\times$ 100. Note that the stability test and estimated uncertainty use different methods to evaluate stability

overseas institutes not using RMNS yet and to contribute to further improvements in global-scale comparability from nutrient measurement data produced by institutes in each country.

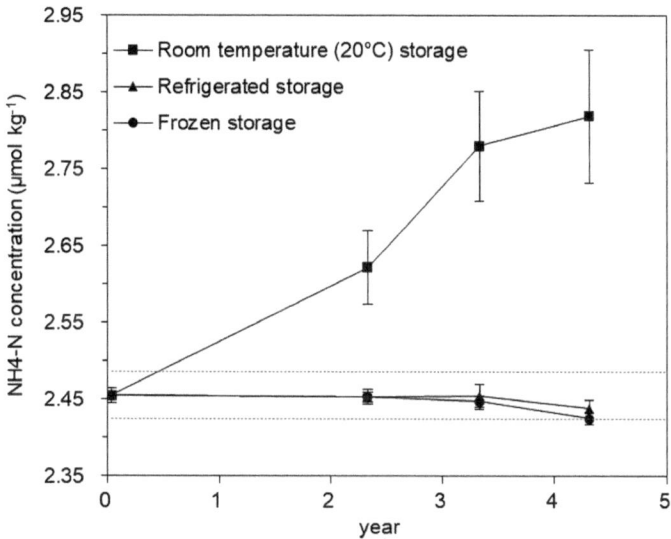

Fig. 2.8 Ammonium concentration changes of the test samples stored under three temperature conditions. The dashed line shows 3 standard deviations (3σ) of the initial value. The error bars indicate 1σ of the measured values ($n = 10$)

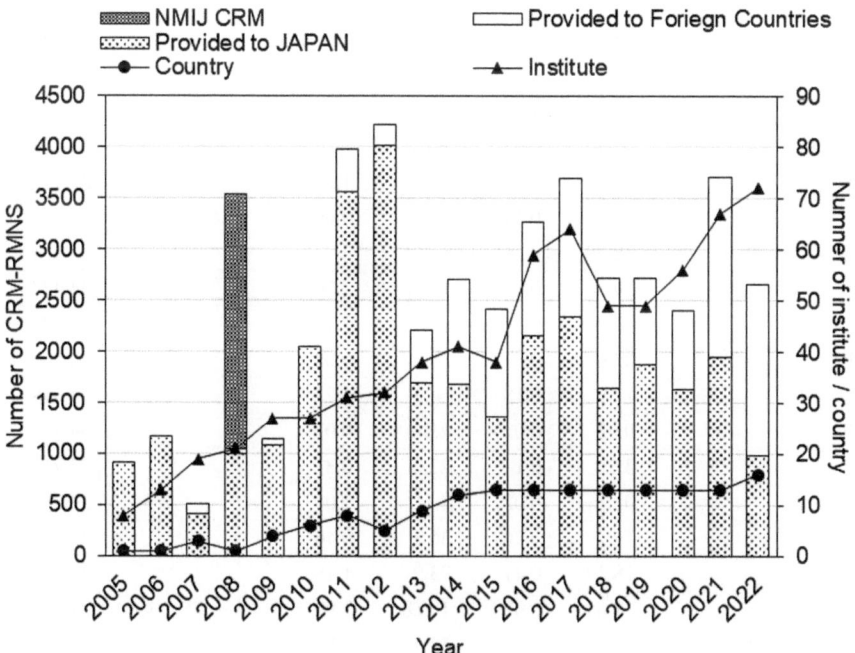

Fig. 2.9 Temporal changes in the number of RMNS bottles distributed (bar graph: open bars for foreign countries, large-meshed bars for within Japan, and fine-meshed bars for NMIJ CRM), the number of foreign institutes (solid triangle), and the number of countries (solid circle)

Table 2.6 RMNS lots produced from 1995 to 2022

Lot. no	Nitrate	Nitrite	Silicate	Phosphate	Ammonium	Salinity
#H2	25.12	1.53	144	1.78		
#L1	0.15	0.1		0.01		
#H3						
#H4	26.92	0.54		2.08		
#L2						
#H5	23.41	1.28	57.57	0.41		
#H5′	0.58	0.05		0.19		
Old yellow	23.5	1.41	58.4	0.4		
A	16.88	0.85	149.6	1.45		
B	7.74	0.6	75.5	0.73		
C	0.11	0.04	2.2	0.1		
A	17.96	0.94	152.4	1.43		
B	8.15	0.67	78	0.66		
C	0.14	0.04	2.4	0.05		
D	22.83	0.08	133.7	1.63		34.711
E	15.86	0.58	72.4	1.19		34.673
F	0.04	0.03	2.4	0.07		34.652
G	0.01	0.02	1.2	0.17		35.344
H	26.28	0.1	131	1.67		34.702
I	20.46	1.57	118.2	1.48		34.721
J	0.08	0.03	2.1	0.26		35.326
K	19.69	0.15	127.6	1.52		34.732
L	6.21	0.17	31	0.69		35.042
M	38.77	0.36	129.6	2.72		34.499
N	39.93	0.09	131.3	2.76		34.473
O	37.82	0.5	131.5	2.78		34.474
P	0.06	0	2.7	0.08		34.764
Q	17.9	0.18	64.7	1.25		34.618
R	0.03	0.02	2.3	0.07		34.757
S	3.2	0.1	22.2	0.3		34.71
T	0.01	0.02	2.1	0.08		34.768
U	3.87	0.24	36.7	0.46		34.784
V	18.17	0.41	93.9	1.48		34.665
W	28.84	0.02	100.8	1.99		34.651
X	7.02		55.4	0.81		34.917
Y	2.61		31.2	0.44		35.158
Z	11.74	0.49	20.9	1.3		35.321

(continued)

Table 2.6 (continued)

Lot. no	Nitrate	Nitrite	Silicate	Phosphate	Ammonium	Salinity
AA	27.91		142.9	1.97		34.661
AB	12.92	0.92	71.5	1.07		34.577
AC	0.04	0	2	0.07		34.549
AD	7.89	0.06	68.5	3.5		35.152
AE	4.49	0.05	28.2	2.21		35.168
AF	1.83	0.05	16.1	0.99		35.267
AG	16.54	0.52	156.8	1.79		34.665
AH	35.49	0	133.4	2.09		34.702
AI	3.87	0.31	53.3	0.55		34.914
AJ	6.46	1.34	92	3.01		35.128
AK	22.28	0	68.3	1.21		34.047
AL	0.1	0	1.4	0.05		34.559
AM	28.97	0.01	91.4	1.31		33.855
AN	7.33	0.03	25.8	0.38		34.386
AO	36.19	0.04	99.7	2.73		34.339
AP	36.05	0.05	99.5	2.65		34.339
AQ	36.17	0.04	99.6	2.67		34.337
AR	0.02	0.04	2.3	0.04		34.592
AS	0.03	0.03	2	0.04		34.587
AT	7.37	0.06	18.2	0.54		34.442
AU	29.67	0.06	67.3	2.12		34.146
AV	33.21	0.1	156.6	2.48		34.332
AW	23.42	0.44	120.1	2.11		34.332
AX	21.39	0.38	59.5	1.59		34.445
AY	5.52	0.68	30.11	0.5		34.627
AZ	41.87	0.01	136.3	3.01		34.448
BA	0.02	0	1.66	0.08		34.617
BB	40.25	0.03	156.6	2.65		34.625
BC	40.54	0.02	159.5	2.72		34.615
BD	29.86	0.04	64.95	2.23		34.274
BE	36.64	0.02	101.2	2.67		34.351
BF	41.28	0.02	151.6	2.79		34.614
BG	35.96	0.06	259.2	2.58		34.336
BH	20.66	0.57	21.9	1.51		34.959
BI	40.44	0.01	149.7	2.6		34.591
BJ	7.45	0.01	29.5	0.59		34.328
BK	19.74	0.62	22.3	1.38		34.951

(continued)

Table 2.6 (continued)

Lot. no	Nitrate	Nitrite	Silicate	Phosphate	Ammonium	Salinity
BL	0.05	0	2.4	0.05		34.268
BM	8.35	1.25	22.2	0.79		35.07
BN	36	0.01	102.3	2.66		34.329
BO	24.51	0.27	99.7	1.78		34.336
BP[b] (NMIJ CRM 7601-a)	0.04	0.01	1.8	0.04		34.7
BQ[b] (NMIJ CRM 7603-a)	43.69	0.01	144.9	3.03		34.449
BR[b] (NMIJ CRM 7602-a)	15.33	0.41	29.6	1.12		34.589
NMIJ CRM 7601-a[c] (BP)	0.021[a]	0.035[a]	1.28	0.02[a]		
NMIJ CRM 7603-a[c] (BQ)	44.27	0.026[a]	146.2	3.03[a]		
NMIJ CRM 7602-a[c] (BR)	15.2	0.41	29.8	1.06[a]		
BS	0.058[a]	0.017[a]	2.411	0.054		34.663
BT	18.15	0.471	42.02	1.296		34.413
BU	3.937	0.072	20.92	0.345		34.538
BV	35.36	0.047	102.2	2.498		34.364
BW	24.59	0.067	60.01	1.541		34.33
BX	43	0.034	136.1	2.906		34.423
BY	0.024[a]	0.019[a]	1.763	0.039[a]		34.692
BZ	43.35	0.215[a]	161	3.056		34.427
CA	19.66	0.063	36.58	1.407		34.376
CB	35.79	0.116[a]	109.2	2.52		34.374
CC	30.88	0.116[a]	86.16	2.08		34.338
CD	5.498	0.018	13.93	0.446		34.484
CE	0.01[a]	0.03	0.06[a]	0.012[a]		35.224
CF	43.4	0.09	159.7	3.06		34.455
CG	23.7	0.06	56.4	1.7		34.32
CH	16.94	0.18	29.84	1.172		34.991
CI	13.77	0.41	8.26	0.948		35.654
CJ	16.2	0.031	38.5	1.19		34.357
CK	0.02[a]	0.011[a]	0.73[a]	0.048		35.211
CL	5.47	0.015[a]	13.8	0.425		34.685
CM	33.2	0.018[a]	100.5	2.38		34.414
CN	43.6	0.010[a]	152.7	2.94		34.536

(continued)

Table 2.6 (continued)

Lot. no	Nitrate	Nitrite	Silicate	Phosphate	Ammonium	Salinity
CO	15.86	0.04	34.72	1.177		34.282
CP	24.8	0.31	61.1	1.753		34.398
CQ	0.06	0.07[a]	2.20[a]	0.030	1.76[a]	34.703
CR	5.46	0.97[a]	14.0	0.394	0.95[a]	34.619

[a] These values were used as guides
[b] These values were measured by KANSO
[c] These values were NMIJ CRM's certified values

2.14 Supplemental Information: Alphanumeric Identification of RMNS Bottles

Each bottle of RMNS has a sticker with an alphanumeric identifier (e.g., "AB-1234"). The alphabetical portion of the identification code represents the sequence number of the production lot. An identifier starting with "A-" designates the first-produced lot. An identifier starting with "Z-" designates the 26th lot produced. The 27th lot produced is designated "AA-;" "AB-" designates the 28th lot produced. The latest lot is designated "CR-." Sometime in the future, there may be bottles carrying a code starting with "AAA-."

The number portion of the identifier represents the bottling sequence in each production lot. The identifier "AB-1234," therefore, indicates the 1234th bottle in lot AB. In this way, each bottle is uniquely identified.

The RMNS bottles used for preliminary experiments or produced as prototypes were identified in the same way. The bottles distributed to the market, therefore, may not necessarily carry successive alphanumeric identifiers.

2.15 Supplemental Information: RMNS Production History

Table 2.6 lists the RMNS lots produced to date, along with their nutrient concentrations (μmol kg^{-1}). More information on the produced RMNS up to lot CR is available in the JAMSTEC database: https://doi.org/10.17596/0002218 (Aoyama et al., 2022).

Competing Interests The authors have no conflicts of interest to declare that are relevant to the content of this chapter.

References

Aoyama, M., Ota, H., Kimura, M., Kitao, T., Mitsuda, H., Murata, A., & Sato, K. (2012). Current status of homogeneity and stability of the reference materials for nutrients in seawater. *Analytical Sciences, 28*, 911–916.

Aoyama, M., Mitsuda, H., Fujii, T., & Nagaya, C. (2022). A dataset of nutrients concentration of certified reference material since 2011 and those in reference material from 1995 to 2010. JAMSTEC. https://doi.org/10.17596/0002218.

International Organization for Standardization. (2016). *General requirements for the competence of reference material producers* (ISO 17034:2016).

Toshiba Lighting and Technology Co. (2003). *Technical report STD NO. F-35G.* Toshiba Lamp. [in Japanese].

Japan Meteorological Agency. (1999). *Manual on oceanographic observation part I* (p. 200) [in Japanese].

Murata, A., Aoyama, M., Cheong, C., Miura, T., Fujii, T., Mitsuda, H., Kitao, T., Sasano, D., Nakano, T., Nagai, N., Kodama, T., Kasai, H., Kiyomoto, Y., Setou, T., Ono, T., Yokogawa, S., Arii, Y., Sone, T., Ishikawa, Y., … Wakita, M. (2020). Current situation and future perspective for environmental standards of seawater: Commencing with certified reference materials (CRMs) for nutrients. *Oceanography in Japan, 29*, 157–159.

Ota, H., Kitao, T., & Kimura, M. (2006). Japan Patent Kokai No. 058080.

Ota, H., Mitsuda, H., Kimura, M., & Kitao, T. (2010). Reference materials for nutrients in seawater: Their development and present homogeneity and stability. In M. Aoyama, A. G. Dickson, D. J. Hydes, A. Murata, J. R. Oh, P. Roose, & E. M. S. Woodward (Eds.), *Comparability of nutrients in the world's ocean* (pp. 11–30). Tsukuba, Japan: MOTHER TANK.

Ota, H. (2003). Japan Patent Kokai No. 214996.

International Organization for Standardization. (2006). *Reference materials–general and statistical principles for certification* (ISO Guide 35:2006).

Chapter 3
Production and Distribution of SCOR-JAMSTEC Certified Reference Materials (CRM) of Nutrients in Seawater

Akihiko Murata, E. Malcolm S. Woodward, Karel Bakker, and Masahide Wakita

Abstract To encourage the worldwide use of available Certified Reference Materials (CRMs) for measurements of nutrients in seawater, five new lots of CRMs were produced and distributed globally as an activity of the Scientific Committee on Oceanic Research (SCOR) Working Group #147. The production process was the same as that used for commercially available CRMs, and Atlantic Ocean seawater was also used, considering the potential usefulness of such CRMs to countries bordering the Atlantic Ocean. Between 2016 and 2021, 4186 CRM samples were provided to 48 institutions as part of a promotional campaign for the wider global use of CRM. Herein we discuss the challenges of preparing these CRMs and sustaining their distribution and accessibility.

Keywords Nutrients · SCOR-JAMSTEC CRM · Atlantic seawater · Essential ocean variable (EOV) · Comparability · Traceability

A. Murata (✉)
Physical and Chemical Oceanography Research Group, Global Ocean Observation Research Center, Japan Agency for Marine-Earth Science and Technology, Yokosuka, Japan
e-mail: murataa@jamstec.go.jp

E. M. S. Woodward
Plymouth Marine Laboratory, Plymouth, UK
e-mail: EMSW@pml.ac.uk

K. Bakker
Department of Ocean Systems, Royal Netherlands Institute for Sea Research, Den Burg, The Netherlands
e-mail: Karel.Bakker@nioz.nl

M. Wakita
Mutsu Institute for Oceanography, Japan Agency for Marine-Earth Science and Technology, Mutsu, Japan
e-mail: mwakita@jamstec.go.jp

© The Author(s) 2025   33
M. Aoyama et al. (eds.), *Chemical Reference Materials for Oceanography*, Springer Oceanography, https://doi.org/10.1007/978-981-96-2520-8_3

3.1 Introduction

Nutrients in seawater are useful indicators for assessing variations and alterations in the ocean's climate and ecosystems. As a result, nutrients are classified by the Global Ocean Observing System (GOOS) as Essential Ocean Variables (EOVs), which are parameters considered necessary to address scientific and social issues related to the ocean. Nutrient concentrations have been measured since the earliest scientific oceanographic observations in the late nineteenth century. Despite the large amount of data collected over time, there have been few studies of environmental changes in the ocean based on nutrient concentrations, and this is because historical nutrient concentration datasets have had limited internal and external consistency. Consequently, it is still difficult to compare data from one laboratory's analysis of a certain part of the ocean with data from another laboratory's analysis for samples taken at a geographically identical part of the ocean (Bindoff et al., 2007). To enhance comparability among nutrient measurements, efforts have been historically made to produce seawater Certified Reference Materials (CRMs) (Clancy & Willie, 2004), and several of these materials were commercially available. One of the commercially available CRMs is the Reference Material for Nutrients in Seawater (RMNS), which is manufactured by KANSO TECHNOS Co., Ltd., in Osaka, Japan (KANSO). Since its introduction in 2004, RMNS has been used by research institutions and laboratories globally as a certified reference to help achieve accurate nutrient measurements. Given the efficacy of RMNS for accurate measurements, it was decided to promote nutrient measurement based on RMNS globally, particularly in countries bordering the Atlantic Ocean, through a collaborative effort between a Scientific Committee on Oceanic Research (SCOR) International working group and the Japan Agency for Marine-Earth Science and Technology (JAMSTEC). The SCOR activity was conducted by SCOR Working Group #147: "Towards comparability of global oceanic nutrient data (COMPONUT)." For this purpose, we developed three specific RMNS based on Atlantic Ocean seawater, and between 2016 and 2021 availability period, we distributed a total of five lots of RMNS linked to the SCOR working group, including those made from Atlantic seawater. We included Atlantic seawater because differences in nutrient characteristics (ratios, concentrations, etc.) between the Pacific and the Atlantic may be one of the reasons why RMNS produced from Pacific seawater has not been used, especially in Atlantic countries.

In this chapter, we discuss our activity conducted under an international framework for the collection of raw seawater and the subsequent handling and preservation processes involved in producing and distributing the processed CRMs. In the Appendix, we also present some techniques relevant to sterilization, which is essential for producing high-concentration nutrient CRMs. Hereafter, the CRMs produced and distributed through this activity will be referred to as the SCOR-JAMSTEC CRMs.

3.2   Collection of Raw Seawater

For the SCOR-JAMSTEC CRMs, we created five CRM lots, three of which were derived from North Atlantic Ocean seawater and two from Pacific seawater (Table 3.1). Lot CG is Pacific Ocean seawater with a moderate concentration of nutrients. Seawater with a similar nutrient concentration was obtained from a facility in Yaizu City, Shizuoka Prefecture, Japan, which routinely pumps seawater from a depth of ~397 m in Suruga Bay for commercial purposes. Under the right conditions, we could easily produce a large amount of lot CG. In contrast, for lot CF, which has high nutrient concentrations for Pacific water, and the highest concentrations in the world ocean (Table 3.1), the source seawater is only present at a depth of around 1,200 m in the subarctic gyre of the western North Pacific. Therefore, we had no alternative but to collect seawater using a vessel with deep-water sampling capability. In October 2015, we collected approximately 1,000 L of seawater from a depth of 1,188 m at 45.8°N, 166.5°E, for the SCOR-JAMSTEC CRM, using R/V *Mirai* of JAMSTEC. Onboard R/V *Mirai*, we conducted a quick-fix sterilization to prevent any decreases in nutrient concentrations prior to the full-scale sterilization conducted at the KANSO facility. Two methods can be used for a quick-fix onboard sterilization: ultraviolet irradiation and heat treatment at around 80 °C. Because the heat treatment method is cumbersome, as described in the Appendix to this chapter (Sect. 3.5.2), we decided to use ultraviolet irradiation in the production of lot CF. Ultraviolet irradiation often causes the decomposition of organic matter, which inadvertently raises nutrient concentrations. Consequently, we used ultraviolet irradiation with an output power of approximately 100 W to inhibit decomposition. This is much less than the ~1 kW output power typically used to break down organic matter. The details of the ultraviolet irradiation method are in the Appendix to this chapter (Sect. 3.5.1).

For lot CE (Table 3.1), we asked the Royal Netherlands Institute for Sea Research (NIOZ) to collect the seawater and ship it to Japan. The NIOZ collected the surface seawater that originated from the Atlantic Ocean east of Sicily in the Eastern Mediterranean Gyre enhanced in salinity through evaporation to a practical salinity of just

Table 3.1 CRM lots provided by the SCOR-JAMSTEC CRM activity

Name	Lot	Seawater source	Bottling date	Certification date
Low in Atlantic	CE	NIOZ[a]	29 Oct. 2015	17 Nov. 2016
Medium in Atlantic	CI	PML[b]	27 Dec. 2017	27 Jul. 2018
Medium in Pacific	CG	JAMSTEC/Yaizu[c]	21 Apr. 2016	24 May 2017
High in Pacific	CF	JAMSTEC[d]	15 Dec. 2015	17 Nov. 2016
High in Atlantic	CH	PML	19 Jan. 2018	27 Jul. 2018

[a] Royal Netherlands Institute for Sea Research (NIOZ), Texel, The Netherlands
[b] Plymouth Marine Laboratory (PML), Plymouth, UK
[c] Yaizu is a city in Shizuoka prefecture, Japan. There is a facility there where seawater of ~397 m depth can be collected
[d] Japan Agency for Marine-Earth Science and Technology

over 38. This surface seawater being fully irradiated by sunlight over a long time was depleted for all nutrients. The seawater was transported in the light in a High-Density PolyEthylene (HDPE) plastic 1000-L container on deck of the vessel back to NIOZ in the Netherlands and kept irradiated using artificial fluorescent light from above the container for two years at 22 °C before being processed. After a few months of being transported by ship, the nutrient concentrations of seawater can be slightly reduced through biological uptake. Nevertheless, the decrease is minimal and even advantageous in terms of preventing biological production after delivery. Finally, the water was diluted with ultra-pure water with a resistivity of 18.2 MΩ (UPW) to a practical salinity of approximately 35 (to match the Atlantic Ocean's salinity) and filtered over Sartorius Sartobran P, type Medicaps pore size combined 0.45 and 0.2 μm before shipping in 20-L plastic containers to Japan. After reaching Japan, the seawater was completely sterilized at KANSO. Subtropical surface seawater such as this is usually collected for its low concentrations of nutrients. Consequently, dealing with seawater for shipping was relatively straightforward; samples of seawater were collected into 20-L plastic containers and could be transported to Japan as regular freight cargo (Fig. 3.1). We asked the Plymouth Marine Laboratory to collect seawater from the Atlantic Ocean and send it to Japan for the production of lots CI and CH, which have medium and high nutrient concentrations in Atlantic seawater, respectively (Table 3.1). In this instance, it was necessary to suspend biological activity to preserve the medium and high nutrient concentrations. The water samples were collected during the RRS *James Cook* (JC150) cruise using a conductivity-temperature-depth (CTD) Rosette system with OTE Niskin bottles fired at the required depths. The medium concentration samples (Lot CI) were taken at a depth of 600 m at the position: Lat.: 22°00'N, Long.: 54°00'W. The deep Atlantic water sample with the higher nutrient concentration (Lot CH) was sampled from a depth of 2500 m at the position Lat.: 22°19.8'N, Long.: 35°52.199'W.

Although it would have been ideal to use ultraviolet irradiation for sterilization and to store sterilized seawater in a large container, this was not feasible because of

Fig. 3.1 Atlantic Ocean surface seawater imported into Japan as regular cargo

Fig. 3.2 A drying oven was installed on RRS *James Cook*. On the left is Dr. Malcolm Woodward, a co-chair of SCOR Working Group #147, and on the right is Akihiko Murata, who is the first author of this chapter

a delay in the construction of such a setup that could be used on a ship. Instead, we sent a large, commercially available drying oven (Fig. 3.2) that had been modified for temporary sterilization to the RRS *James Cook*, from which the mid- and deep-Atlantic seawater was collected. The samples were pasteurized for 36 h at about 90 °C with the drying oven. As with the low concentration CRM (Lot CE) based on Atlantic seawater, 20-L plastic containers filled with sterilized seawater were shipped to KANSO in Japan at the end of the research cruise for final autoclaving to produce Lots CI and CH.

3.3 Distribution of SCOR-JAMSTEC CRMs

SCOR-JAMSTEC CRMs were produced on a commission basis by KANSO but sold and distributed by JAMSTEC. A website available with lots of CRMs was launched on JAMSTEC's homepage (Fig. 3.3), allowing customers to purchase and receive CRMs on demand, with the price fixed at JPY 6700 per bottle (~90 mL). Distribution began in January 2017 and ended in March 2022. Over this period, we provided 4,186 bottles to 48 institutions and laboratories. The SCOR-JAMSTEC

SCOR-JAMSTEC CRM

> Scientific background
> (lifted from the SCOR WG#147
> proposal)

> The SCOR WG#147

> IOCCP nutrients web page

> Previous inter-laboratory
> calibration exercises

> General Property

> Available Labs

> Price

> Order

> Storage/Handling

> Quality Control

> CRM production procedure

> References

> Chronological records

> Contact

> Other available CRMs

SCOR-JAMSTEC CRM

Japan Agency for Marine-Earth Science and Technology (JAMSTEC) is happy to start to provide nutrient CRM's with the SCOR-JAMSTEC logo, with a new cost structure making them more accessible for the global science and research community collaborating with the SCOR WG#147 "Towards comparability of global oceanic nutrient data (COMPONUT) ".

The SCOR-JAMSTEC CRMs are primarily intended to support the community of open ocean research aiming for improving the knowledge of trends and variability of nutrient fields. The SCOR International Working Group #147, and JAMSTEC, request that users of the SCOR-JAMSTEC CRMs submit their final nutrient data to an internationally recognized, public available, data repository. It is important that the associated metadata should include a description of how the SCOR-JAMSTEC CRM's were used.

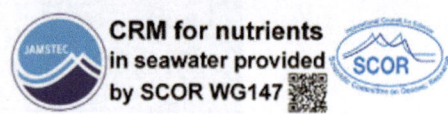

CRM for nutrients in seawater provided by SCOR WG147

This CRM is produced by KANSO on a commission basis and distributed by JAMSTEC based on a framework of SCOR WG147, COMPONUT.
Business contact: crm_nutrients@jamstec.go.jp
Scientific issues: Michio Aoyama. r706@ipc.fukushima-u.ac.jp
and Malcolm Woodward. m.woodward@pml.ac.uk

Fig. 3.3 A screen capture of the CRM distribution website that was accessible from JAMSTEC's homepage (http://www.jamstec.go.jp/scor/). The website is no longer active because sale of these CRMs has ended

CRMs were widely used in countries specifically surrounding the Atlantic Ocean (Fig. 3.4), and we conclude that the aim of the project which was the promotion of nutrient measurements based on CRMs, particularly in Atlantic Ocean countries, was achieved.

3.4 Future Challenges

We encouraged the measurement of nutrients using CRMs through the work of SCOR Working Group #147. This is because nutrient levels can be used to evaluate the marine environment, ocean productivity, and ocean climate change, as long as the nutrient concentrations are measured with a high level of consistency. We consider our CRMs to be widely accepted standards in the oceanic community. This will have little impact, however, if the distribution of the CRMs is not sustained. In this section, we examine the problems that must be addressed to move forward.

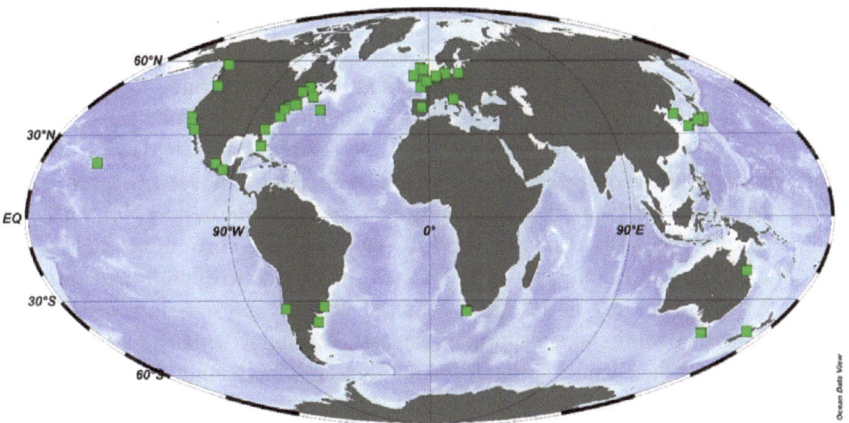

Fig. 3.4 Locations of the main institutions from which SCOR-JAMSTEC CRMs were ordered. This figure is reproduced from Fig. 3.7 in Murata et al. (2020) with permission from the Oceanography Society of Japan

We initiated the creation of nutrient CRMs in collaboration with a private firm so that the CRMs developed could be widely utilized in the oceanic community, similar to the reagents provided by chemical companies. This system ensures the ongoing availability of CRMs if they are marketable. In reality, there are no issues with the availability of CRMs with low or medium concentrations of nutrients because raw seawater for those CRMs can be easily sourced from the oceans near Japan. This implies that a private firm could acquire raw seawater at a low cost. Nevertheless, CRMs for high nutrient concentrations, such as CF (Table 3.1), must be extracted from deep waters (~1,200 m) in the western North Pacific, where the highest concentrations of nutrients are found. Additionally, it is necessary to use a specialized instrument like the CTD water sampling system to acquire deep seawater. Furthermore, the seawater collected must be sterilized onboard using a dependable method (see Sects. 3.5.1 and 3.5.2) as soon as possible after collection because the high nutrient concentrations must remain unaltered until the collected seawater is delivered to a facility where complete sterilization can be conducted. Thus, equipment typically found on a research vessel is required for the production of high-concentration CRMs, which incurs a substantial expense that a private company probably cannot cover. For the SCOR-JAMSTEC CRM, JAMSTEC supplied raw seawater and performed a quick-fix sterilization onboard R/V *Mirai*. We could do this because the distribution of SCOR-JAMSTEC CRM is solely for academic purposes. Generally, it is difficult to provide academic organizational resources to private businesses. It is therefore a challenge to maintain a steady supply of CRMs with high nutrient concentrations. The fact that users of high-concentration nutrient CRMs are mainly located in countries around the North Pacific may improve the cost-effectiveness. We are now experimenting with a simpler technique for producing high-concentration nutrient CRMs (see Sect. 3.5.3).

A separate issue exists for CRMs with low to moderate nutrient concentrations. Those who usually observe the Atlantic Ocean are wanting to use CRMs created from Atlantic seawater. This request is understandable considering the variations in seawater composition between the Pacific and Atlantic Oceans, which can influence the accuracy of the nutrient data produced. It is preferable to use CRMs with compositions similar to those of the samples to be analyzed, but it is not always practical to do so on a commercial basis; e.g., collecting seawater from the depths of the Atlantic Ocean and transporting it to Japan to be fully sterilized in a customized facility.

At present, there are no technical issues in the manufacture of high-concentration nutrient CRMs. The only issues are associated with an appropriate framework of acquiring and providing the appropriate seawater. Nutrient CRMs should be consistently available to the oceanic community into the foreseeable future if at all possible as highly accurate nutrient data are essential for assessing and diagnosing changes in the ocean. We hope that our initiative for the production and distribution of SCOR-JAMSTEC CRMs will be a useful reference point for future efforts.

Continuing nutrient CRM production is imperative because CRMs are the only accurate way of having continuing global inter-comparability between nutrient data sets over short- and long-term time scales, which will allow for studies of changes to global nutrient concentrations that will be affected by current and future climate change.

Competing Interests The authors have no conflicts of interest to declare that are relevant to the content of this chapter.

Appendix

Quick-Fix Sterilization for Lot CF

This section of the Appendix describes the process of sterilizing seawater to produce CRM lot CF, which had high nutrient concentrations of Pacific Ocean seawater, and the UV sterilization system used for quick-fix sterilization.

During the return voyage of R/V *Mirai* from the Arctic Ocean to Japan (cruise ID: MR15-03), we collected seawater from the subsurface layers in the western North Pacific (Table 3.2) and passed it through a UV sterilization system into a 1,000-L plastic bag (SHOWA PAXXS, Tokyo, Japan) that had been pre-sterilized on land with gamma irradiation at 15 kGy.

A flow diagram of the UV sterilization system is shown in Fig. 3.5. This system was designed by Prof. M. Aoyama (JAMSTEC/Fukushima University, affiliation at that time). Before use, components of the system, such as hoses and connectors, were pasteurized by placing them in a bag for sterilization and keeping them in a dry oven

Table 3.2 Sampling positions, dates, times, and depths

Cast	Latitude	Longitude	Time	Depth
	(degrees)	(degrees)	(UTC)	(dbar)
1	45.7670°N	166.4743°E	10/14/2015 18:54	1,200
2	45.7747°N	166.4704°E	10/14/2015 21:25	1,200
3	45.7990°N	166.4942°E	10/14/2015 23:45	1,200

Note The above information is from Aoyama et al. (2015)

at a temperature above 70 °C for at least 5 h. A two-layer cartridge filter (pore sizes 0.45 and 0.2 μm) was included in the setup to eliminate particles and bacteria. The collected seawater was circulated for a period of 24 h. Tables 3.3 and 3.4 show the nutrient concentrations before and after sterilization, respectively.

Table 3.3 Nutrient concentrations ($\mu mol\ kg^{-1}$) before UV sterilization (Aoyama et al., 2015)

Cast	Bottle	Nitrate	Nitrite	Silicate	Phosphate	Ammonium
1	1	43.61	0.01	159.56	3.125	0.06
1	10	43.63	0.01	159.48	3.124	0.06
1	19	43.68	0.01	160.11	3.128	0.06
1	28	43.58	0.01	159.78	3.126	0.05
2	4	43.55	0.01	160.01	3.120	0.03
2	13	43.49	0.01	159.87	3.119	0.03
2	22	43.50	0.01	159.99	3.119	0.02
2	31	43.48	0.01	160.00	3.120	0.02
3	7	43.63	0.01	159.95	3.132	0.01
3	16	43.59	0.01	159.46	3.126	0.02
3	25	43.62	0.01	159.92	3.130	0.02
3	34	43.66	0.01	160.19	3.135	0.03

The seawater taken from each Niskin bottle (Table 3.2) was collected into a 1000-L plastic bag and sterilized by UV exposure for 24 h. The seawater was then sampled into 5 plastic tubes for analysis

Table 3.4 Nutrient concentrations ($\mu mol\ kg^{-1}$) after UV sterilization (Aoyama et al., 2015)

Nitrate	Nitrite	Silicate	Phosphate	Ammonium
43.26	0.18	159.48	3.116	0.01
43.26	0.25	159.81	3.119	0.01
43.21	0.27	159.88	3.121	0.02
43.22	0.26	159.88	3.122	0.01
43.20	0.27	159.85	3.122	0.02

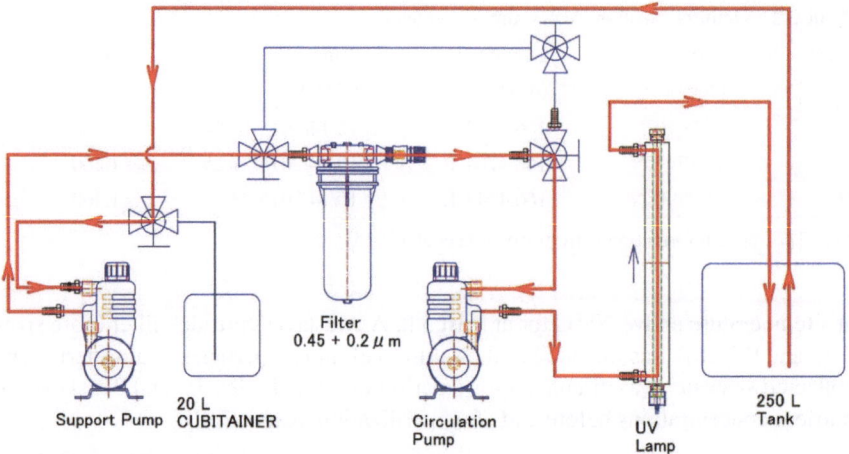

Fig. 3.5 Flow diagram of the UV sterilization system. For the Lot CF, 1,000-L plastic bag has been used instead of 250-L tank

Heat Treatment Sterilization Method

We have often performed quick-fix sterilization for high-concentration nutrient CRMs using the equipment aboard R/V *Mirai*. This section of the Appendix describes the actual sterilization process.

The seawater is transferred from the CTD Niskin bottles to 20-L plastic containers using drawing tubes. We use plastic containers that were rinsed with HCl, H_2O_2, for example, or sterilized by gamma rays. The drawing tubes are sterilized by low-temperature pasteurization. The Niskin bottles are washed with non-potable water from the ship before sample collection to prevent contamination of sampled seawater

Fig. 3.6 Seawater in the sauna of R/V *Mirai*. Seawater in a plastic container is placed in a colored crate. The piled crates are supported by poles to prevent shifting because of the ship's movement

by seawater adhering to the surface of the Niskin bottles. The volume of seawater collected by the CTD is usually greater than 200 L, so 10 plastic containers can be easily filled.

To halt biological activity in seawater collected from a cold, dark environment, the seawater collected in the plastic containers is rapidly heated by submerging the containers in a water bath at >60 °C for 2 h. Because of the large number of plastic containers, we had to improvise and use a bathtub in a bathroom onboard R/V *Mirai*. After the treatment in the water bath, the containers of seawater are placed in a sauna onboard R/V *Mirai* (Fig. 3.6). Seawater is kept in the sauna at >80 °C for 5 h. However, there are usually vertical temperature variations in the sauna, with higher temperatures at higher positions and lower temperatures at lower positions. The plastic containers that reach temperatures >80 °C for 5 h are removed from the sauna, and those in the lower-temperature tier are moved to the tier that reached 80 °C. After heat treatment, the seawater is cooled to room temperature.

Preservation Experiment for High Concentrations of Nutrients in Seawater: Preliminary Results

This section of the Appendix presents the results of a preservation experiment for seawater with high nutrient concentrations, i.e., ~45 μmol kg^{-1}, 3 μmol kg^{-1}, and 160 μmol kg^{-1} for nitrate, phosphate, and silicate, respectively. These concentrations are the highest known in the entire ocean and are only observed around 1,500 m depth near time-series stations KNOT (155°E, 44°N) and K2 (160°E, 47°N) in the northwestern North Pacific Ocean. Therefore, to produce a high-concentration lot of nutrient CRM, it is necessary to collect a large volume of water (500 L) using a water sampler and to minimize concentration changes due to biological activity. R/V *Mirai* has a sauna that has been used for pasteurization, and a small ultraviolet (UV) sterilizer. However, the sauna pasteurization requires heavy work onboard and the facilities soon will no longer be available because R/V *Mirai* will be decommissioned in 2025. Also, the UV sterilizer does not provide sufficient sterilization for 500 L. For this reason, we conducted an experiment to see if there was a simpler method for preservation.

Sampling and Handling of Seawater

The plastic containers (10 L) were cleaned on land before the research cruise. They were washed with ~3% H_2O_2, rinsed with tap- and Milli-Q water, and pasteurized at 80 °C for 5 h. Water sampling was conducted during leg 2 of cruise MR23-05 of R/V *Mirai*. The seawater used for this experiment was collected from the deep layers (~1,200 m) at station K2 on 7 August 2023. Seawater sterilized with a simple made-to-order UV equipment (Shiko Giken Co., Ltd., Awaji, Japan) was collected from each of two Niskin bottles (12 L) into each of two plastic containers. Seawater without sterilization was also collected as a control for the experiment. Experimental storage conditions were as follows: (i) storage at room temperature with UV sterilization; (ii) storage at refrigerated temperature (~4 °C) with UV sterilization; (iii) storage at room temperature without UV sterilization; and (iv) storage at refrigerated temperature without UV sterilization. All samples used in the preservation experiments were transported after the cruise under refrigeration (~4 h, ~4 °C) from Shimizu port to the Yokosuka headquarters of JAMSTEC. The room temperature samples were stored in an air-conditioned laboratory until measurements, and the refrigerated samples were stored in a refrigerator at ~4 °C until measurements.

Nutrient Measurements of the Stored Samples

Nutrient concentrations were measured three times aboard R/V *Mirai* and once on land. Temporal variations of nutrient concentrations in stored samples are shown in Fig. 3.7, with the day of the first measurement onboard R/V *Mirai* (8 August) as day 0. Nutrients were also measured onboard on 11 and 14 August (day 3 and 6, respectively), and in a laboratory on land after the cruise on 25 August (day 17). Nutrient concentrations were measured with a QuAAtro 39-J system (BLTEC K. K., Tokyo, Japan) both onboard and on land following Hydes et al. (2010) with minor changes. The results onboard are presented as the average of two measurements, and the results on land are the average of five measurements.

As shown in Fig. 3.7, refrigerated samples with or without UV sterilization can withstand storage for more than two weeks. The results are promising for the production of nutrient CRMs of high concentration because the operations onboard a ship can be simplified (see Sects. 3.5.1 and 3.5.2). However, the effectiveness of this method has only been confirmed for 10-L plastic containers; for CRM batches > 500-L, 50 plastic 10-L containers would be needed to see if similar results could be obtained. Space for storing such a large number of plastic containers onboard would also be required. In the case of R/V *Mirai*, there is a large refrigerator onboard, and if such a facility were available, it would be feasible to prepare larger volumes of CRMs. Although normal research vessels do not usually have such a large refrigerator available for scientific purposes, it should be possible to provide refrigerated containers (which could be rented) for the same purpose.

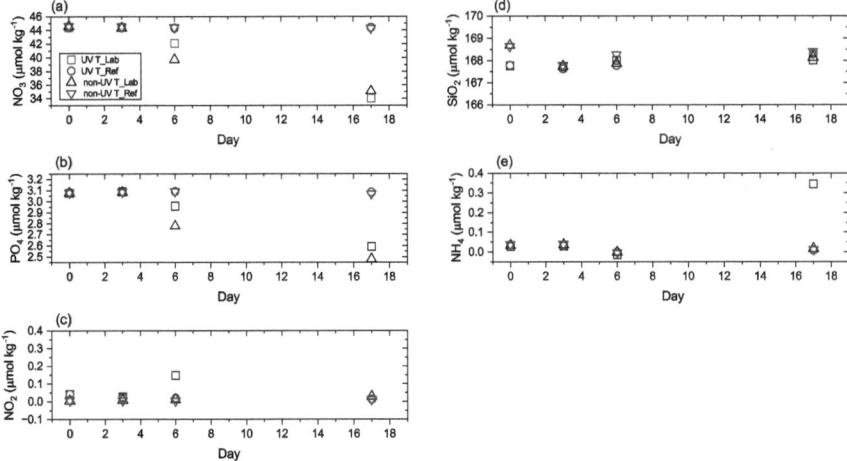

Fig. 3.7 Temporal changes in nutrient concentrations after collecting seawater onboard R/V *Mirai*. **a** nitrate, **b** phosphate, **c** nitrite, **d** silicate, and **e** ammonium. Squares and circles indicate results of storage at room temperature and refrigerated temperature (~4 °C), respectively, both with UV sterilization, while triangles and inverted triangles indicate results of storage at room temperature and refrigerated temperature, respectively, both without UV sterilization

References

Aoyama, M., Nishino, S., Arii, I., Sone, Y., Hayashi, E., Yokogawa, S., & Enoki, M. (2015). Nutrients In: *R/V mirai cruise report MR15–03*. https://doi.org/10.17596/0002842.

Bindoff, N. L., Willebrand, J., Artale, V., Cazenave, A., Gregory, J., Gulev, S., Hanawa, K., Le Quéré, C., Levitus, S., Nojiri, Y., Shum, C. K., Talley, L. D., & Unnikrishnan, A. (2007). Observations: Oceanic climate change and sea level. In S. Solomon, D. Qin, M. Manning, Z. Chen, M. Marquis, K. B. Averyt, M. Tignor, & H. L. Miller (Eds.), *Climate change 2007: The physical science basis. contribution of working group i to the fourth assessment report of the intergovernmental panel on climate change* (pp. 385–433). Cambridge University Press.

Clancy, V., & Willie, S. (2004). Preparation and certification of a reference material for the determination of nutrients in seawater. *Analytical Bioanalytical Chemistry, 378*, 1239–1242.

Hydes, D. J., Aoyama, M., Aminot, A., Bakker, K., Becker, S., Coverly, S., Daniel, A., Dickson, A. G., Grosso, O., Kerouel, R., van Ooijen, J., Sato, K., Tanhua, T., Woodward, E. M. S., & Zhang, J.-Z. (2010). Determination of dissolved nutrients (N, P, Si) in seawater with high precision and inter-comparability using gas-segmented continuous flow analysers. In: *GO-SHIP repeat hydrography manual: A collection of expert reports and guidelines*. IOCCP Report No. 14, ICPO Publication Series No 134.

Murata, A., Aoyama, M., Cheong, C., Miura, T., Fujii, T., Mituda, H., Kitao, T., Sasano, D., Nakano, T., Nagai, N., Kodama, T., Kasai, H., Kiyomoto, Y., Setou, T., Ono, T., Yokogawa, S., Arii, Y., Sone, T., Ishikawa, Y., & Wakita, M. (2020). Current situation and future perspective for environmental standards of seawater: Commencing with certified reference materials (CRMs) for nutrients. *Oceanography in Japan, 29*, 153–187. (in Japanese with English abstract).

Chapter 4
Maintaining Inter-Comparability of Nutrient Measurements

Karel Bakker, Sharyn Ossebaar, Rob Middag, and Matthew P. Humphreys

Abstract Measurements of marine nutrient concentrations are necessary to investigate global biogeochemical cycles, including in the context of climate change, as nutrients are closely coupled to the carbon cycle. To study these phenomena on a global scale, measurements must be internationally comparable between laboratories and calibrated to a metrologically traceable reference standard. In this chapter, we describe how well-trained chemical analysts following Good Laboratory Practice can meet these requirements. Inter-laboratory comparability is not possible without first achieving consistent and reproducible measurements within a single laboratory. First, the measurement reproducibility must be monitored and kept constant within each session of analysis. Second, this criterion should be maintained from one analysis session to the next, for example by using an internal laboratory reference standard and/or conducting duplicate measurements across different sessions. Third, an internationally accepted Certified Reference Material (CRM) should be used to assess the comparability of the data with other laboratories worldwide. Regular participation in laboratory inter-comparison exercises can also reveal systematic bias, thus helping to improve general data quality. Improving analytical quality involves many aspects, such as calibrating volumetric flasks and pipettes gravimetrically to reduce systematic errors, but also a number of seemingly small procedural improvements that can noticeably improve measurement precision and consistency. Each step in the

K. Bakker (✉) · S. Ossebaar · R. Middag · M. P. Humphreys
Department of Ocean Systems, Royal Netherlands Institute for Sea Research, Den Burg, The Netherlands
e-mail: Karel.Bakker@nioz.nl

S. Ossebaar
e-mail: sharyn.ossebaar@nioz.nl

R. Middag
e-mail: rob.middag@nioz.nl

M. P. Humphreys
e-mail: matthew.humphreys@nioz.nl

R. Middag
Center for Isotope Research-Oceans, University of Groningen, Groningen, The Netherlands

© The Author(s) 2025
M. Aoyama et al. (eds.), *Chemical Reference Materials for Oceanography*, Springer
Oceanography, https://doi.org/10.1007/978-981-96-2520-8_4

overall procedure from sampling, calibration and analysis through to data reporting contributes to the precision and accuracy of the results.

Keywords CFA · Precision · Reproducibility · Consistency · Bias · International comparability · CRM · Normalisation

4.1 Introduction

Dissolved inorganic nutrients are central to the functioning of the global marine ecosystem and the carbon cycle. The main marine "macronutrients" (those with interior ocean concentrations typically in the μM range include nitrate (NO_3^-), nitrite (NO_2^-), ammonium (NH_4^+), phosphate (PO_4^-) and silicate (Si)). Near the ocean surface, these nutrients are often the limiting factor controlling the growth of phytoplankton at the base of the marine food web (Moore et al., 2013) although micronutrients (trace metals) play an important role too (Anderson, 2020). Deeper down in the water column, the chemical signature of nutrient concentrations can be used to infer biogeochemical processes such as organic matter remineralisation (e.g. Humphreys et al., 2022) and to distinguish water masses in combination with physical parameters (Tomczak, 1981). Accurate measurements of nutrient concentrations are therefore essential to understand global biogeochemical cycles and to observe changes in the marine ecosystem, for example in response to climate change.

However, it is challenging to obtain nutrient measurements sufficiently accurate to investigate these processes as highlighted in the 2007 Intergovernmental Panel on Climate Change report: "Uncertainties in deep ocean nutrient observations may be responsible for the lack of coherence in the nutrient changes. Sources of inaccuracy include the limited number of observations and the lack of compatibility between measurements from different laboratories at different times" (Bindoff et al., 2007). It is estimated that differences in nutrient measurements between laboratories and expeditions lead to discrepancies of 1–3%, (Aoyama et al., 2018) and the goal set by the Global Ocean Ship-based Hydrographic Investigations Program (GO-SHIP) for the near future is to be comparable within 1% through improvements of analytical and calibration procedures, including the growing use of international certified reference material (CRM) (Hydes et al., 2010; Becker et al., 2020).

Consistency of existing nutrient measurements in the deep open ocean can also be assessed and improved by using cross-over analysis. For example, data compilation projects such as CARINA (Carbon Dioxed in the Atlantic Ocean; Tanhua et al., 2010) and its successor GLODAP (Global Ocean Data Analysis Project; Lauvset et al., 2022) compare deep ocean nutrient measurements from different oceanographic expeditions at close-by sampling stations and apply adjustments to ensure consistency when significant offsets are found. However, while this approach can help to maximise the scientific value of existing datasets, it should not be considered as the primary route to consistency as the adjustment process is somewhat subjective and could lead to real trends in deep ocean nutrient concentrations being hidden. It should also be noted that these data compilations are primarily focussed on marine carbonate system measurements, with the nutrients included as useful auxiliary variables. Instead, our focus moving forwards should be on producing accurate nutrient measurements that do not require adjustments.

During the seventies and until the early nineties, the only way of knowing the quality of a laboratory's data was by joining a comparison exercise of the International Council for the Exploration of the Sea (ICES) and measuring unknown samples that would be compared against results from many participants. The World Ocean Circulation Experiment (WOCE) program from 1990 set a data quality goal of an overall reproducibility of 1–3% with a precision of better than 0.4% for nutrients depending on the species analysed. At the time, accuracy was not mentioned as CRMs were not available (Joyce et al., 1994). Unfortunately, the WOCE goal was not always achieved, but nevertheless, it was a good starting point. Nowadays, most oceanographic laboratories are capable of analysing nutrients with data quality close to, or within those goals set by WOCE in 1991.

In this chapter, we discuss the steps required for producing precise, accurate and internationally comparable nutrient measurements. We begin with precision, illustrating various points of methodology that lead to precise, repeatable measurements within individual sessions of analysis. We then consider how consistent measurements can be produced across separate analysis runs within the same laboratory, illustrating the approach with a case study of nutrient measurements from several research expeditions to the Weddell Sea. Finally, we discuss how inter-laboratory comparison exercises (e.g. Aoyama et al., 2010, 2013, 2016, 2018) and the recent advent of widely available and reliable nutrient CRMs can deliver the internationally consistent measurements demanded by marine biogeochemical research at the present time of profound environmental change.

4.2 Internal Precision Using a Tracking Standard

Before 2000, overall precision reported for nutrients was typically referred to as "% of *full-scale* values," where full scale denotes the highest point of the used calibration, e.g. 0.4% of the full scale within a run and 0.6% in-between analytical runs. However, this precision is not representative for mid-to-lower concentrations in the calibration range. Nowadays, precision is usually calculated for at least three concentrations from at least five replicate samples as the Relative Standard Deviation (RSD) in percentage or Coefficient of Variation at that concentration (Taylor, 1987).

Prior to the introduction of Certified Reference Materials (CRMs) in 2012, some analysts used an internal Quality Check by monitoring a deep-water sample (e.g. 20 L of seawater from deeper than 1500 m), collected at the start of an expedition as a reference sample for tracing any drift during an expedition. However, the deep-water samples monitored during different expeditions often had stability issues with values drifting over time. Possible causes are biological activity that changes the nutrient concentration or evaporation of the deep-water sample during the expedition (Hydes

Fig. 4.1 NIOZ Silicate tracking standard (NIOZ standard) and a reference material for nutrients in seawater (RMNS batch AZ) as tracking standards to monitor analytical performance in a series of analytical runs

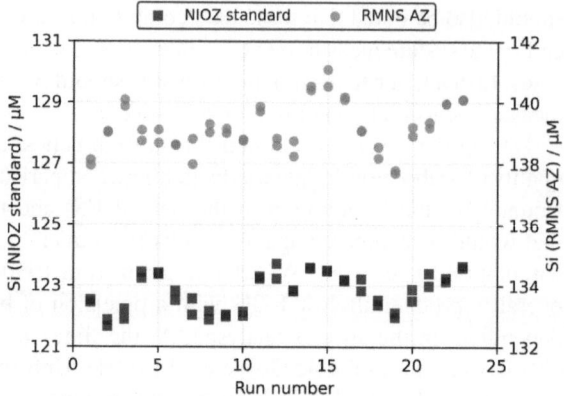

et al., 2010). However, it is possible to produce stable internal tracking standards as described below and this can help improve quality controls.

The idea is that by monitoring the concentration of an internal tracking standard, the daily measured values can be plotted in a so-called Shewhart chart (Fig. 4.1) or similar. Outliers in this dataset can be detected using upper and lower limits of confidence and possible problems solved before an expedition has been completed. Moreover, by analysing such a stable tracking standard containing the major nutrients of interest in every analytical run, it is possible to normalise data based on the average value of this standard to produce a dataset that is more internally consistent.

As an example of this procedure, the nutrient laboratory from the Royal Netherlands Institute for Sea Research (NIOZ) started using an independent internal lab reference as a tracking standard in 1996 to normalise data and improve internal consistency. The idea originated in 1992 on a Weddell Sea expedition with a week-long algae growth experiment. Specifically, an in-situ bio-assay experiment under iron limitation was conducted and samples needed to be analysed for NO_3^- every day to monitor nutrient drawdown. Analysis of such samples that are expected to have similar concentrations are ideally measured in a single run to avoid differences in accuracy or precision between runs. However, this was not possible as a microbial growth would go on during storage in a fridge as the temperature would be similar or even warmer than the environmental conditions. Therefore, the samples were measured every day in combination with a lab tracking standard which was independent from the calibration, to see if it was conceivable to back-correct or normalise to this tracking standard and in this way eliminate the influence of any day-to-day variability as much as possible. The use of this tracking standard at a concentration close to the concentration of the experimental samples seemed to be successful (Table 4.1) as the corrected concentration either remained similar between days or decreased, whereas the raw data also showed an increase between days (Fig. 4.2), which is not likely unless there was contamination. Overall, one cannot conclude the absolute values were better, but consistency in-between days seemingly improved by using this normalisation.

Table 4.1 Nitrate concentrations during a shipboard incubation experiment in year 1992 as published some years later (Van Leeuwe et al., 1997). The measured concentrations for the day-to-day sample analysis are given, followed by the concentration of the tracking standard in the next column, measured in the same analytical run. Based on the average value of the tracking standard, a correction factor was calculated and in bold the corrected data as used in Fig. 4.2 in light blue is reported

Day no.	Measured data (NO$_3$ µM/L)	Tracking standard (NO$_3$ µM/L)	Correction factor	Corrected data (NO$_3$ µM/L)
0	28.8	27.4	0.992	28.6
1	28.5	27.2	0.999	28.5
2	28.5	27.2	0.999	28.5
3	28.4	27.1	1.003	28.5
5	27.9	26.8	1.014	28.3
6	28.0	27.2	0.999	28.0
7	27.7	27.4	0.992	27.5
Average		27.19		
St. Dev		0.20		
C.V. %		0.75		

Note St. Dev. Indicates standard deviation; C.V. indicates coefficient of variation

Fig. 4.2 The nitrate concentrations during the shipboard incubation experiment in year 1992 experiment with uncorrected raw measurements in dark blue and the corrected values in light blue. This was part of the experiment II in a suite of six experiments II through VI (Van Leeuwe et al., 1997)

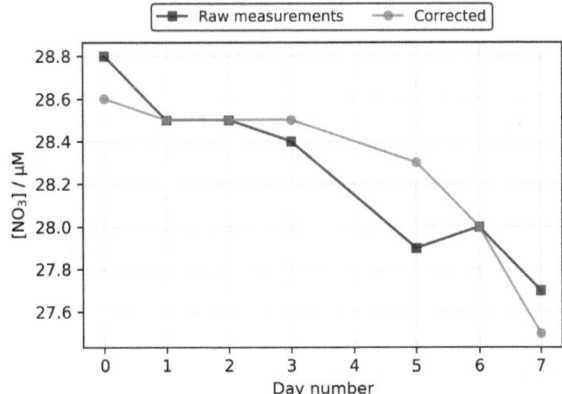

The same approach of normalisation described above was later used along an oceanographic transect on the R.V. Polarstern (Expedition ANT-XIII/4) in the Weddell Sea in 1996.

4.3   Case Study of Nutrient Measurements in the Weddell Sea

Silicate concentrations in the Atlantic Ocean typically range from zero near the surface to around 40 μM or higher in Antarctic Bottom Water. In a system with such a large gradient, a reproducibility between stations of 1% is sufficient to discern the overall spatial distribution and differentiate between different water masses, mixing and biogeochemical processes. However, in the Weddell Sea, macro-nutrient concentrations vary by less than 0.8% over a depth range of 3000 m. Silicate concentrations vary by only ± 1 μM at a concentration of 127 μM (i.e. <0.8%), over a depth range from 1000 to 4000 m (Fig. 4.3). As such, the Weddell Sea is a challenge for analysts, with the small variation in nutrients below the surface layer requiring the highest possible reproducibility and accuracy to distinguish trends over time (Hoppema et al., 2015). From our experience, to produce a reliable transect plot over this depth range, the precision needs to be better than 0.4% in each single run, allowing a reproducibility better than the aim of 0.6% between multiple analytical runs. However, such between-run reproducibility is only achievable by normalising to an independent standard containing the major nutrients PO_4, NO_3 and Si.

An independent lab tracking standard was produced for the 1996 Weddell Sea expedition ANT-XIII/4 aboard R.V. Polarstern. This standard contained a known mixture of PO_4, Si and NO_3 at high concentrations and was sterilised using mercuric (II) chloride ($HgCl_2$). For every analytical run, this high-concentration stock mixture was diluted in Low Nutrient Seawater (LNSW) to the same concentration (near the local deep ocean concentrations) and measured in triplicate. After completion of the section through the Weddell Sea, the average value of the tracking standard was calculated (after removing outliers that deviated more than 2 standard deviations from the average) and the deviation from this average was calculated as a factor for each run, which was used to normalise the data from the different analytical runs. To verify that this indeed reduced run-to-run variation, in every run a near bottom duplicate sample from a previous station was re-analysed in consecutive runs resulting in a series of duplicate samples from every station (except the last station of the expedition) that were analysed in separate runs. Reproducibility between these duplicates was compared before and after the correction factor was applied. As an example, for the Si data from this expedition the root-mean-square deviation of all duplicates decreased from 0.60 to 0.35%. This indicates that the correction based on the tracking standard improved the consistency of the data by decreasing the effect of run-to-run variability. This was also visible in the contour plots of the section data by showing smoother distributions and fewer concentration anomalies for the 1996 expedition (Fig. 4.3). To maintain longer term consistency within a lab, the same tracking standard should ideally be used indefinitely (Hoppema et al., 2015): however, this is not necessarily feasible, as in practice, such a standard would run out eventually. Instead, any new tracking standard should be carefully calibrated against a previous one by measuring them together over multiple analytical runs.

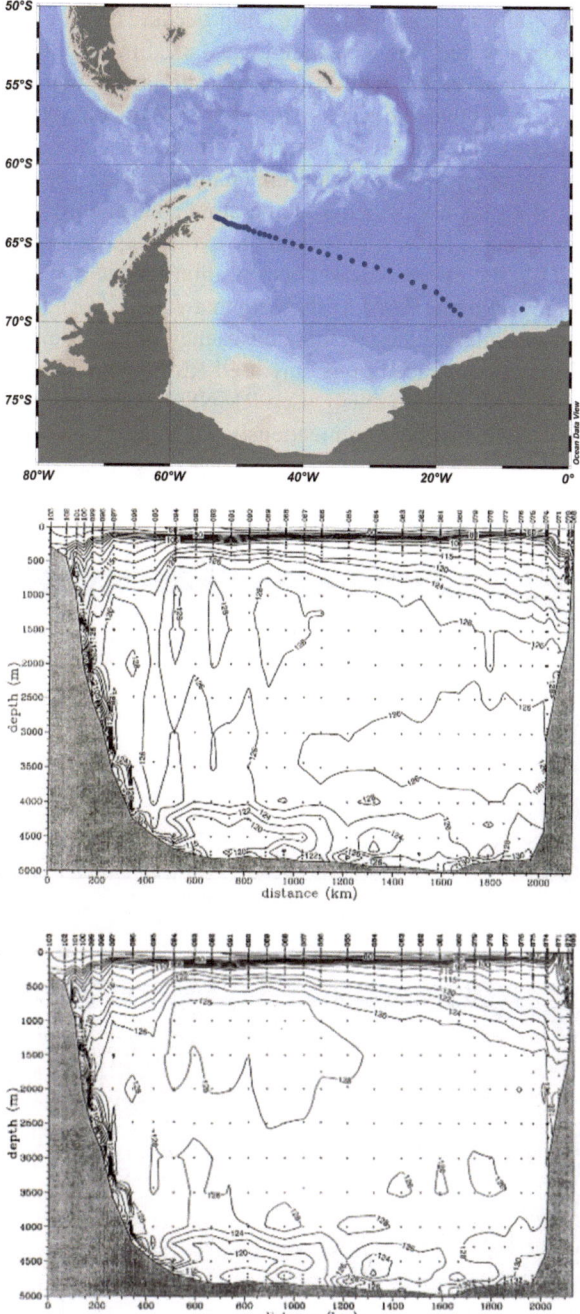

Fig. 4.3 Silicate concentrations (μM/L) of a Weddell Sea hydrographic section from 1996, expedition ANT-XIII/4 with R.V. Polarstern. Upper image shows the uncorrected silicate data and the lower image shows the corrected silicate data using the external tracking standard. Plots made by A. Wisotzki in 1996 on request during the expedition

Given the successful application in 1996, this procedure became common practice in the NIOZ laboratory and was for example also used in 2008 for a repeat occupation of the same transect, but with Reference Material for Nutrients in Seawater (RMNS) batch AZ rather than the previously used NIOZ tracking standard (Fig. 4.4). Additionally, within-analytical run consistency of uncorrected data has improved in our lab over the years as well (compare upper images from Fig. 4.3 uncorrected silicate with Fig. 4.4), mainly by better temperature control during sample analysis and covering open sample tubes with parafilm to avoid evaporation. For the Weddell Sea section of 2008, the differences in concentrations of the duplicate samples are also sorted from low to high (Fig. 4.5) showing that when correctly used, an RMNS or CRM can indeed improve the internal consistency of the data. To also improve the comparability between expeditions, any used RMNS should be the same, or at least be calibrated against a previously used RMNS. Besides the use of an RMNS, inter-comparisons have also been tremendously valuable to achieve (international) comparability.

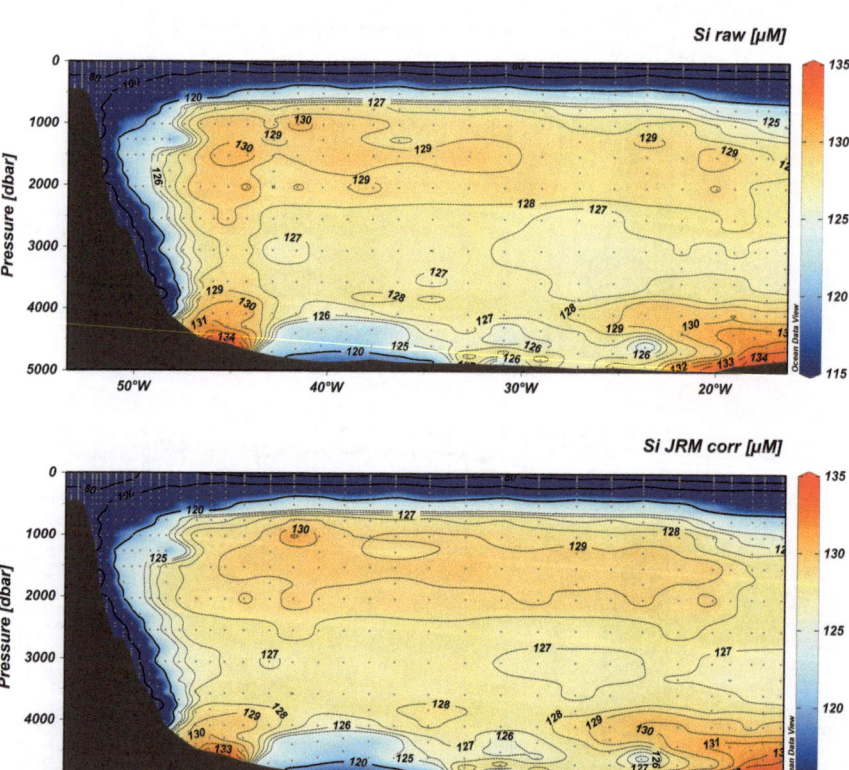

Fig. 4.4 Silicate concentrations (μM/L) of a Weddell Sea Hydrographic Section from 2008, expedition ANT-XXIV/3 with R.V. Polarstern. Upper image shows the uncorrected silicate data and the lower image shows the corrected silicate data using the external RMNS (Batch AZ)

Fig. 4.5 Concentration differences between analysis of duplicate samples in successive runs, sorted from low too high for both uncorrected raw data (dark blue) and data corrected using the external tracking standard RMNS, batch AZ (light blue)

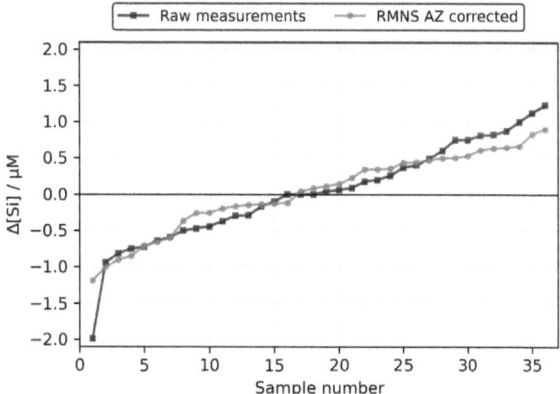

4.4   Inter-Comparison Exercises and Certified Reference Material

The Meteorological Research Institute of Japan (MRI) organised inter-comparisons (ICs) in time periods of 3 years (Aoyama et al., 2010, 2013). During these ICs, the participants' data on unknown, homogenous and sterile seawater samples were published together with the names and addresses of the participating labs. The participation of marine laboratories in ICs, like Quality Assurance of Information for Marine Environmental Monitoring in Europe and the IOCCP-JAMSTEC organised in Japan that have been going on for years, have undoubtedly improved the comparability of data. Notably, ICs provide participants the opportunity to compare their data with laboratories and detect possible bias from the mean value. Moreover, the exchange of information on methods and calibration procedures can lead to a generally accepted methodology that can also improve the comparability of marine nutrient concentration data. For example, NIOZ in combination with Plymouth Marine Laboratory (PML) organised a workshop in 2012 focussing on problems encountered with PO_4 analysis where a selection of 10 laboratories worked together to establish best possible practices for analysing PO_4 (Aoyama et al., 2015). Subsequently, the following IOCCP-JAMSTEC inter-comparisons (Aoyama et al., 2016, 2018) showed better consistency of the PO_4 data produced than obtained in previous ICs (see Table 4.2), underlining both the success and importance of such workshops.

Overall, due to the open mindset from the nutrient community participants in the ICs, a lot of progress has been made. Participants have been willing to improve their techniques and thereby over the years resulting in less bias for PO_4. Silicate RSD has remained the same which has been somewhat surprising as Si is usually analysed without many issues or difficulties. There are several potential reasons for the lower accuracy for silicate. Silicate standards are mostly made by weighing in sodium hexafluorosilicate (Na_2SiF_6) with a purity better than 97% and these salts are hard to completely dissolve, potentially leading to uncertainty. Additionally, different choices of source material between labs, such as quartz material or alkaline NIST

Table 4.2 Comparing 2008 and 2018 results for the IC study with consensus S.D. expressed as RSD showing a clear improvement in PO_4 analysis after the joint workshop

	IC 2008 (n = 31)			IC 2018 (n = 54)		
	Mean (µM/kg)	S.D (µM/kg)	RSD (%)	Mean (µM/kg)	S.D (µM/kg)	RSD (%)
NO_3	21	0.33	1.6	23.7	0.38	1.6
PO_4	1.6	0.04	2.5	1.7	0.03	1.8
Si	59	0.84	1.4	56	0.78	1.4

Note S.D. indicates standard deviation; RSD indicates relative S.D

standards that first need to be neutralised before being used in LNSW, could introduce further biases. Recently, ongoing efforts have been made by the international community to prepare an internationally accepted concentrated silicate stock solution which would be available for the nutrient community to overcome this lower accuracy issue. Although the RSD for NO_3 also remained the same, more investigation within the nutrient community is needed to attempt to improve this value.

The unknown, homogeneous and sterile seawater samples used over the years in the MRI and IOCCP-JAMSTEC ICs were in fact different CRMs produced. The production and use of internationally accepted CRMs for marine nutrients and the reporting of the analysed CRMs with its assigned value as metadata together with the nutrient dataset makes it now possible to compare nutrient data worldwide. CRMs are scarce and expensive and hence at NIOZ we still use an internal standard daily, in addition to fewer measurements of the CRM. It should be noted that even though CRMs are indeed the key to compare data, the internal precision of nutrient data produced by a laboratory should be optimal before CRMs are used for comparing. Using an internationally accepted CRM is the best tool to achieve international comparability where it should be noted that this is most useful after underlying causes for bias between groups have been identified and addressed as described in the next section.

4.5 Achieving High Precision and Some Practical Considerations

The overall procedure from the preparation of standards to sampling, analysis and data quality control, determines the precision of the final measurements. Generally, the aim is to produce comparable, precise and accurate data without bias. We strongly recommend using the methods published in the GO-SHIP Repeat Hydrography Nutrient Manual (Becker et al., 2020) as the basic approach, with some important notes being mentioned here of processes that in our experience positively contributed to improved precision and accuracy in the NIOZ nutrient lab, while acknowledging that this list is not exhaustive and that laboratories can achieve good precision in different ways.

When preparing for an analytical run, one should consider the calibration range which should be based on the expected values of nutrients in the research area (e.g. previously measured concentrations can be found in Geochemical Ocean Sections or eWOCE databases). Theoretically, the best approach is to analyse samples in-between 20–70% of the used calibration range, but this is of course not always possible when encountering nutrient-depleted waters. Diluting stock standards with pipettes and volumetric flasks may introduce systematic errors. Therefore, we recommend that when preparing the calibration standards, to use the same set of pipettes and individual numbered plastic pre-calibrated volumetric flasks. Systematic errors for calibrated flasks are small, around ±0.05% (accuracy from calibration on a balance), and precision of a micropipette is about 0.2% but sometimes with only 0.8% accuracy. Therefore, consistently using the same combination of volumetric flasks and pipettes during an expedition results in an uncertainty of approximately 0.2% for the calibration points whereas any bias stays the same for all, where such bias could be corrected by the use of a CRM straight away for use without dilution.

Sampling seawater in the open ocean is usually done using a Conductivity-Temperature-Depth (CTD) profiler mounted to a rosette frame equipped with 24 or 36 so-called Niskin samplers, but of course many other sampling techniques and sampler types exist (e.g. Middag et al., 2023). Niskin bottles or similar samplers are lowered down into the sea and during the upcast the samplers are closed at a chosen depth. At this chosen depth, it is recommended that the winch is stopped to make sure that the actual water from that specific depth is collected (Swift, 2010). Given that Niskins have a relatively small opening diameter relative to the total diameter of the bottle, we found that a waiting time of 2 min is required after stopping the winch before closing the Niskin to obtain consistent nutrient concentrations from sequentially closed bottles at the same depth, notably when there are steep gradients observed in the water column (Paver et al., 2020). To improve the flushing of the samplers, NIOZ developed its own samplers (Rijkenberg et al., 2015), but nevertheless still observes the 2-min waiting period. Additionally, differences in the size and construction of the frame holding the samplers may also cause water to be dragged up from deeper depths affecting the collection of water. This wake effect combined with the hampered flushing of regular Niskin samplers can be clearly identified in analysed samples such as salinity or oxygen for which the values can be compared to the sensor data. This is especially seen when the CTD passes through strongly stratified water. Paver et al. (2020) showed that data from water collected from a sampler closed without waiting were consistent with sensor data from 8 ± 2 m deeper.

Preferably at the start of an expedition and if time permits, it is recommended to close all CTD-Rosette bottles at one depth and take samples from all bottles to measure the nutrient concentrations as a check for malfunctioning or contamination of samplers. Such a run can also give a good impression of the precision for all the nutrients being analysed. As described in section (*Case study of nutrient measurements in the Weddell Sea*) to observe in-between run reproducibility, re-analysing a duplicate sample from a previous station (e.g. the deepest sampled depth) in the subsequent analytical run is good practice, especially when done in combination with an independent lab reference or better a CRM.

After collection, samples should be left to condition in the dark for approximately 2 hours to reach the lab temperature prior to analysis. It is crucial to analyse samples at the same temperature as the calibration standards to ensure the same reaction conditions and to avoid thermal expansion or contraction at different temperatures in the sampler. This temperature should also be noted together with the salinity of the sample to be able to convert concentrations that are measured in $\mu M/l$ to units of $\mu M/kg$.

Finally, during lab tests, we observed that evaporation due to extremely dry environments can notably change the concentration of a sample or calibrant while it awaits analysis when left open and uncovered (AWI, 1981–2000, pages 38–39). In principle this effect can be corrected for using a sensitivity drift standard during the analysis run, however, it doesn't necessarily correct for samples that may be open for longer periods in such environments and it is better to avoid the need for such a correction. It is advisable to carry out such tests and if necessary to cover the Continuous Flow Analysis (CFA)-cups using highly stretched parafilm to avoid evaporation as much as possible and where a sharpened sampler needle can still penetrate through the stretched covering.

Overall, as said at the start of this section, there are many steps and considerations involved in obtaining high-quality data for which we refer the reader to more detailed descriptions (Becker et al., 2020), with some additional suggestions provided here.

4.6 Summary

Producing high-quality marine nutrient concentration data that provides insights into (changing) biogeochemical cycles and that is inter-comparable between labs and oceanographic research expeditions starts with high-precision measurements in individual labs. Subsequently, the consistency needs to be optimised to minimise variability between analytical runs and any drift over time. Only when internal precision is sufficient and between-run variability has been minimised, inter-laboratory comparisons become useful. Recent inter-comparisons organised in Japan in 2018 with participants from mainly oceanographic institutes showed a good agreement within 2% for most nutrient species of the average mean. However, deviations of as much as 10% still exist between some labs, especially for phosphate and silicate (Aoyama et al., 2018). By organising workshops on specific methods and reporting the outcome to the community, improvements can be obtained, underlining the importance of such exercises. Besides inter-comparisons, the use of CRMs has resulted in an important step in the journey to inter-laboratory comparability. Nevertheless, despite all the progress that has been made, there is still room for further improvements, notably for silicate.

Acknowledgements All recommendations in this chapter resulted from workshops, visits and conversations with many colleagues worldwide. We are very thankful to our former colleague Rob de Vries for starting the nutrient facility using CFAs and his advice to always be critical of all analytical steps, including how final data output in general is obtained. Together with our NIOZ colleagues Jan van Ooijen and Malcolm Woodward from PML, two international workshops were organised that led to a published report from which the nutrient community could benefit and we thank all participants for their contributions. We also thank Andreas Wisotzki for producing Fig. 4.3 back in 1996 to be able to visualise the effects of normalisation of data while onboard. Finally, many thanks to Stephen Coverly who was always willing to answer questions and issues related to segmented flow chemistry.

Competing Interests The authors have no conflicts of interest to declare that are relevant to the content of this chapter.

References

Anderson, R. F. (2020). GEOTRACES: Accelerating research on the marine biogeochemical cycles of trace elements and their isotopes. *Annual Review of Marine Science, 12*(1), 49–85. https://doi.org/10.1146/annurev-marine-010318-095123.

Aoyama, M., Anstey, C., Barwell-Clarke, J., Baurand, F., Becker, S., Blum, M., Coverly, S. C., Czobik, E., d'Amico, F., Dahllöf, I., Dai, M., Dobson, J., Pierre-Duplessix, O., Duval, M., Engelke, C., Gong, G.-C., Grosso, O., Hirayama, A., Inoue, H., Ishida, Y., & Zhang, J.-Z. (2010). *2008 Inter-laboratory comparison study for reference material for nutrients in seawater.* Technical Reports of the Meteorological Research Institute (No. 60, p. 134), Tsukuba, Japan. https://doi.org/10.11483/mritechrepo.60.

Aoyama, M., Carignan, M., Molina, L. D. A., Terhell, D., Prove, G., Knockaert, M., Bell, S., de Santis Braga, E., Paranhos, R., Anstey, C., Payne, C., Barwell-Clarke, J., Richardson, W., Monteiro, I., Dai, M., Zhao, Z., Hu, Y., Sun, J., Larsen, M. M., Jensen, D. W., Kerouel, R., & Márquez, A. (2013). *2012 inter-laboratory comparison study of a reference material for nutrients in seawater.* Technical Report of the Meteorological Research Institute, Japan Meteorological Agency, Tsukuba-city, Ibaraki, Japan. https://doi.org/10.13140/RG.2.1.4058.1207.

Aoyama, M., Bakker, K., van Ooijen, J., Ossebaar, S., & Woodward, E. M. S. (2015). *Report from an international nutrient workshop focusing on phosphate analysis.* Yang Yang Publisher, Fukushima, Japan. ISBN: 978-4-908583-01-8.

Aoyama, M., Abad, M., Anstey, C., Ashraf P, M., Bakir, A., Becker, S., Bell, S., Berdalet, E., Blum, M., Briggs, R., Garadec, F., Cariou, T., Church, M., Coppola, L., Crump, M., Curless, S., Dai, M., Daniel, A., Davis, C., & Zhang, J.-Z. (2016). *IOCCP-JAMSTEC 2015 inter-laboratory calibration exercise of a certified reference material for nutrients in seawater.* IOCCP Report Number 1/2016. ISBN 978-4-901833-23-3.

Aoyama, M., Abda, M., Aguilar-Islas, A., Ashraf P, M., Azetsu-Scott, K., Bakir, A., Becker, S., Benoit-Cattin-Breton, A., Berdalet, E., Björkman, K., Blum, M., de Santis Braga, E., Caradec, F., Cariou, T., Chiozzini, V. G., Collin, K., Coppola, L., Crump, M., Dai, M., & Zhang, J.-Z. (2018). *IOCCP-JAMSTEC 2018 inter-laboratory calibration exercise of a certified reference material for nutrients in seawater.* Yokosuka, Japan, International Ocean Carbon Coordination Project/Japan Agency for Marine-Earth Science and Technology (JAMSTEC), p. 214. (IOCCP Report Number 1/2018). https://doi.org/10.25607/OBP-429.

AWI. (1981–2000). Berichte zur Polarfoschung. ISSN 0176 – 5027, 001–376F.

Becker, S., Aoyama, M., Woodward, E. M. S., Bakker, K., Coverly, S., Mahaffey, C., & Tanhua, T. (2020). GO-SHIP repeat hydrography nutrient manual: The precise and accurate determination of dissolved inorganic nutrients in seawater, using continuous flow analysis methods. *Frontiers in Marine Science, 7*, 581790. https://doi.org/10.3389/fmars.2020.581790.

Bindoff, N. L., Willebrand, J., Artale, V., Cazenave, A., Gregory, J., Gulev, S., Hanawa, K., Le Quéré, C., Levitus, S., Nojiri, Y., Shum, C. K., Talley, L. D., & Unnikrishnan, A. (2007).

Observations: Oceanic climate change and sea level. In S. Solomon, D. Qin, M. Manning, Z. Chen, M. Marquis, K. B. Averyt, M. Tignor, & H. L. Miller (Eds.), *Climate change 2007: The physical science Basis. Contribution of working group I to the fourth assessment report of the intergovernmental panel on climate change* (pp. 385–433). Cambridge University Press.

Hoppema, M., Bakker, K., van Heuven, S. M., van Ooijen, J. C., & de Baar, H. J. (2015). Distributions, trends and inter-annual variability of nutrients along a repeat section through the Weddell Sea (1996–2011). *Marine Chemistry, 177*, 545–553. https://doi.org/10.1016/j.marchem.2015.08.007.

Humphreys, M. P., Meesters, E. H., de Haas, H., Karancz, S., Delaigue, L., Bakker, K., Duineveld, G., de Goeyse, S., Haas, A. F., Mienis, F., Ossebaar, S., & van Duyl, F. C. (2022). Dissolution of a submarine carbonate platform by a submerged lake of acidic seawater. *Biogeosciences, 19*, 347–358. https://doi.org/10.5194/bg-19-347-2022.

Hydes, D. J., Aoyama, M., Aminot, A., Bakker, K., Becker, S., Coverly, S., Daniel, A., Dickson, A. G., Grosso, O., Kerouel, R., van Ooijen, J., Sato, K., Tanhua, T., Woodward, E. M. S., & Zhang, J. Z. (2010). Determination of dissolved nutrients (N, P, SI) in seawater with high precision and inter-comparability using gas-segmented continuous flow analysers. In*: The GO-SHIP repeat hydrography manual: A collection of expert reports and guidelines.* IOCCP Report Number 14, ICPO Publication Series Number 134. https://doi.org/10.25607/OBP-15. Version 1, 2010.

Joyce, T., & Corry, C. (1994). *Requirements for WOCE hydrographic programme data reporting.* WHPO Publication 90–1, Revision 2, WOCE Hydrographic Programme Office, WHOI, Woods Hole, MA 02543, USA, p. 144.

Lauvset, S. K., Lange, N., Tanhua, T., Bittig, H. C., Olsen, A., Kozyr, A., Alin, S., Álvarez, M., Azetsu-Scott, K., Barbero, L., Becker, S., Brown, P. J., Carter, B. R., da Cunha, L. C., Feely, R. A., Hoppema, M., Humphreys, M. P., Ishii, M., Jeansson, E., & Key, R. M. (2022). GLODAPv2.2022: The latest version of the global interior ocean biogeochemical data product. *Earth System Science Data, 14*, 5543–5572. https://doi.org/10.5194/essd-14-5543-2022.

Middag, R., Zitoun, R., & Conway, T. M. (2023). Trace metals. In J. Blasco, & A. Tovar-Sanchez (Eds.), *Marine analytical chemistry*. Cham, Switzerland: Springer. https://doi.org/10.1007/978-3-031-14486-8_3.

Moore, C. M., Mills, M. M., Arrigo, K. R., Berman-Frank, I., Bopp, L., Boyd, P. W., Galbraith, E. D., Geider, R. J., Guieu, C., Jaccard, S. L., Jickells, T. D., La Roche, J., Lenton, T. M., Mahowald, N. M., Marañón, E., Marinov, I., Moore, J. K., Nakatsuka, T., Oschlies, A., & Ulloa, O. (2013). Processes and patterns of oceanic nutrient limitation. *Nature Geoscience, 6*, 701–710. https://doi.org/10.1038/ngeo1765.

Murphy, J., & Riley, J. P. (1962). A modified single solution method for the determination of Phosphate in natural waters. *Analytica Chimica Acta, 27*, 31–36. https://doi.org/10.1016/S0003-2670(00)88444-5.

Paver, C. R., Codispoti, L. A., Coles, V. J., & Cooper, L. W. (2020). Sampling errors arising from carousel entrainment and insufficient flushing of oceanographic sampling bottles. *Limnology and Oceanography, Methods, 18*, 311–326. https://doi.org/10.1002/lom3.10368.

Rijkenberg, M. J. A., de Baar, H. J. W., Bakker, K., Gerringa, L. J. A., Keijzer, E., Laan, M., Laan, P., Middag, R., Ober, S., van Ooijen, J., Ossebaar, S., van Weerlee, E. M. & Smit, M. G. (2015). "PRISTINE", a new high-volume sampler for ultraclean sampling of trace metals and isotopes. *Marine Chemistry, 177*, Part 3, 501–509. https://doi.org/10.1016/j.marchem.2015.07.001.

Snyder, L.R. (1976). *The prediction and control of sample dispersion in continuous flow-analysis.* Advances in Automated Analysis: Technicon International Congress, Vol. 1, pp. 76–81. Mediad, Incorporated, 1977.

Swift, J. H. (2010). Quality water sample data: Notes on acquisition, record keeping, and evaluation. In E. M. Hood, C. L. Sabine, & B. M. Sloyan (Eds.), *The GO-SHIP repeat hydrography manual: A collection of expert reports and guidelines* (Version 1, 38 pp.). https://doi.org/10.25607/OBP-1346.

Tanhua, T., van Heuven, S., Key, R. M., Velo, A., Olsen, A., & Schirnick, C. (2010). Quality control procedures and methods of the CARINA database. *Earth System. Science Data, 2*, 35–49. https:// doi.org/10.5194/essd-2-35-2010.

Taylor, J.K. (1987). *Quality assurance of chemical measurements*. Lewis Publishers Inc., Chelsea, MI. ISBN: 0-87371-097-5. https://doi.org/10.1201/9780203741610.

Tomczak, M. (1981). A multi-parameter extension of temperature/salinity diagram techniques for the analysis of non-isopycnal mixing. *Progress in Oceanography, 10*, 147–171. https://doi.org/ 10.1016/0079-6611(81)90010-0.

Van Leeuwe, M. A., Scharek, R., De De Baar, H. J. W., Jong, J. T. M., & Goeyens, L. (1997). Iron enrichment experiments in the Southern Ocean: Physiological responses of plankton communities. *Deep Sea Research Part II: Topical Studies in Oceanography, 44*(1–2), 189–207. https:// doi.org/10.1016/S0967-0645(96)00069-0.

Zhang, J.-Z., Fischer, C. J., & Ortner, P. B. (1999). Optimization of performance and minimization of silicate interference in continuous flow phosphate analysis. *Talanta, 49*, 293–304. https://doi. org/10.1016/S0039-9140(98)00377-4.

Chapter 5
Certification of Reference Materials of Nutrients in Seawater, NMIJ CRM 7601-a, 7602-a, and 7603-a

Chikako Cheong and Tsutomu Miura

Abstract The National Metrology Institute of Japan (NMIJ) has developed three levels of certified reference materials (CRMs) (NMIJ CRM 7601-a, 7602-a, and 7603-a) for nutrient analysis to provide SI traceability of analysis values of nutrients (nitrate, nitrite, phosphate, and dissolved silica) in oceanographic observations and to enable the comparison of these values across time and space. These property values were determined using the analytical results of multiple analytical methods and standard solutions of inorganic ions and elements as the SI traceability source. Depending on their characteristics, each property value was finally distinguished as a certified value, indicative value, or information. The homogeneity and long-term stability of the CRMs were evaluated on the basis of the International Organization for Standardization (ISO) Guide 35 using the results of continuous flow analysis. The certified values of nitrate and dissolved silica in 7603-a, from the deep Pacific Ocean, could be certified with uncertainties as low as approximately 1%, which is required for the observations. The values for nitrite and phosphate, as well as for relatively low nutrient concentrations, met the minimum requirements for the oceanographic observations, although the uncertainties were slightly higher. In terms of direct distribution from the National Research Institute of Metrology, these CRMs are the first CRMs in the world that can cover a wide nutrient concentration range from that of surface water to that of the deep Pacific Ocean. This development is expected to make the metrological accuracy of nutrient analyses in the observations more robust.

Keywords Nutrients · Certified reference materials · International System of Units traceable · Continuous flow analysis · Ion chromatography · ICP–MS · Seawater

C. Cheong (✉) · T. Miura
Research Institute for Material and Chemical Measurement/National Metrology Institute of Japan, Tsukuba, Ibaraki, Japan
e-mail: c-kato@aist.go.jp

T. Miura
e-mail: t.miura@aist.go.jp

© The Author(s) 2025
M. Aoyama et al. (eds.), *Chemical Reference Materials for Oceanography*, Springer Oceanography, https://doi.org/10.1007/978-981-96-2520-8_5

5.1 Introduction

Nutrients (nitrate, nitrite, phosphate, and dissolved silica) in seawater are considered major items in global oceanographic observations. They are used to assess and predict the effects of material recycling in the ocean, climate change, and global warming, and are becoming increasingly important from the perspective of global environment and climate change monitoring (Knap et al., 1996). However, their temporospatial variations are often slight despite their large impact, and reliable reference materials are now required to achieve precise comparisons (Aoyama et al., 2012).

In recent years, with the globalized and borderless nature of the market, traceability requirements have increased for chemical standards in general. The oceanographic community has a strong interest in the traceability of measurements of nutrient concentrations in seawater, and there is a demand for the supply of certified reference materials (CRMs) with measurement traceability that can meet the requirements of the International Mutual Recognition Agreement (Global MRA) for measurement standards. So far, the MOOS series (currently MOOS-3; (Clancy et al., 2014)) of the National Research Council (NRC) Canada is the only one supplied for nutrients in seawater. However, MOOS has only one concentration level for North Atlantic seawater. There is thus a growing demand, both in Japan and abroad, for a CRM with a higher concentration level that can be used in the extreme layers of the Pacific Ocean and a so-called zero seawater CRM with a concentration level as close to zero as possible. Therefore, the National Metrology Institute of Japan (NMIJ) of the National Institute of Advanced Industrial Science and Technology (AIST) developed three levels of seawater CRMs containing high (Pacific equivalent), medium (Atlantic equivalent), and very low (near zero) concentrations of nutrients to cover a wide range of nutrient concentrations, including four components (nitrite ion, nitrate ion, phosphate ion, and dissolved silica) in the candidate standard materials. NMIJ developed a method for the determination of four nutrient components (nitrite, nitrate, phosphate, and dissolved silica) in three levels of seawater candidate reference materials containing very low (near zero) concentrations of nutrients. This chapter describes the details of the development of the CRMs supplied by NMIJ.

5.2 Production of Candidate CRMs

The production of seawater CRMs was contracted to KANSO Technos Co., LTD. Raw seawater was filtered through a Millipore membrane filter, sterilized in an autoclave, and bottled in small batches of approximately 100 mL each and sealed in aluminum laminated bags. All operations were performed in a clean room (details of the production process are described in Chap. 2). Table 5.1 shows the raw seawater for the three seawater CRM levels.

Of the raw seawater samples, extremely low-concentration seawater (collected on November 18, 2009), high-concentration seawater (collected on October 26, 2008),

Table 5.1 Raw seawater for the candidate CRMs

Concentration level of nutrients	NMIJ CRM No.	Abbreviation	Source of materials
Extremely low	7601-a	ELSW	Seawater from the surface layer in the Pacific Ocean
Middle	7602-a	MSW	Blended seawater from (1) a 690-m depth in the Arctic Sea, (2) a 1500-m depth in the Atlantic Ocean, and (3) a 397-m depth in the Suruga-wan Bay
High	7603-a	HSW	Seawater from the nutrient maximum layer (3000-m depth) in the Pacific Ocean

and Arctic seawater (collected on October 7, 2010) were obtained with the cooperation of the Meteorological Research Institute and the Japan Agency for Marine-Earth Science and Technology. The Atlantic seawater (collected on July 21, 2010) was collected with the cooperation of the Meteorological Research Institute and Plymouth Marine Laboratory (UK), whereas the Suruga Bay seawater (collected on November 8, 2010) was collected with the cooperation of KANSO Technos Co., LTD.

The Atlantic seawater was prepared by mixing an appropriate amount of Suruga Bay seawater (from a 397 m depth) and Arctic Ocean seawater (from a 690 m depth) because of an unintended decrease in concentration during transportation after collection.

5.3   Analytical Strategy for Each Nutrient Component

The property values for the three CRM levels, four nutrient components for each and 12 items in total, were determined on the basis of the results of the quantitative analysis using five analytical methods:

(1) Continuous flow analysis (CFA)
(2) Ion chromatography (IC) direct method (IC direct)
(3) IC desalination method (IC desalination)
(4) Ion exclusion chromatography (IEC) postcolumn absorption photometric method (IEC postcolumn)
(5) IEC isotopic dilution inductively coupled plasma mass spectrometry (IEC–ID–ICP–MS).

The analytical methods used to determine the property values are listed in Table 5.2. The property values were determined using two or more analytical methods to ensure the reliability of the values, and items for which quantitative values were obtained using only one method were treated as "indicative values" or "information." The details of each analytical method are described in the following sections.

Table 5.2 Analytical methods for determining the property values of the candidate CRMs

	(1) CFA	(2) IC direct	(3) IC desalination	(4) IEC postcolumn	(5) IEC–ID–ICP–MS
NO_2^-	○	● (middle)	–	–	–
NO_3^-	○	● (middle and high)	● (middle and high)	–	–
PO_4^{3-}	○	×	–	–	–
Dissolved silica (as Si)	○	–	–	○	○

Here, ○ indicates that the method could be applied to all three concentration levels of CRMs, ● shows the method could be applied to only some concentration levels, and × represents that no determination value could be obtained because it was below the detection limit

5.4 Chemicals

All reagents were of analytical grade, available from FUJIFILM Wako Pure Chemical Corp. (Osaka, Japan) or Kanto Chemical Co. Inc. (Tokyo, Japan) unless otherwise specified. Water was purified using a Milli-Q SP.

ICP–MS system (Merck, Darmstadt, Germany) or a Milli-Q Integral Q- POD Element system (Merck, Darmstadt, Germany).

The silicon standard solution used was a Standard Reference Material (SRM) 3150 supplied by the National Institute of Standards and Technology (NIST, Gaithersburg, USA). The standard solutions of nitrate, nitrite, and phosphate ions used in this study were equivalent to NMIJ CRM 3805-a, 3506-a, and 3808-a, respectively; they were prepared using the gravimetric method by dissolving potassium nitrate, sodium nitrite, and potassium dihydrogen phosphate in water, respectively. The purity of each salt was assayed through coulometric titration and/or gravimetric analysis (Hioki et al., 1990, 1994).

Hereinafter, the concentrations of nitrite, nitrate, phosphate, and dissolved silica were identified on the basis of the mass fractions ($mg\ kg^{-1}$) of nitrite (NO_2^-, molecular weight 46), nitrate (NO_3^-, molecular weight 62), phosphate (PO_4^{3-}, molecular weight 95), and silicon (Si, atomic weight 28), respectively. All preparations and measurements were conducted in a laboratory at temperatures between 24 and 26 °C.

5.5 Continuous Flow Analysis

CFA is widely used for nutrients in hydrographic observations. Therefore, it was used as the basis for the determination of the certified candidate CRMs. This section describes the nutrient determination by CFA, which was used to determine four nutrient components (nitrite, nitrate, phosphate, and dissolved silica). When measuring samples containing a high salt concentration matrix, such as seawater, the standard addition method is usually employed. This extrapolates the obtained calibration curve to calculate the determination values. Basically, the linearity of the calibration curve must be analyzed accurately. However, the linearity of the calibration curve is controversial in CFA (Bakker et al., 2010). Therefore, the bracketing method (Cheong et al., 2020), where the calibration curve is narrowed to the range where linear regression is possible, was employed for the analysis of the four nutrient components of each of the three candidate CRMs.

5.5.1 *Apparatus*

A continuous flow analyzer AACS-V (BLTEC K. K., Osaka, Japan), which automatically conducts coloring reactions followed by photometric measurements, was employed for the measurements of seawater nutrients in a CFA mode. The sample solutions were injected using an autosampler, where a sample aliquot of 1.8 mL was stored in each of the vials covered by polytetrafluoroethylene (PTFE) septa. For the measurement of nitrate, a reduction column (Glastron, Inc., NJ, USA) was used to convert nitrate to nitrite, which was made of a coiled hollow cylindrical cadmium (Zhang et al., 2000). Three reduction columns were incorporated in the CFA reaction line; the columns were treated before use with a copper sulfate solution. Complete conversion was allowed under the three-column condition (Kato & Hioki, 2009). The measuring flasks, beakers, and spoons used in the preparation of the measuring solutions and reagents were all made from polymethylpentene, polyethylene, or PTFE. We refer to the solution just before mixing with the reagents as the measuring solution. Storage bottles for the measuring solutions and reagents were made from polypropylene.

5.5.2 *Artificial Seawater*

In this study, original artificial seawater (ASW) was prepared following the reference "Protocols for JGOFS Core Measurements" (Knap et al., 1996) by dissolving 64.3 g of sodium chloride (Merck, Darmstadt, Germany; superpure grade or NMIJ CRM 3008-a, NMIJ/AIST), 14.3 g of magnesium sulfate heptahydrate (Kanto), and 0.34 g of sodium hydrogen carbonate (Kanto) in an adequate amount of water and diluting to 2000 mL with water at 25 °C. The major ionic compositions in both the average of the world ocean and the original ASW used in this study are listed in Table 5.3. Meanwhile, the salinity varied, approximately from 34 to 36 g kg^{-1}, depending on the ocean area, seasons, and depth (Yoshiyuki, 1997). The concentration of each ion contained in the original ASW was like that in the average of the world ocean within ~10%, except for that of magnesium.

Table 5.3 Major ionic compositions in seawater and ASW

	Average of the world's oceans (g kg^{-1})	ASW (g kg^{-1})
Cl$^-$	19.35	18.75
Na$^+$	10.78	12.20
Mg^{2+}	1.28	0.68
SO$_4{}^{2-}$	2.69	2.67
HCO$_3{}^-$	0.14	0.12
Salinity	34–36	34.7
Cl$^-$	19.35	18.75

The nutrient levels in water and ASW were confirmed by IC (for nitrite and nitrate), IC–ICP–MS (Cheong et al., 2021) (for phosphate), and atomic absorption spectrophotometry (Cheong et al., 2014) (for dissolved silica), all of which were below the detection limit (DL). The DLs for nitrite, nitrate, phosphate, and dissolved silica were 0.009 mg kg^{-1} (2 μmol kg^{-1}), 0.005 mg kg^{-1} (0.09 μmol kg^{-1}), 0.038 μg kg^{-1} (0.0004 μmol kg^{-1}), and 0.12 μg kg^{-1} (0.004 μmol kg^{-1}), respectively.

Hereinafter, the measuring solutions were identified using the mass fractions of ASW and seawater, respectively. They were referred to, for instance, as 0.9 g g^{-1} ASW and 0.5 g g^{-1} seawater. The ASW or seawater mass fraction indicated in this chapter was always calculated after adding a standard solution of an analyte.

5.5.3 Coloring Reagents

All the coloring reagents were prepared as described in the literature (Kawano & Uchida, 2007).

The Griess method (Griess, 1879; Huygen, 1970; Miranda et al., 2001) was used for the measurement of nitrite and nitrate, with nitrate reduced to nitrite before measurement. The following three solutions were prepared for the Griess method: (1) 0.09 mol L^{-1} imidazole solution (containing 0.02 mol L^{-1} sulfuric acid and 0.1% TritonX-100) as a catalyst, (2) 0.06 mol L^{-1} sulfanilamide solution (containing 1.2-mol L^{-1} hydrochloric acid and 0.1% TritonX-100) as a modifier, and (3) 0.004 mol L^{-1} N-1-naphthylethylenediamine dihydrochloride (1-NED) solution (containing 0.12 mol L^{-1} hydrochloric acid) as a coupling agent.

The phosphomolybdenum blue method (Murphy & Riley, 1962; Namiki, 1964) was used to determine phosphate. First, 0.02-mol L^{-1} stock molybdate solution was prepared by dissolving 5.6 g of disodium molybdate dihydrate and 0.12 g of potassium antimonyl tartrate in ~500 mL of water containing 35 mL of sulfuric acid, diluting to 1000 mL with water, and adding 5 mL of a 15% solution of sodium dodecyl sulfate. Finally, the following two solutions were prepared: (1) as a mixed coloring reagent, 1.1 g of L-ascorbic acid dissolved in 200 mL of the 0.02-mol L^{-1}

stock molybdate solution, mixed with 4 mL of a 15% solution of sodium dodecyl sulfate and (2) 0.04-mol L^{-1} sulfuric acid containing 0.075% sodium dodecyl sulfate as a pH-controlling agent.

The silicomolybdenum blue method (Grasshoff, 1964; Kato, 1976; Ramachandran & Gupta, 1985) was used to determine dissolved silica. In the method using CFA, a molybdate solution of 0.06-mol L^{-1} (as molybdenum) as a complexing agent, prepared by dissolving 15.0 g of disodium molybdate dihydrate (Wako Pure Chemical Ind., Osaka, Japan) in ~300 mL of water containing 6 mL of sulfuric acid (Wako) and diluting to 1000 mL with water. A 0.40-mol L^{-1} oxalic acid solution as a phosphate masking agent was prepared by dissolving 25.0 g oxalic acid dihydrate (Kanto Chemical Co., Inc., Tokyo, Japan) in ~300 mL of water and diluting to 500 mL with water. A 0.14-mol L^{-1} L-ascorbic acid solution as a reducing agent was prepared by dissolving 2.5 g of L-ascorbic acid (Wako) in ~80 mL of water and diluting to 100 mL with water.

5.5.4 Measurement Procedure

A flow diagram of CFA is shown in Fig. 5.1, and its supporting explanation is in Table 5.4. All reactions and detections were performed on a single continuous flow. The sample solutions (i.e., the measuring solutions) were injected using an autosampler; when no sample solution flowed, a washing solution, prepared by diluting ASW with water in the same ratio as that of the seawater in the measuring solutions, flowed instead. A measuring solution was mixed with all reaction reagents in a flow path. The flow rates are expressed in Fig. 5.1. The absorption of the complex formed by the reaction was detected at 820 nm for dissolved silica, 550 nm for nitrite and nitrate, and 880 nm for phosphate. In the case of the nitrate measurement, the absorption was provided as the sum of the values for nitrite and nitrate because nitrate was reduced, as mentioned above. The amount of substance content (μmol kg^{-1}) of nitrate in a sample was calculated by subtracting that of nitrite from the total amount of substance content of nitrite and nitrate. The measuring solution injection time was optimized to 90 s so that a stable peak plateau was obtained. The washing time of 300 s (the maximum setting time of the instrument used) was applied to improve data accuracy. In a series of measurements, the measuring solutions (90 s) and the washing solution (300 s) were delivered continuously and alternately. The peak area was obtained by accumulating ten intensities with 1-s intervals for a 10-s duration from 70 to 80 s after the peak start. At the beginning of each series of measurements, the washing solution put in a sample vial was analyzed in place of a measuring solution; its absorption was always not distinguishable from the fluctuation of the baseline. Although it was not necessary to bring the absorption signals down to the reagent-blank level between each pair of measuring solutions by prolonged washing times according to "Protocols for JGOFS" (Knap et al., 1996), the washing time of 300 s (the maximum setting time of the instrument used) was applied in this study to improve data accuracy.

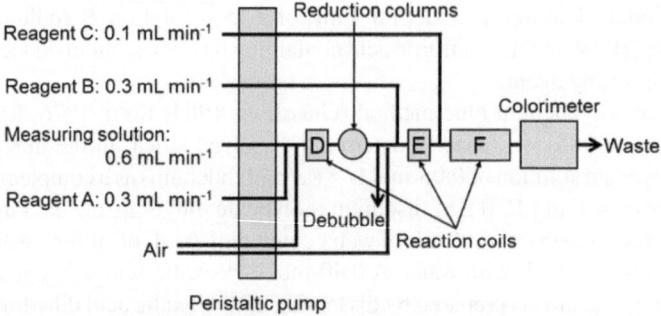

Fig. 5.1 Flow diagram for CFA (Modified from Cheong et al., 2024)

Table 5.4 Detail of each item in the flow diagram

Analyte	Nitrite	Nitrite + Nitrate	Phosphate	Dissolved silica
Reagent A[*1]	Sulfanilamide[*3]	Imidazole[*3]	Sulfuric acid[*3]	Molybdate solution[*3]
Reagent B[*1]	1-NED[*3]	Sulfanilamide[*3]	Mixed coloring reagent[*3]	Oxalic acid[*3]
Reagent C[*1]	–	1-NED[*3]	–	Ascorbic acid[*3]
Reaction coil D[*1]	5 turns	5 turns	20 turns	20 turns
Reaction coil E[*1]	5 turns	5 turns	55 turns	5 turns
Reaction coil F[*1]	15 turns	15 turns	–	50 turns
Reduction columns[*2]	None	Inserted	None	None
Wavelength of the colorimeter	820 nm	550 nm	880 nm	550 nm

[*1] The symbols A–F correspond to those in Fig. 5.1
[*2] See Fig. 5.1
[*3] For each exact composition, see the text

A detailed description of the general operation of the CFA instrument and a calculation of the addition calibration method were given in a paper on dissolved silica in seawater (Cheong et al., 2014).

For the determination of the candidate CRMs, 10 bottles of samples were selected through stratified random sampling. The candidate CRMs were diluted to match the absorbance range measurable by the CFA instrument, and 10 bottles of measuring samples were prepared. The dilution was done with the following seawater-containing rates: 0.04 $g\,g^{-1}$ for measurements of nitrate in HSW, 0.50 $g\,g^{-1}$ for nitrate in MSW, and 0.75 $g\,g^{-1}$ for nitrate in ELSW, for nitrite in MSW, and phosphate in HSW. Standard solutions with concentrations around 0.95 and 1.05 times the concentration of the target component in the seawater sample were prepared using the mass ratio mixing method (designated as low- and high-concentration standard solutions, respectively), and the peak areas of the two points were used to determine the concentrations of the target components in the seawater samples. The two standard solutions used were added with the ASW to have the same content as the seawater samples to match the matrix with the seawater samples.

The analyte concentration in seawater, x_{sample} (mg kg^{-1}), was determined as follows:

$$x_{sample} = \frac{\left(\frac{(y_{sample} - y_{low})}{(y_{high} - y_{low})} + x_{low} \right)}{D} \tag{5.1}$$

Here, y_{sample}, y_{high}, and y_{low} are the peak areas of the seawater samples, high-concentration standard solution, and low-concentration standard solution, respectively; x_{high} and x_{low} are the concentrations of the target nutrient components in the high-concentration standard solution and low-concentration standard solution, respectively (all units in mg kg^{-1}); and D is the seawater content ratio in the seawater sample (= mass of added seawater (g)/total seawater sample after dilution (g)).

5.5.5 Analytical Results by the CFA Analysis

Before the analysis of the candidate CRMs, five addition–recovery tests were conducted by diluting three seawater samples with the same seawater-containing rate as that during the determination of the candidate CRMs and adding a standard solution of known concentration. The measured solutions before and after addition were measured using the CFA bracket method. The difference in the experimental addition mass fraction was compared with the addition mass fraction in the gravimetric method preparation. The results are shown in Table 5.5. Although deviations beyond reproducibility were observed for nitrate with the equivalent of ELSW, nitrite, and phosphate, they were deemed acceptable considering the uncertainty required for oceanographic observations. They were thus included in the uncertainty of the analytical results as matrix differences between the sample and standard solutions (Cheong et al., 2020).

On the basis of the above considerations, candidate CRMs were analyzed using the bracket method. The analytical values by the CFA are listed in Table 5.6. Here, determination is the process of determining the concentration when using one analytical method by calculation based on the signal intensity obtained from the measurement, and the value obtained is called the quantitation value. The determination value is calculated for each analytical method. Finally, the property value is obtained based on multiple determination values for each component. Measurement was performed twice (or four times) for each seawater sample. The average values of 10 × 2 (or 10 × 4) $x_{samples}$ obtained in this way were used as the determination values of the CRMs.

Table 5.5 Results of recovery tests by CFA

Nutrient	Base Seawater	Seawater-containing rate	Added w_{pre} (mg kg^{-1})	Found[*] w_{dat} (mg kg^{-1})	Recovery[*] (w_{pre}/w_{dat})
NO$_3^-$	HSW	0.04	0.1046	0.1046 ± 0.0004	100.0% ± 0.4%
	MSW	0.50	0.0459	0.0458 ± 0.0002	99.9% ± 0.5%
	ELSW	0.75	0.00010	0.00008 ± 0.00001	75.2% ± 16.8%
NO$_2^-$	MSW	0.75	0.00159	0.00164 ± 0.00002	102.7% ± 1.3%
PO$_4^{3-}$	HSW	0.75	0.0208	0.0202 ± 0.0002	97.3% ± 0.8%

[*]Each value after "±" indicates the U ($k = 2$) due to the bracketing method including the repeatability

Table 5.6 Analytical values of the three NMIJ CRMs by CFA

	HSW/mg kg^{-1}	MSW/mg kg^{-1}	ELSW/mg kg^{-1}
NO$_2^-$	0.00123 ± 0.00013 (10.5%)	0.0183 ± 0.0007 (3.64%)	0.00165 ± 0.00012 (7.4%)
NO$_3^-$	2.7528 ± 0.0071 (0.26%)	0.9565 ± 0.0020 (0.21%)	0.0013 ± 0.0003 (23%)
PO$_4^{3-}$	0.2876 ± 0.0010 (0.36%)	0.1013 ± 0.0010 (0.94%)	0.0015 ± 0.0002 (11%)
Dissolved silica (as Si)	4.1122 ± 0.0069 (0.17%)	0.8500 ± 0.0018 (0.22%)	0.0366 ± 0.0001 (0.33%)

[*]Each value after "±" indicates U ($k = 2$), and the figures in parentheses in the lower row represent the relative values of U

The combined standard uncertainty u_c of the determination values comprised the following: (1) standard uncertainty derived from the determination of a single bottle $s_r/\sqrt{n}$ ($n = 2$ or 4), the same as the measurement repeatability, obtained from the standard deviation s_r obtained by the analysis of variance (ANOVA) of $x_{samples}$ 10×2 (or 10×4); (2) standard uncertainty u_{mtrx} derived from the difference in absorbance resulting from the difference in the matrix between the seawater and the standard solution in ASW; and (3) standard uncertainty, u_{std}, of the standard solution used. The u_c values were estimated using the square root of the sum of squares of these relative values:

$$u_c = \sqrt{\left(s_r/\sqrt{n}\right)^2 + u_{mtrx}^2 + u_{std}^2} \qquad (5.2)$$

The relative combined standard uncertainty of nitrate ($u_{c(NO3)}$) was calculated from the relative combined standard uncertainty of nitrate ($u_{c(NO3)}$) and the relative combined standard uncertainty of the total concentration of nitrate and nitrite ($u_{c(NO2+NO3)}$) as follows:

$$u_{c(NO3)} = \frac{\sqrt{\left(u_{c(NO2+NO3)} \times [NO_2 + NO_3]\right)^2 + \left(u_{c(NO2)} \times [NO_2]\right)^2}}{[NO_3]} \qquad (5.3)$$

Here, the units for $[NO_2 + NO_3]$, $[NO_2]$, and $[NO_3]$ are $\mu mol\ kg^{-1}$.

For nitrate and dissolved silica, analytical values were obtained with the expanded uncertainties U ($k = 2$), generally below the target of 1%. Although the uncertainties for nitrite and phosphate were a bit large, they were within acceptable limits considering the concentration levels contained in the candidate CRMs.

5.6 Ion Chromatography (Direct and Desalination Methods)

Nitrite, nitrate, and phosphate were analyzed using the standard addition method with IC. Although IC is generally an analytical method not affected by the sample matrix, whether the seawater matrix does not truly affect the analysis is important in this certification. Thus, the halogen removal method was also conducted for the analysis of nitrate ions to evaluate the effect of the matrix precisely, where the seawater matrix was removed from samples using offline columns and then introduced into the IC. The difference from the direct method without an off-line column was discussed.

5.6.1 Apparatus and Measurement Procedure

A DX-500 series ion chromatograph (Dionex, Sunnyvale, USA) consisting of a gradient pump GP-40, a chromatography oven LC-30, a UV–visible absorbance detector AD-20, an autosampler AS with a 50-μL sample loop, and a self-regeneration suppressor SRS-300 was used for the analysis. Three anion exchange columns were used to separate analyte anions from the seawater matrix: (1) Dionex IonPac AG10 (4 × 50 mm) and AS10 (4 × 200 mm), (2) Dionex IonPac AG23 (4 × 50 mm) and AS23 (4 × 200 mm), and (3) IonPac AG12A (4 × 50 mm) and AS12A (4 × 200 mm). Suppressors were used to suppress the conductivity derived from the eluent.

For the nitrate analysis, the analytical conditions used are shown in Table 5.7. In considering the separation analyte and the seawater matrix, three conditions were studied. The candidate CRMs were diluted to 1.05 times with water after adding 0–6 mg kg^{-1} of nitrate. These were used as the measuring samples, and the amount of nitrate was determined using the standard addition method. For some nitrate analysis, seawater salts (chloride, bromide, and iodide) in the samples were removed using OnGuard II Ag (2.5 mL), OnGuard II Ba (2.5 mL), and OnGuard II Na (2.5 mL) before the measurements with IC.

Table 5.7 Analytical conditions for nitrate by IC

	Condition A	Condition B	Condition C
Guard column	IonPac AG10 (Dionex/Thermo)	IonPac AG23 (Dionex/Thermo)	IonPac AG12A (Dionex/Thermo)
Separation column	IonPac AS10 (Dionex/Thermo)	IonPac AS23 (Dionex/Thermo)	IonPac AS12A (Dionex/Thermo)
Eluent	80 mmol L^{-1} HCl	4.9 mmol L^{-1} Na_2CO_3 + 0.5 mmol L^{-1} $NaHCO_3$	2.7 mmol L^{-1} Na_2CO_3 + 0.3 mmol L^{-1} $NaHCO_3$
Flow rate (mL min^{-1})	1.0	1.0	1.5
Colum oven temperature (°C)	30	30	30
Suppressor current (mA)	–	50	50
UV detector wavelength (nm)	210 or 225	210	210

For the nitrite analysis, the candidate CRMs were diluted five times with water after adding 0–0.3 mg kg^{-1} of nitrite. The 40 mmol L^{-1} sodium chloride solution eluent flowed at 1.0 mL min^{-1}, and absorption at 210 nm of the measuring solutions was observed.

For the phosphate analysis, the candidate CRMs were diluted 3.4 times with water after adding 0–1.0 mg kg^{-1} of phosphate. Electric conductivities were observed using 4.9 mmol L^{-1} of sodium carbonate with 0.5 mmol L^{-1} sodium hydrogen carbonate solution as the eluent at a flow rate of 1.0 mL min^{-1}.

5.6.2 Analytical Results of the IC Analysis

The determination results for nitrate by IC with each condition are shown in Table 5.8. For nitrate in HSW and MSW, the determination values for each condition agreed with the standard uncertainty, regardless of the different measurement conditions, such as the pretreatment of the seawater samples and the separation and detection methods. The results suggest that halogens coexisting in the seawater matrix do not interfere with the nitrate readings in IC. Meanwhile, ELSW was measured under Condition A and was below the DL (<0.005 mg kg^{-1}). The direct method without pretreatment (A and B) and the desalination method using OnGuard II columns (D, F, and H) were treated as different analytical methods. Weighted means by the weights of the inverse of the standard uncertainties, respectively, were adopted as the analytical values for each method. For nitrite and phosphate, only the results of the direct method are obtained. For nitrite, the concentration was low, and only MSW could be analyzed. The results of the direct and desalination methods are shown in Tables 5.9 and 5.10, respectively.

Table 5.8 Determination results for nitrate by IC

	Pretreatment[*1]	Condition	Detection (wavelength [nm])	HSW[*2] (mg kg^{-1})	MSW[*2] (mg kg^{-1})
A	None	(A)	UV (225)	2.748 ± 0.010 ($n = 5$)	0.934 ± 0.005 ($n = 4$)
B	None	(B)	UV	2.748 ± 0.021 ($n = 1$)	0.937 ± 0.009 ($n = 1$)
D	Desalination-1	(A)	UV (210)	2.751 ± 0.026 ($n = 2$)	0.936 ± 0.028 ($n = 1$)
F	Desalination-1	(C)	ECD	2.759 ± 0.024 ($n = 1$)	0.931 ± 0.012 ($n = 1$)
H	Desalination-2	(C)	ECD	2.701 ± 0.019 ($n = 1$)	–

[*1] Desalination-1 means desalination for Cl^-, Br^-, and I^- by the OnGuard II Ag cartridge. Desalination-2 means desalination for Cl^-, Br^-, I^-, and SO_4^{2-} by both the OnGuard II Ag and the OnGuard II Ba cartridges
[*2] Each value after "±" indicates combined standard uncertainty due to the standard addition method and standard solutions
[*3] ELSW could not be determined because it was below the DL (0.005 mg kg^{-1})

Table 5.9 Analytical values of the three NMIJ CRMs by the IC direct method

	HSW (mg kg^{-1})	MSW (mg kg^{-1})	ELSW (mg kg^{-1})
NO_2^-	<0.009	0.0202 ± 0.0015 (7.3%)	<0.009
NO_3^-	2.748 ± 0.010 (0.35%)	0.936 ± 0.005 (0.51%)	<0.005
PO_4^{3-}	0.243 ± 0.040 (16.3%)	0.109 ± 0.027 (25.0%)	0.019 ± 0.008 (41.2%)

[*] Each value after "±" indicates combined standard uncertainty, and the figures in parentheses in the lower row represent the relative values of the combined standard uncertainty

Table 5.10 Analytical values of the three NMIJ CRMs by the IC desalination method

	HSW (mg kg^{-1})	MSW (mg kg^{-1})	ELSW (mg kg^{-1})
NO_3^-	2.735 ± 0.013 (0.47%)	0.933 ± 0.010 (1.12%)	–

[*] Each value after "±" indicates combined standard uncertainty, and the figures in parentheses in the lower row represent the relative values of the combined standard uncertainty

A recovery test was conducted to validate the method: 2.908 mg kg^{-1} (equivalent to HSW) of nitrate was added to ELSW, and the absorption at 225 nm was analyzed using Condition A. The measured concentration of 2.919 mg kg^{-1} agreed with the added concentration by preparation within the expanded uncertainty, U, with a coverage factor of 2, 0.018 mg kg^{-1}. Therefore, the IC methods were validated.

5.7 Ion Exclusion Chromatography Postcolumn Absorption Photometric Method

The ion exchange column used in the previous section is difficult to apply for dissolved silica. Thus, in this section, for the determination of dissolved silica in the three candidate CRMs, dissolved silica was separated from the seawater matrix using an ion exclusion column, followed by an online coloring reaction. Then, the concentration of dissolved silica was determined by measuring the absorption of the yellow complex. The molybdenum yellow method was used for the coloring reaction to detect dissolved silica. This method had a lower molar absorption coefficient than that of the molybdenum blue method used in CFA, which was disadvantageous in terms of color intensity. Meanwhile, the online operation was easy because only one coloring reagent was needed. Complicated operations might affect the accuracy of the measurements. Therefore, we decided to give priority to the latter advantage and adopt the molybdenum yellow method. Details of this method are described in the literature (Cheong et al., 2018).

5.7.1 Apparatus and Measurement Procedure

An IEC postcolumn absorption photometric system is shown in Fig. 5.2. An ICS-5000 series ion chromatograph (Dionex) consisting of an autosampler AS with a 100 μL sample loop and a UV–visible absorbance detector VWD was used for the analysis. The separation of dissolved silica from the seawater matrix was achieved when a Dionex IonPac ICE-AS1 (9 × 250 mm) was applied. The measurement conditions were as follows: the column oven temperature was 30 °C. The flow rate of the eluent was 1.0 mL min^{-1}, and that of the postcolumn reagent was 0.5 mL min^{-1}. A reaction coil of 2250 μL was adopted. The absorption of the produced silicomolybdenum yellow was measured at a wavelength of 410 nm.

The eluents were prepared by diluting hydrochloric acid (HCl) in water to 5.0 mmol L^{-1}. The postcolumn coloring reagent was prepared by dissolving disodium molybdate dehydrate, nitric acid, and sodium lauryl sulfate in water: 20-mmol L^{-1} sodium molybdate containing 0.2-mol L^{-1} sodium lauryl sulfate.

The candidate CRMs were analyzed using the standard addition method. The ELSW was diluted to 0.5 g g^{-1}, whereas the MSW and HSW were diluted to 0.2 g g^{-1} with pure water. Five levels of standard addition samples were prepared by adding a Si standard solution and measured using the IEC postcolumn absorption photometric method.

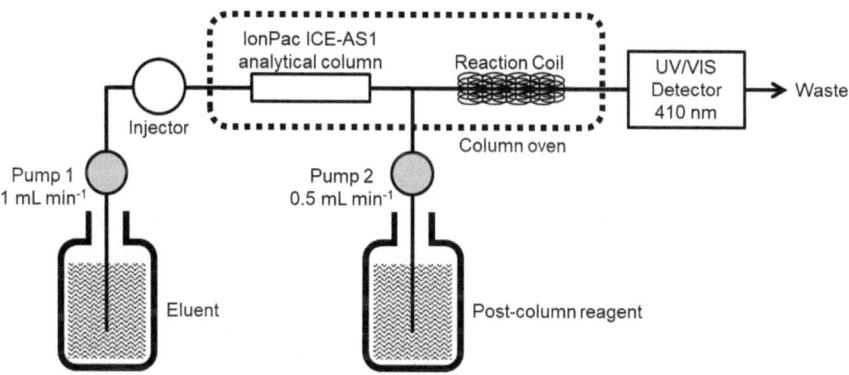

Fig. 5.2 Schematic diagram of the IEC postcolumn absorption photometric method (Modified from Cheong et al., 2018)

5.7.2 Analytical Results of the IEC Postcolumn Analysis

Four bottles of each of the three candidate CRMs were selected through stratified random sampling and analyzed in parallel; the average of the four measurements was used as the analytical value. The analytical results are shown in Table 5.11. The correlation coefficient r of the calibration curve for the standard addition method ranged from 0.9999 to 1.0000 for the HSW and MSW analyses, which was sufficiently linear considering the required uncertainty. The relative combined standard uncertainty for each analysis was evaluated from the relative standard uncertainty u_{SA} due to the standard addition method and the relative standard uncertainty of the silicon standard solution used in the analyses (0.15%). u_{SA} was calculated from the following equation (De Beer et al., 2007; Miller, 1991).

$$u_{SA} = \frac{s}{a} \sqrt{\frac{1}{n} + \frac{\bar{y}^2}{a^2 \sum_i (x_i - \bar{x})^2}} \qquad (5.4)$$

where, a is the slope of the standard addition calibration curve, s is the residual standard deviation from the calibration curve, n is the total number of measured quantity values on which the standard addition calibration curve is based, x_i is the

Table 5.11 Analytical values of the three NMIJ CRMs by the IEC postcolumn

	HSW (mg kg^{-1})	MSW (mg kg^{-1})	ELSW (mg kg^{-1})
Dissolved silica (as Si)	4.132 ± 0.010 (0.24%)	0.830 ± 0.003 (0.30%)	0.033 ± 0.001 (2.7%)

* Each value after "±" indicates combined standard uncertainty, and the figures in parentheses in the lower row represent the relative values of the combined standard uncertainty

concentration of dissolved silica added to the sample solution at each point i of the standard addition calibration curve, $\bar{x}$ is the average of all x_i values, and $\bar{y}$ is the average of all measured quantity values.

A recovery test was conducted to validate the method: an artificial sample of 4.17 mg kg^{-1} (equivalent to HSW) of dissolved silica was added to ASW (described in Sect. 2.6) and determined as the same as that for the candidate standard; the amount of dissolved silica in ASW was estimated to be 0.12 µg kg^{-1} (Cheong et al., 2014). The mean value of the six measurements was 4.17 mg kg^{-1}, with an expanded uncertainty U ($k = 2$) of 0.08 mg kg^{-1} (1.8%, relative), which was calculated from the standard deviation of six measurements (0.91%) and the standard uncertainty of the standard solution used (0.15%). The method was validated because the concentrations agreed with the prepared concentrations within U ($k = 2$). The DL of the method was 0.003 mg kg^{-1}, which was sufficient for the determination of dissolved silica in seawater.

5.8   Ion Exclusion Chromatography–Isotopic Dilution Inductively Coupled Plasma Mass Spectrometry

As silicon has multiple stable isotopes, isotope dilution mass spectrometry (IDMS), which is a primary method of measurement (Richter, 1997) with high metrological reliability, can be applied. Therefore, dissolved silica in the candidate CRMs was determined using IEC–ID–ICP–MS. IDMS is an analytical method where a known amount of enriched isotope of the same element with a different isotopic composition as the sample is added as a spike solution. Then, the amount of the substance in the sample is determined from the change in isotope ratio before and after addition. The silicon concentration of the spike solution used for ID was precisely determined through reverse ID using two different silicon standard solutions. Silicon derived from dissolved silica was separated in an ion exclusion column to avoid instrumental contamination by the seawater matrix, and the eluate was introduced online into an inductively coupled plasma mass spectrometer. Details of this method were described in Nonose et al. (2014).

5.8.1   Apparatus

The IEC–ID–ICP–MS system is shown in Fig. 5.3. The ion exclusion chromatograph was a Dionex ICS-3000 with an autosampler AS with a 100 µL sample loop and an IonPac ICE-AS1 ion exclusion column for separation. An inductively coupled plasma mass spectrometer Element2 (Thermo Fisher Scientific Inc., Bremen, Germany) was used for isotope ratio measurements. The resolution of the mass spectrometer ($m/\Delta m$) was set to 4000 to avoid polyatomic interferences. A quartz-free introduction system (Pt injector, Pt cones, PFA double-pass spray chamber) was used to reduce the

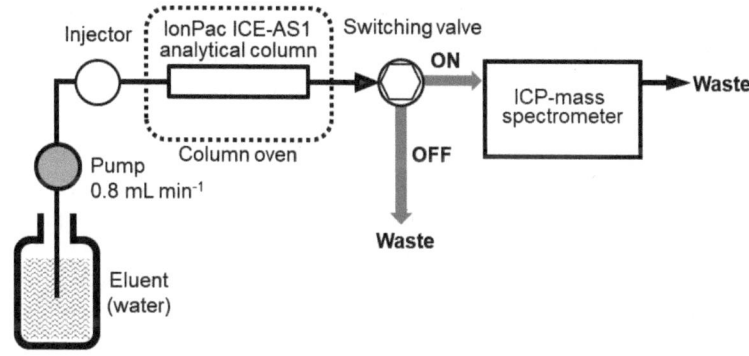

Fig. 5.3 Schematic diagram of IEC–ID–ICP–MS

instrumental background effect. The chromatograph and ICP–MS were connected by a six-port switching valve.

5.8.2 Preparing the Spike Solution and Measuring Solutions

Ultrapur grade water (Kanto) was used to prepare sample solutions for ID and reverse ID. Meanwhile, the silicon spike solution was prepared through the alkaline fusion of Oak Ridge National Laboratory ^{29}Si oxide powder (95.65% enrichment) with Suprapur grade (>99.999%) sodium carbonate (Kanto), followed by dilution with water to approximately 500 mg kg^{-1}.

Three ID measurement solutions for ID were prepared by mixing 0.91 g g^{-1} of the candidate standard with ^{29}Si spike solutions (3.6 mg kg^{-1} for HSW, 0.78 mg kg^{-1} for MSW, and 0.033 mg kg^{-1} for ELSW).

Two kinds of reverse ID measurement solutions were prepared to accurately determine the silicon concentration in the ^{29}Si spike solution: a silicon standard solution and a ^{29}Si spike solution containing the same concentration of silicon as that of the ^{29}Si spike in the ID measurement solutions.

For isotope ratio correction, a silicon standard solution containing the same concentration of silicon as that of the reverse ID measurement solutions was prepared using NIST SRM 3150.

5.8.3 Measurement Procedure

The prepared measuring solutions were introduced into the IEC–ID–ICP–MS apparatus to measure silicon at mass numbers 28 and 29 and obtain a chromatogram. The eluent for IEC was water, its flow rate was 0.8 mL min^{-1}, and the measurement time for one sample was set to 14 min. One example of the chromatogram for the candidate CRM (HSW) is shown in Fig. 5.4. There was a difference in retention time of ~70 s between the peaks derived from chloride and sulfate (retention time 460 s), the major components of the seawater matrix, and the peak derived from dissolved silica (retention time 530 s). The switching valve was switched at 500 s to remove the major components of the seawater matrix from the sample introduction system of the ICP mass spectrometer, and only dissolved silica was introduced to the apparatus.

The isotope ratio of ^{28}Si and ^{29}Si was calculated from the peak area of the chromatogram, and the concentration of silicon in the ^{29}Si spike solution C_y and the concentration of dissolved silica in seawater C_x were determined according to the following equations:

$$C_y = C_z \cdot \frac{m_z}{m'_y} \cdot \frac{A_f(\text{mass A, z})\text{-}K_{b'} \cdot R_{b'} \cdot A_f(\text{mass B, z})}{K_{b'} \cdot R_{b'} \cdot A_f(\text{mass B, y})\text{-}A_f(\text{mass A, y})} \tag{5.5}$$

$$C_x = C_y \cdot \frac{m_y}{m_z} \cdot \frac{K_b \cdot R_b \cdot A_f(\text{mass B, y})\text{-}A_f(\text{mass A, y})}{A_f(\text{mass A, z})\text{-}K_b \cdot R_b \cdot A_f(\text{mass B, z})} \tag{5.6}$$

The subscripts $x, y, z, b,$ and b' indicate seawater, ^{29}Si spike solutions, silicon standard solution, ID measurement solutions (mixture of sample and spike), and reverse ID measurement solutions (mixture of spike and metal standard solution), respectively. C is the amount of material per kilogram, m is the mass, R is the measured isotope ratio with detector dead time correction, K is the correction factor for the isotope ratio calculated from the internal standardization method, and A_f (mass X, Y) is the isotopic abundance of the mass X isotope in Y (sample or spike). The values for mass A and mass B are 28 and 29 for the measurement of silicon. The sample concentration C_x was calculated by combining the uncertainties due to the spike concentration.

The isotopic abundance, the atomic weight of silicon, and the uncertainty of silicon were taken from the IUPAC Inorganic Chemistry Division, CIAAW (Berglund & Wieser, 2011).

Fig. 5.4 Chromatogram of HSW with the addition of the ^{29}Si spike and after diluting by 100 times (Modified from Nonose et al., 2014)

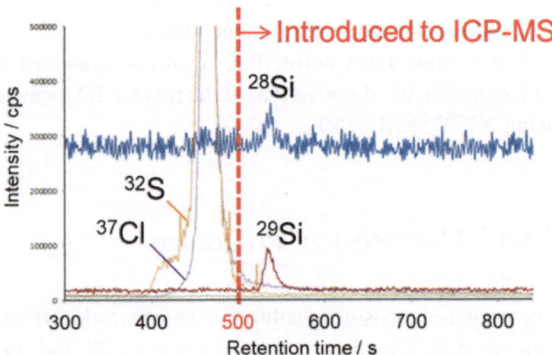

Table 5.12 Analytical values of the three NMIJ CRMs by the IEC-ID-ICP-MS

	HSW (mg kg^{-1})	MSW (mg kg^{-1})	ELSW (mg kg^{-1})
Dissolved silica (as Si)	4.075 ± 0.021 (0.53%)	0.830 ± 0.004 (0.46%)	0.0386 ± 0.0015 (4.0%)

[*]Each value after "$\pm$" indicates combined standard uncertainty, and the figures in parentheses in the lower row represent the relative values of the combined standard uncertainty

5.8.4 Analytical Results by IEC–ID–ICP–MS

The results of the analysis of dissolved silica in three levels of candidate RMs by IEC–ID–ICP–MS are shown in Table 5.12. The uncertainty of the sample concentration consisted of the uncertainty of the spike concentration obtained by inverse ID and the uncertainty obtained by ID (not including the uncertainty of the spike concentration) and was calculated assuming no correlation between the two. The uncertainty derived from weighing was not considered because it was small compared with the other factors. See the literature (Nonose et al., 2014) for details on uncertainty calculations.

For method validation, the seawater CRM produced by NRC Canada (MOOS-2) (the certified value of dissolved silica was 0.809 ± 0.028 mg kg^{-1}, where the value following $\pm$ represented U ($k = 2$)) was analyzed similarly as the analysis of candidate CRMs. The analytical result (0.789 ± 0.007 mg kg^{-1}) agreed with the certified value within U ($k = 2$), thus confirming the validity of the analysis using this method.

5.9 Homogeneity of the CRMs

The homogeneities of the seawater samples (candidate NMIJ CRMs) were estimated using the analytical result by CFA. According to the International Organization for Standardization (ISO) guide 35:2006, ANOVA was applied to the determined values of CFA. Then, the SD of between-bottle homogeneity (s_{bb}) and the standard uncertainty due to between-bottle homogeneity (u_{bb}) were obtained. The larger value of s_{bb} and u_{bb} was adopted as the standard uncertainty due to the homogeneity of the seawater samples. All uncertainties due to homogeneity are shown in Table 5.13. Although the homogeneities of the ELSW and a part of HSW were poorer than others due to the extremely low mass fractions of the analyte in those samples, these amounts of substance components were conclusively decided as indicative values or information as described in Sect. 5.3.

Table 5.13 Standard uncertainty due to homogeneity of the candidate CRMs

	HSW (mg kg^{-1})	MSW (mg kg^{-1})	ELSW (mg kg^{-1})
NO_2^-	0.00020 (15.9%)	0.00007 (0.36%)	0.00017 (10.3%)
NO_3^-	0.0082 (0.30%)	0.0007 (0.07%)	0.0001 (6.2%)
PO_4^{3-}	0.0007 (0.23%)	0.0002 (0.22%)	0.0007 (42%)
Dissolved silica (as Si)	0.0017 (0.04%)	0.0005 (0.06%)	0.00005 (0.15%)

[*]Each value in parentheses indicates the relative value to the property value

5.10 Stability of the CRMs

The long-term stabilities of the seawater samples (candidate NMIJ CRMs) were estimated on the basis of the results of the stability monitoring up to almost 1500 days by CFA. According to ISO Guide 35 (ISO, 2017), each mass fraction was trended against the number of days elapsed. If the following equation held, the slope of the regression line was not significant, and no instability was observed.

$$|b_1| < t_{(0.95, n-2)} \times s_{(b_1)} \qquad (5.7)$$

Here, the slope of the regression line was b_1, the uncertainty of the slope was $s_{(b1)}$, the degrees of freedom were $n - 2$, and the Student's t coefficient at the 95% confidence level was $t_{(0.95, n-2)}$. The results of the trend analysis showed that the slope was not significant for any of the components, and instability was not observed for ~4 years since the determination of the certified value. Therefore, the expiration date (T) was set to 10 years after certification, and the standard uncertainty due to the long-term stability u_{lts} was estimated according to ISO guide 35 using the following equation:

$$u_{lts} = s_{(b_1)} \times T \qquad (5.8)$$

All the standard uncertainties due to long-term stability are shown in Table 5.14. For the components that were finally used as "indicative values" or "information," no estimation of long-term stability was made and only the analytical results at the time of certification were obtained.

5.11 Determination of Certified Values for the CRMs

As the results of this study, six kinds of analytical methods were applied to analyze seawater samples. For nitrate, the results were obtained using CFA, IC direct method, and IC without desalination, as shown in Fig. 5.5. For nitrite, the results were obtained using CFA and the IC direct method, as shown in Fig. 5.6. For phosphate, the results

Table 5.14 Standard uncertainties due to the long-term stability of the candidate CRMs

	HSW (mg kg^{-1})	MSW (mg kg^{-1})	ELSW (mg kg^{-1})
NO$_2^-$	–	0.0002 (0.90%)	–
NO$_3^-$	0.0221 (0.81%)	0.0098 (1.0%)	–
PO$_4^{3-}$	0.0022 (0.75%)	0.0021 (2.1%)	–
Dissolved silica (as Si)	0.0078 (0.19%)	0.0045 (0.53%)	0.00052 (1.5%)

*Each value in parentheses indicates the relative value to the property value

were obtained using CFA, IC direct method, and IC–ICP–MS, as shown in Fig. 5.7. For dissolved silica, the results were obtained using CFA, IEC–ID–ICP–MS and IEC postcolumn absorption spectrometry method, and IC without desalination, as shown in Fig. 5.8. Among the total 12 items (three concentration levels × four nutrients), items for which the analytical results were obtained by more than one method were evaluated as certified values, whereas items for which the analytical result was obtained by only one method were evaluated as indicative values. Here among the items that should have been evaluated as indicative values, four items (i.e., nitrite, nitrate, and phosphate in ELSW, and nitrite in HSW) were evaluated just as information because it was found that those measurement SD and uncertainty due to homogeneity and/or stability were large and that the analytical results lacked reliability. Note that the analytical results by IC–ICP–MS were not used for determining the property values because the analytical results had not been obtained yet when the CRMs were certified.

Each property value was determined as follows: for items for which only one available analytical result was obtained, the analytical result was adopted as the property value. For items for which available analytical results were obtained by multiple methods, the arithmetic mean of those results was adopted as the property value. However, some items varied with different analytical methods, which could not be explained by examinations up to now. In particular, the results of CFA had relatively small uncertainties than those of other analytical methods. Thus, an uncertainty factor of type B that was not estimated might have existed. Therefore, the analytical results by each method were regarded as equivalent, and the property value was calculated using the arithmetic mean of those results. The observed differences between the methods were estimated by assuming a rectangular distribution of the maximum difference between the property value and the analytical results of each method.

The following uncertainty factors (1)–(4) were considered standard uncertainties attached to property values.

(1) Relative standard uncertainty due to each method (standard uncertainty of the arithmetic mean). This was estimated using the following formula:

Fig. 5.5 Analytical results for nitrate using the developed methods and the finally determined property values. **a** HSW, **b** MSW, and **c** ELSW were determined. Each error bar means U ($k = 2$). The uncertainty of **c** ELSW was not given because the property value was finally provided as "information." (Modified from Cheong, 2020)

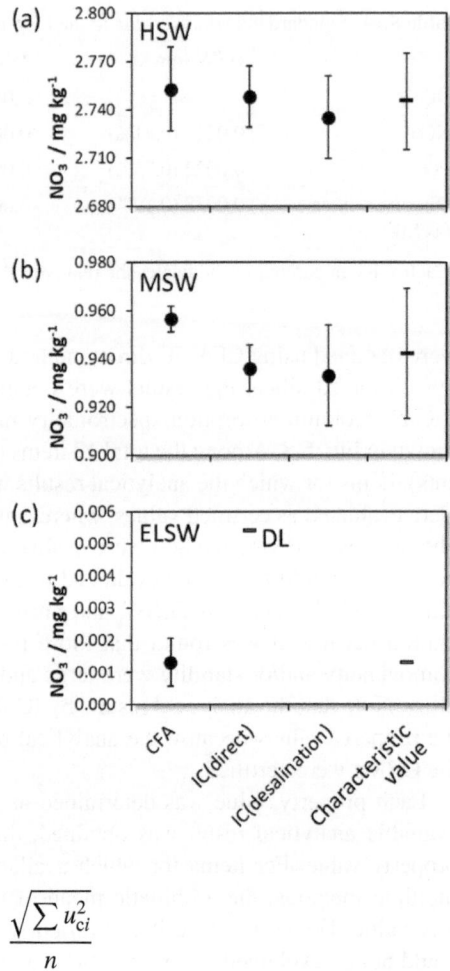

$$\frac{\sqrt{\sum u_{ci}^2}}{n}$$

Here, u_{ci} means the relative standard uncertainty due to each analytical method, and n represents the number of methods.

(2) Relative standard uncertainty due to the difference between the analytical methods. This was estimated using the following equation (as mentioned above):

$$\frac{\frac{\text{maximum deviation between} x_{\text{char}} \text{and} x_{\text{meas}}}{\sqrt{3}}}{x_{\text{char}}}$$

Here, x_{char} refers to a property value, whereas x_{meas} indicates the analytical results of each analytical method.

(3) Relative standard uncertainty due to homogeneity. All relevant items are shown in Table 5.13.

Fig. 5.6 Analytical results for nitrite using the developed methods and the finally determined property values. **a** HSW, **b** MSW, and **c** ELSW were determined. In the case when the result was not obtained, the DL was shown in the figure. Each error bar means U ($k = 2$). The uncertainties of **a** HSW and **c** ELSW were not given because their property values were finally provided as "information." (Modified from Cheong, 2020)

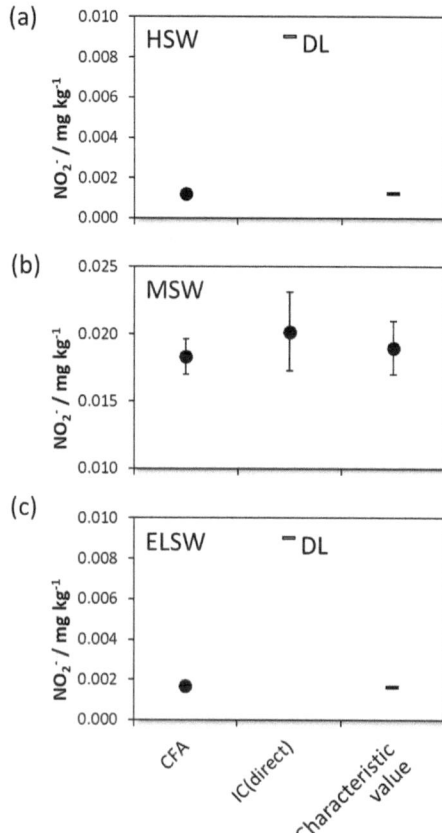

(4) Relative standard uncertainty due to long-term stability. All relevant items are shown in Table 5.14.

The combined standard uncertainty was calculated from the square root of the sum of squares of these standard uncertainties. Certified values, indicative values, and property values as information are summarized in Table 5.15. The magnitudes of the uncertainties of nutrient mass fraction in the developed CRMs fulfilled minimum requirements in the oceanography community at least at relatively high concentrations. The NMIJ CRMs, developed in this way, had a wide nutrient concentration range from high nutrient concentrations in the deep Pacific Ocean to near zero on the surface water, for example, used for blank evaluation. They are the world's first nutrient CRMs useful for the accurate control of nutrient measurements in actual seawater, including ocean observations.

Fig. 5.7 Analytical results for phosphate using the developed methods and the finally determined property values. **a** HSW, **b** MSW, and **c** ELSW were determined. The plots of ● were used for calculating the property values. Those of ○ were not used; the values obtained by the IC direct method gave large uncertainties, and the values obtained by IC–ICP–MS were obtained after the CRMs were certified. In the case when the result was not obtained, the DL was shown in the figure. Each error bar means U ($k = 2$). The uncertainty of **c** ELSW was not given because the property value was finally provided as "information." (Modified from Cheong, 2020)

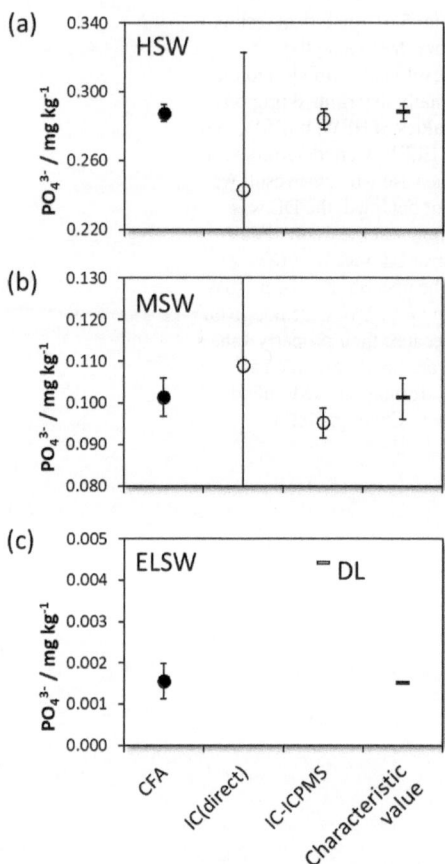

5.12   Conclusions

Seawater CRMs for nutrient analysis were developed using the analytical results of three seawater samples (candidate CRMs). The analytical results of three candidate CRMs using the developed methods in this study were aggregated. On this basis, homogeneity and long-term stability were evaluated and the concentrations of the nutrients were characterized. Three kinds of seawater CRMs for nutrient analysis were certified by NMIJ, with property values according to their properties (certified values, indicative values, and information). The developed CRMs were given uncertainties attached to property values and the range of nutrient concentrations that fulfilled the minimum requirements of the oceanography community. The seawater CRMs were SI traceable and reliable CRMs characterized on the basis of SI traceable national primary standard solutions using multiple validated analytical methods as possible. In addition, the CRMs were developed by a national metrology institute on the basis of MRA. They are the world's first CRMs that can cover a wide range

Fig. 5.8 Analytical results for dissolved silica using the developed methods and the finally determined property values. **a** HSW, **b** MSW, and **c** ELSW were determined. The plots of ● were used for calculating the property values, whereas those of ○ were not used because of the same determination fundamentals. Each error bar means U ($k = 2$). (Modified from Cheong, 2020)

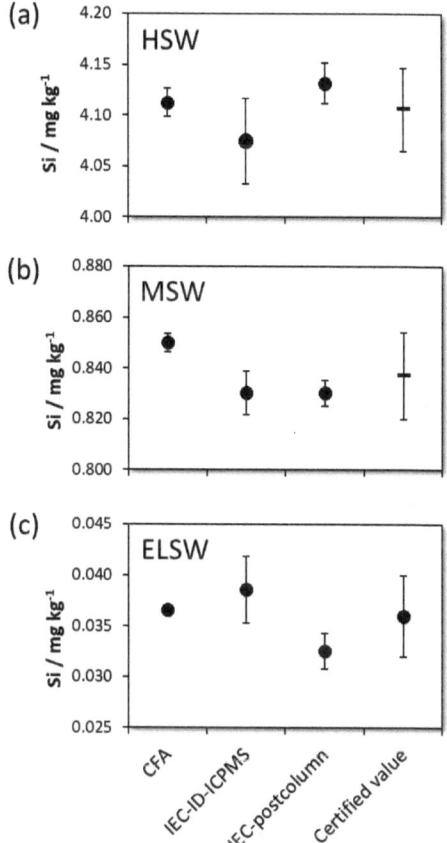

Table 5.15 Certified values of the three NMIJ CRMs (HSW, MSW, and ELSW)

	HSW (mg kg^{-1})	MSW (mg kg^{-1})	ELSW (mg kg^{-1})
NO_2^-	0.0012[*3]	0.019 ± 0.002	0.0016[*3]
NO_3^-	2.745 ± 0.030	0.942 ± 0.027	0.0013[*3]
PO_4^{3-}	0.288 ± 0.005[*2]	0.101 ± 0.005[*2]	0.002[*3]
Dissolved silica (as Si)	4.106 ± 0.041	0.837 ± 0.017	0.036 ± 0.004

[*1] Each value after "±" indicates U ($k = 2$)

[*2] Indicative values

[*3] Provided as information, which means a determination value in time using only one analytical method. Because the final uncertainty was large and the reliability as a property value was not sufficient, these values were provided as such

of nutrient concentrations from the surface water level to that of the deep Pacific Ocean. The use of these CRMs for hydrographic observations, among various applications, could improve the reliability and accuracy of observations and could lead to ensuring the comparability of observation values. Currently, the first key comparison of seawater nutrients is being conducted by the Consultative Committee for Amount of Substances. Thus, the international equivalence of nutrient analyses in different countries is confirmed and is expected to become stronger in the future.

Competing Interests The authors have no conflicts of interest to declare that are relevant to the content of this chapter.

References

Aoyama, M., Ota, H., Kimura, M., Kitao, T., Mitsuda, H., Murata, A., & Sato, K. (2012). Current status of homogeneity and stability of the reference materials for nutrients in seawater. *Analytical Sciences, 28*(9), 911–916. https://doi.org/10.2116/analsci.28.911

Bakker, K., Ooijen, J., Weerlee, E., & Epping, E. (2010). A Comparison of Silicate Standards in Nutrient Analyses. In M. Aoyama, A. Dickson, D. Hydes, A. Murata, P. Roose, & E. Woodward (Eds.), *Comparability of nutrients in the world's ocean* (pp. 127–139). Mother Tank.

Berglund, M., & Wieser, M. E. (2011). Isotopic compositions of the elements 2009 (IUPAC Technical Report). *Pure and Applied Chemistry, 83*(2), 397–410. https://doi.org/10.1351/pac-rep-10-06-02

Cheong, C. (2020). *A study on accurate analysis of seawater nutrients.* Doctoral dissertation, University of Tsukuba. Retrieved from https://tsukuba.repo.nii.ac.jp/records/55415

Cheong, C., Miura, T., & Nonose, N. (2021). Determination of phosphate in seawater by ion chromatography inductively coupled plasma sector field mass spectrometry. *Limnology and Oceanography: Methods, 19*(10), 682–691. https://doi.org/10.1002/lom3.10453

Cheong, C., Nonose, N., Miura, T., & Hioki, A. (2014). Improved accuracy of determination of dissolved silicate in seawater using absorption spectrometry. *Accreditation and Quality Assurance, 19*(1), 31–40. https://doi.org/10.1007/s00769-013-1024-5

Cheong, C., Sakaguchi, A., Sueki, K., & Ohata, M. (2020). Evaluation of the calibration method for accurate analysis of dissolved silica by continuous flow analysis. *Analytical Sciences, 36*(2), 247–253. https://doi.org/10.2116/analsci.19P291

Cheong, C., Suzuki, T., Miura, T., & Hioki, A. (2024). Comparison of continuous flow analysis and ion chromatography for determinations of nitrate, nitrite and phosphate ions in seawater and development of related seawater certified reference materials. *Accreditation and Quality Assurance, 29*(3), 243–251. https://doi.org/10.1007/s00769-024-01586-x

Cheong, C., Yamauchi, Y., & Miura, T. (2018). Determination of dissolved silica in seawater by ion-exclusion chromatography with post-column derivatization/silicomolybdenum yellow detection. *Analytical Sciences, 34*(4), 477–481. https://doi.org/10.2116/analsci.17P508

Clancy, V., Gedara, I. P., Grinberg, P., Meija, J., Mester, Z., Pagliano, E., Willie, S., & Yang, L. (2014). MOOS-3: Seawater certified reference material for nutrients. *Retrieved from.* https://doi.org/10.4224/crm.2014.moos-3

De Beer, J. O., De Beer, T. R., & Goeyens, L. (2007). Assessment of quality performance parameters for straight line calibration curves related to the spread of the abscissa values around their mean. *Analytica Chimica Acta, 584*(1), 57–65. https://doi.org/10.1016/j.aca.2006.11.032

Grasshoff, K. (1964). On the determination of silica in sea water. *Deep Sea Research and Oceanographic Abstracts, 11*(4), 597–604. https://doi.org/10.1016/0011-7471(64)90004-x

Griess, P. (1879). Bemerkungen zu der abhandlung der HH. Weselsky und benedikt ueber einige azoverbindungen. *Berichte der deutschen Chemischen Gesellschaft, 12*, 426–428.

Hioki, A., Kokubun, A., & Kubota, M. (1994). Accuracy of the measurement of purity of sulfamic acid by coulometric titration. *The Analyst, 119*(8). https://doi.org/10.1039/an9941901879

Hioki, A., Watanabe, T., Terajima, K., Fudagawa, N., Kubota, M., & Kawase, A. (1990). Accuracy in gravimetric-determination of nitrate and nitrite as nitron nitrate. *Analytical Sciences, 6*(5), 757–762. https://doi.org/10.2116/analsci.6.757

Huygen, I. C. (1970). Reaction of nitrogen dioxide with Griess type reagents. *Analytical Chemistry, 42*(3), 407–409. https://doi.org/10.1021/ac60285a018

International Organization for Standardization. (2017). *Reference materials—General and statistical principles for certification* (ISO Guide 35:2017).

Kato, C., & Hioki, A. (2009). Evaluation of a reduction rate of nitrate in continuous flow analysis. *Bunseki Kagaku, 58*(8), 723–729. https://doi.org/10.2116/bunsekikagaku.58.723

Kato, K. (1976). Spectrophotometric determination of dissolved silica based on alfa-molybdosilicic acid formation. *Analytica Chimica Acta, 82*, 401–408.

Kawano, T., & Uchida, H. (2007). *WHP P10 Revisit Data Book*. JAMSTEC. Retrieved from https://doi.org/10.17596/0000035

Knap, A., Michaels, A., Close, A., Ducklow, H., & Dickson, A. (Eds.). (1996). *Protocols for the Joint Global Ocean Flux Study (JGOFS) Core Measurement, JGOFS Report Nr. 19*. Woods Hole, MA: Scientific Committee on Oceanic Research.

Miller, J. N. (1991). Basic statistical methods for analytical chemistry. Part 2. Calibration and regression methods. A review. *The Analyst, 116*(1), 3–14. https://doi.org/10.1039/an9911600003

Miranda, K. M., Espey, M. G., & Wink, D. A. (2001). A rapid, simple spectrophotometric method for simultaneous detection of nitrate and nitrite. *Nitric Oxide, 5*(1), 62–71. https://doi.org/10.1006/niox.2000.0319

Murphy, J., & Riley, P. J. (1962). A modified single solution method for the determination of phosphate in natural waters. *Analytica Chimica Acta, 27*, 31–36. https://doi.org/10.1016/S0003-2670(00)88444-5

Namiki, H. (1964). Fundamental conditions and reaction mechanisms in formation of phosphomolybdenum blue. *Bulletin of the Chemical Society of Japan, 37*(4), 484–491. https://doi.org/10.1246/bcsj.37.484

Nonose, N., Cheong, C., Ishizawa, Y., Miura, T., & Hioki, A. (2014). Precise determination of dissolved silica in seawater by ion-exclusion chromatography isotope dilution inductively coupled plasma mass spectrometry. *Analytica Chimica Acta, 840*, 10–19. https://doi.org/10.1016/j.aca.2014.06.018

Ramachandran, R., & Gupta, P. K. (1985). An improved spectrophotometric determination of silicate in water based on molybdenum blue. *Analytica Chimica Acta, 172*, 307–311. https://doi.org/10.1016/S0003-2670(00)82621-5

Richter, W. (1997). Primary methods of measurement in chemical analysis. *Accreditation and Quality Assurance, 2*(8), 354–359. https://doi.org/10.1007/s007690050165

Yoshiyuki, N. (1997). Updated (1996 version) table of the elements in seawater and its remarks. *Bulletin of the Chemical Society of Japan, 51*, 302–308.

Zhang, J.-Z., Fischer, C. J., & Ortner, P. B. (2000). Comparison of open tubular cadmium reactor and packed cadmium column in automated gas-segmented continuous flow nitrate analysis. *International Journal of Environmental Analytical Chemistry, 76*, 99–113. https://doi.org/10.1080/03067310008034123

Chapter 6
Development of Primary Inorganic Standard Solution of Si, NMIJ CRM 3645-a

Tsutomu Miura and Chikako Cheong

Abstract The need to develop Si standard solutions as national standards has been expanded, since the importance of Si measurement for evaluating industrial material and environmental analysis is increasing. For developing a Si standard solution, the purity assignment method for high-purity SiO_2 using the gravimetric analysis and gravimetric preparation method for Si standard solution was established. Finally, an NMIJ CRM 3645-a Si standard solution was developed. Thirty-three elements in the high-purity SiO_2 were measured for impurity evaluation. As a result, the amounts of impurities in SiO_2 were low enough not to affect the analytical value obtained by the gravimetric analysis. The purity value of SiO_2 was 99.648%. The relative standard deviation (RSD) of the mean value (RSD/$\sqrt{n}$) was considered the standard uncertainty in the gravimetric analysis, so the final purity value was 99.648% $\pm$ 0.074% ($k = 2$). The stability and homogeneity of the prepared Si standard solution were evaluated by inductively coupled plasma optical emission spectrometry and a continuous flow analyzer. Based on the variability and homogeneity, the standard uncertainties were 0.13% and 0.017%, respectively. Finally, the expanded uncertainty of NMIJ CRM 3645-a Si standard solution was 0.28%.

Keywords Inorganic standard solution · Certified reference material · Silicon · Traceability · Gravimetric analysis · ICP-OES · Continuous flow analysis · Concentration ratio

T. Miura (✉) · C. Cheong
National Institute of Advanced Industrial Science and Technology, National Metrology Institute of Japan, Tsukuba, Ibaraki, Japan
e-mail: t.miura@aist.go.jp

C. Cheong
e-mail: c-kato@aist.go.jp

© The Author(s) 2025
M. Aoyama et al. (eds.), *Chemical Reference Materials for Oceanography*, Springer Oceanography, https://doi.org/10.1007/978-981-96-2520-8_6

6.1   Introduction

Generally, the signal intensities of analytical instruments for inorganic analyses, such as atomic absorption spectrometry (AAS), inductively coupled plasma optical emission spectrometry (ICP-OES), and inductively coupled plasma mass spectrometry (ICP-MS), must be calibrated using elemental standard solutions for the target elements (Bièvre & Taylor, 1997; Choquette et al., 2020; Milton & Quinn, 2001; Quinn, 1997; Röthke et al., 2020; Vogl, 2018; Vogl et al., 2018). The National Metrology Institute of Japan (NMIJ) is responsible for developing certified reference materials and establishing the traceability of SI (The International System of Units) on chemical metrology in Japan (Hioki, 2008; Miura & Wada, 2022; Nonose et al., 2017; Suzuki et al., 2007). To establish SI traceability, the primary measurement method should be applied to characterize certified reference materials. The metrological traceability of inorganic elemental analysis to SI (The International System of Units) is established using purity-assayed high-purity metals and inorganic compounds. The analytical instrument is calibrated using an elemental standard solution prepared gravimetrically from the purity-assayed inorganic material. This calibration procedure provides traceability to SI.

Si has a wide range of industrial applications, and it has been used in high-temperature structural materials, refractory materials, heat-resistant insulating materials, and cutting tools in glass, ceramics, and fine ceramics. In the steel industry, Si is added as an oxygen absorber to electric furnaces. Additionally, Si steel plates are used in transformers to reduce losses owing to eddy currents. Si alloys are also used in the aluminum industry. The demand for high-purity single crystals as semiconductor devices is increasing. Consequently, the applications of amorphous and polycrystalline Si in liquid-crystal displays and solar cells are also increasing. Si has also been subjected to chemical analysis in air dust, asbestos, suspended particulate matter, soil, sediment, land water, and seawater. Dissolved silicate in seawater is crucial for nutrient analysis in global warming evaluation and modeling. The importance of Si measurement for evaluating industrial materials and environmental analysis is increasing, with an increasing need to develop Si standard solutions as national standards. Therefore, the NMIJ has developed a certified reference material (CRM), NMIJ CRM 3645-a Si standard solution as a national standard in Japan.

6.2   Materials and Methods

6.2.1   Instruments and Apparatus

A semi-microbalance (Mettler Toledo XP-205) was used to weigh the samples. The balance was calibrated using the Japan Calibration Service System (JCSS). The balances were installed in a balance room controlled at 25 °C ± 2 °C and 50% ± 10% relative humidity. All weighing was performed in the balance room. A platinum crucible (nominal volume: 40 mL) was used as the container for Na_2CO_3 fusion and the gravimetric analysis of Si. A bead and fuse-sample TK-4100 and program controller TK-5910 (AmenaTech, Yokohama, Japan) were used for Na_2CO_3 fusion of SiO_2, gravimetric analysis of Si, and preparation of the Si standard solution. Water was purified using a Merck Millipore Milli-Q Advantage A10 instrument with a Q-pod element purification system. An ICP-OES PerkinElmer Optima 4300DV (PerkinElmer Japan, Yokohama, Japan) was used to determine the unprecipitated Si content in the filtrate and the washing solution for SiO_2 hydrate precipitation and impurity analysis of SiO_2 as a raw material. A continuous flow analyzer AACS-V (BLTEC K. K., Osaka, Japan) was used to evaluate the preparation variability of the Si standard solution.

6.2.2 Reagents

High-purity SiO_2 (10 kg, Norwegian Crystallites AS, powder, Lot No. 2045, informative purity value: 99.99%) was purchased from FUJI FILM Wako Pure Chemicals Corporation (Osaka, Japan). The SiO_2 was divided into 100 bottles, each containing 100 g of SiO_2. Ultrapur-100^{TM} grade HCl, HNO_3, HF, and H_2SO_4 were purchased from Kanto Chemicals Co. Inc. (Tokyo, Japan). Na_2CO_3 (Suprapur, Merck) was used for SiO_2 fusion. A 1% tetramethyl ammonium hydroxide (TMAH) solution was prepared by diluting 25% TMAH solution (TAMAPURE-AA TMAH, TAMA CHEMICALS Co., LTD, Tokyo, Japan) in water. A NIST SRM 3150 Si standard solution was used to determine the Si content in the filtrate and washing solution. A Ga solution, intended as an internal standard for Si determination using ICP-OES, was prepared by diluting the JCSS standard solution (Ga: 1000 mg dm^{-3}, Kanto Chemicals Co. Inc.). NIST traceable standard solutions of SPEX XSTC-331 were used for impurity analysis using ICP-OES. The preparation variability of the Si standard solution was evaluated using a continuous flow analyzer (CFA) and the Si molybdenum blue method (Cheong et al., 2014, 2020). A 0.06 mol L^{-1} (as molybdenum) molybdate solution, as a complexing agent, was prepared by dissolving 15.0 g of disodium molybdate dihydrate (FUJIFILM Wako Pure Chemical Co., Osaka, Japan) in 300 mL of water containing 6 mL of H_2SO_4 (FUJIFILM Wako Pure Chemical Co.) and diluting the resultant solution to 1000 mL with water. As a phosphate masking agent, a 0.40 mol L^{-1} oxalic acid solution was prepared by dissolving 25.0 g of oxalic acid dihydrate (Kanto Chemical Co., Inc., Tokyo, Japan) in 300 mL of water and diluting the solution to 500 mL with water. As a reducing agent, a 0.14 mol L^{-1} L-ascorbic acid solution was prepared by dissolving 2.5 g of L-ascorbic acid (FUJIFILM Wako Pure Chemical Co.) in 80 mL of water and diluting the solution to 100 mL with water.

6.2.3 Measurement Procedure of Metallic Impurities Using ICP-OES

Approximately 2.3, 3.6, 2.1, and 2.5 g of SiO_2 were weighed for impurity analysis. The samples were dissolved in HNO_3 and HF in polytetrafluoroethylene (PTFE) beakers. Subsequently, the solution was slowly heated until it had dried via evaporation. After evaporation, 30 mL of 1.3% HNO_3 was added to the sample. The metallic impurities in the high-purity SiO_2 were measured using ICP-OES (PerkinElmer Optima 4300DV). The measurement conditions for the ICP-OES are summarized in Table 6.1.

Table 6.1 Instrumental conditions of the PerkinElmer Optima 4300DV ICP-OES measurement of the impurities in high-purity SiO_2

	ICP source operation parameters
RF power	1300 W
Plasma gas flow	Ar 15 L min^{-1}
Auxiliary gas flow	Ar 0.2 L min^{-1}
Nebulizer gas flow	Ar 0.7 L min^{-1}
Nebulizer	PEEK MiraMist nebulizer
Spray chamber	Borosilicate glass cyclonic spray chamber
Sample uptake	1 mL min^{-1}
	Spectrometer operating parameters
Viewing direction	Axial
Signal measurement mode	Peak area (3 points)
Background correction	Manually selected, 2 points interpolations
Measurement time	Auto
Measurement replications	10 times
Measured emission line (nm)	Ag; 338.289, Al; 396.153, As; 228.812, Ba; 413.065, Be; 234.861, Bi; 223.061, Ca; 393.366, Cd; 228.802, Co; 228.616, Cr; 267.716, Cu; 324.752, Fe; 238.204, Ga; 294.364, Hf; 277.336, In; 325.609, K; 766.490, Li; 670.784, Mg;285.213, Mn; 257.610, Na; 589.592, Nb; 309.418, Ni; 221.648, Pb; 220.353, Rb; 780.023, Se; 196.026, Sr; 421.552, Ta; 240.063, Ti; 334.940, Tl; 276.787, V; 290.880, W; 224.876, Zn; 213.857, Zr; 343.823

6.2.4 Analytical Procedures of Gravimetric Analysis of Si for Purity Determination

The gravimetric analysis of the Si in the high-purity SiO_2 determined the purity of the SiO_2. Figure 6.1 shows the procedure for the gravimetric analysis of Si.

In a clean quartz crucible, 1 g of the original SiO_2 was weighed and ignited at 800 °C in an electric furnace for 1 h. After cooling to room temperature in a silica gel desiccator for 1 h in the balance room, 1 g of SiO_2 was precisely weighed in a clean 40 mL Pt crucible. Na_2CO_3 (1.8 g) was added and mixed with the SiO_2 in the Pt crucible. The SiO_2 was fused with the Na_2CO_3 to obtain Na_2SiO_3 using a bead and fuse sampler. The operating conditions of the bead and fuse sampler are summarized in Table 6.2.

Subsequently, the fused sample in the Pt crucible was cooled to ambient temperature. After cooling, 25 mL of water was added to the fused sample, followed by heating to 100 °C on a hot plate until the fused Na_2SiO_3 had dissolved completely. After dissolution, the sample solution was completely transferred to a 100 mL Pt dish.

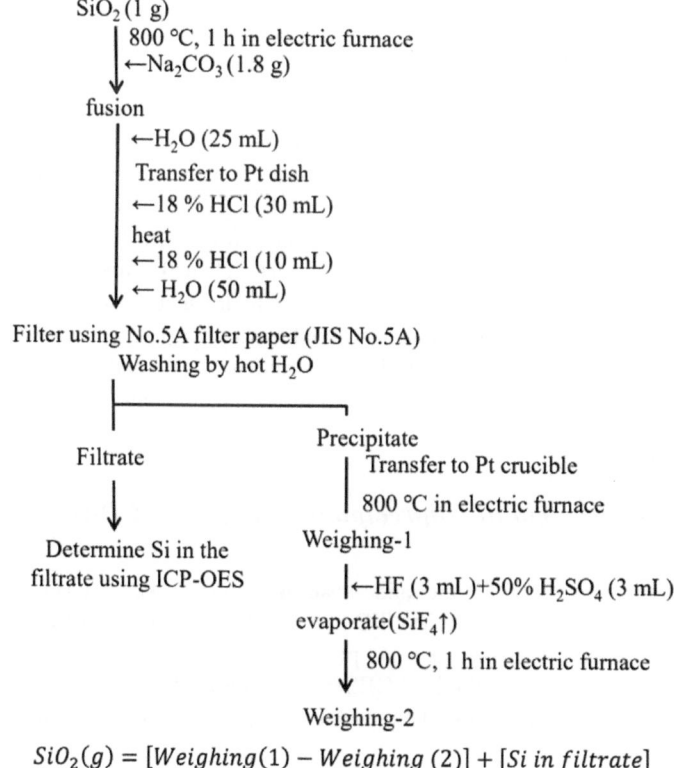

Fig. 6.1 Flow diagram for the gravimetric analysis of the Si in the original high-purity SiO_2

Table 6.2 Operating conditions of an Amenatech bead and fuse-sampler during the fusion of SiO$_2$ with Na$_2$CO$_3$

Step No	Set temperature, °C	Time, s	
1	0 → 600	200	
2	600 → 700	200	
3	700 → 800	200	
4	800 → 900	200	
5	900 → 1000	200	
6	1100 → 1200	200	
7	1200	400	Agitation

To the sample solution, 30 mL of 18% HCl solution was added followed by heating to 95 °C to precipitate SiO$_2$ hydrate. After dehydrating the precipitate, 10 mL of 18% HCl and 50 mL of water were added to dissolve the soluble salt. Subsequently, the solution was filtered through filter paper (No. 5A, diameter 9 cm, ADVANTEC, Tokyo, Japan). The precipitate on the filter paper was washed with hot water. The Si eluted in the filtrate was determined using ICP-OES and used for gravimetric analysis. The filtered precipitate was transferred to a weighed Pt crucible using filter paper. The filter paper was charred to ashes in the Pt crucible using a burner, and the resultant precipitate was ignited at 800 °C for 1 h in an electric furnace. The mixture was allowed to cool for 1 h in a silica gel desiccator in the balance room and weighed until the mass remained constant (weighing 1). This study used a variation of less than 0.1 mg as the criterion for constant mass. The filtered SiO$_2$ hydrate was converted to SiO$_2$ during ignition. After reaching a constant mass, the SiO$_2$ in the Pt crucible was decomposed using 3 mL each of HF and 48% H$_2$SO$_4$. The Pt crucible was heated to 200 °C on a hot plate to eliminate Si as SiF$_4$. The Pt crucible was heated to 350 °C on a hot plate until the excess H$_2$SO$_4$ had evaporated completely. After evaporation, the Pt crucible was transferred to an electric furnace. The sample was ignited at 800°C for 1 h, allowed to cool for 1 h in a silica gel desiccator in the balance room, and weighed until the mass remained constant (weighing 2). The mass difference between weighing 1 and 2 was used as the mass of SiO$_2$.

6.2.5 Measurement of Unprecipitated Si Using ICP-OES

Unprecipitated Si in the filtrate and washing solution was measured using a PerkinElmer Optima 4300DV ICP-OES to evaluate the amount of Si lost. Known amounts of Ga were added to the sample solutions as an internal standard. The measurement conditions for the ICP-OES are summarized in Table 6.3. The analytical results were calculated from the mass of SiO$_2$ and the amount of unprecipitated Si present in the filtrate.

Table 6.3 Instrumental conditions of PerkinElmer Optima 4300DV ICP-OES measurement of Si in the filtrate solution using gravimetric analysis

	ICP source operation parameters
RF power	1300 W
Plasma gas flow	Ar 15 L min^{-1}
Auxiliary gas flow	Ar 0.2 L min^{-1}
Nebulizer gas flow	Ar 0.7 L min^{-1}
Nebulizer	PFA* nebulizer
Spray chamber	PTFE** cyclonic spray chamber
Sample uptake	1 mL min^{-1}
	Spectrometer operating parameters
Viewing direction	Axial
Signal measurement mode	Peak area (3 points)
Background correction	Manually selected, 2 points interpolations
Measurement time	Auto
Measurement replications	10 times
Measured emission lines	Si: 251.611 nm, Ga: 417.206 nm
Mass fractions of Ga as internal standard	Ga 20 mg kg^{-1}

*Perfluoroalkoxy alkanes; **Polytetrafluoroethylene

6.2.6 Preparation of the NMIJ CRM 3645-a Si Standard Solution

NMIJ CRM 3645-a was prepared using a gravimetric method. The detailed preparation procedure is described below.

The high-purity SiO_2 was weighed in a clean quartz crucible and ignited at 800 °C in an electric furnace for 1 h. After cooling to room temperature in a silica gel desiccator for 1 h in the balance room, 2.139 g of SiO_2 was precisely weighed in a clean 40 mL Pt crucible. Subsequently, 3.845 g of Na_2CO_3 was added and mixed with the SiO_2 in the Pt crucible. The SiO_2 was fused with the Na_2CO_3 to obtain Na_2SiO_3 using a bead and fuse sampler. The operating conditions of the bead and fuse sampler are summarized in Table 6.2. After cooling to room temperature, approximately 25 mL of pure water was added and slowly heated on a hot plate for the dissolution of Na_2SiO_3. The solution was washed with pure water and transferred to a PFA beaker (250 mL capacity). Subsequently, the solution was washed with water, transferred into a 1 L narrow-necked high-density polyethylene bottle, and diluted to 1 kg. The temperature and air pressure in the balance room were measured at the time of weighing, and buoyancy was corrected. The mass fraction of the NMIJ CRM 3645-a Si standard solution was determined using the following equation:

$$Si\left(mg\,kg^{-1}\right) = \frac{m_{sio_2} \times p_{sio_2} \times f_{sio_2}}{m_{soln} \times f_{soln}} \times \frac{AW_{si}}{AW_{si} + 2 \times AW_O} \times 10^6,$$

Where, m_{SiO_2} is the mass of weighed SiO_2 as the raw material, p_{SiO_2} is the determined purity value of the SiO_2, f_{SiO_2} is the buoyancy correction factor of the mass of weighed SiO_2, m_{soln} is the mass of weighed standard solution, f_{soln} is the buoyancy correction factor of the mass of weighed standard solution, and AW_{Si} and AW_O are the atomic weights of Si and O, respectively.

The prepared standard solution in a 1 kg bottle was divided into nine 100-mL high-density polyethylene bottles (narrow-necked, with a medium lid). The lids were tightened with a torque of 2.0 Nm. Each sample was weighed and recorded.

6.2.7 ICP-OES Measurement of Si for Evaluating the Stability and Homogeneity of the NMIJ CRM 3645-A Si Standard Solution

The homogeneity and stability of Si in the NMIJ CRM 3645-a Si standard solution were evaluated using the values obtained via precise ICP-OES measurements. The operating conditions of ICP-OES are summarized in Table 6.4.

Table 6.4 Instrumental conditions of the PerkinElmer Optima 4300DV ICP-OES measurement of Si for evaluating the homogeneity and stability of Si in the NMIJ CRM 3645-a Si standard solution

	ICP source operation parameters
RF power	1300 W
Plasma gas flow	Ar 15 L min^{-1}
Auxiliary gas flow	Ar 0.2 L min^{-1}
Nebulizer gas flow	Ar 0.7 L min^{-1}
Nebulizer	PEEK* Mira Mist nebulizer
Spray chamber	Borosilicate glass cyclonic spray chamber
Sample uptake	1 mL min^{-1}
	Spectrometer operating parameters
Viewing direction	Radial
Signal measurement mode	Peak area (5 points)
Background correction	Manually selected, 2 points interpolations
Measurement time	10 s
Measurement replications	10 times
Measured emission lines	Si: 251.611 nm, Ga: 417.206 nm
Mass fractions of Ga as internal standard	Si: 25 mg/kg, Ga: 20 mg/kg as internal standard

*Polyether ether ketone

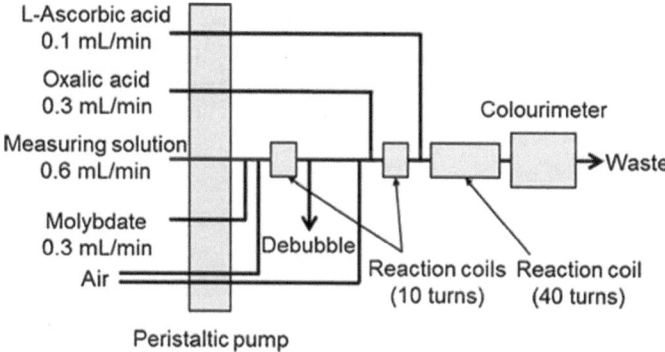

Fig. 6.2   Flow diagram of a CFA for Si measurement (Cheong et al., 2014, 2020)

6.2.8   CFA Measurement of Si in Si Standard Solution for Evaluating Preparation Variability

The measurement conditions of Si using CFA were similar to those described in previous papers (Cheong et al., 2014, 2020). Figure 6.2 shows the flow diagram of the measurement using the CFA, and all reactions and detection were performed on a continuous flow. During the CFA measurement, a measuring solution containing silicate and a $0.06 \, mol \, L^{-1}$ (as molybdenum) molybdate solution were mixed to form a molybdate complex, followed by the additions of 0.40 and $0.14 \, mol \, L^{-1}$ oxalic acid and L-ascorbic acid solutions, respectively. The flow rates were 0.6, 0.3, and 0.1 $mL \, min^{-1}$ for the measuring, molybdate, oxalic acid, and L-ascorbic acid solutions, respectively. The absorption of the complex was measured at a wavelength of 820 nm. A measuring solution injection time of 75 s was determined via optimization to obtain a stable peak plateau. When the washing time was 75 s, 0.11% of a peak was carried over to the next peak.

6.3   Results and Discussion

6.3.1   Analytical Results of Impurities

Thirty-three elements in the high-purity SiO_2 were measured using impurity analysis. The analytical results of the impurities are summarized in Table 6.5.

The impurity analysis results in the high-purity SiO_2 confirmed that the amounts of impurities in the SiO_2 did not affect the value obtained from the gravimetric analysis.

Table 6.5 Analytical results of the impurities in the high-purity SiO_2

Impurity elements	Measured emission line, nm	Measured value, mg/kg	Impurity elements	Measured emission line, nm	Measured value, mg/kg
Ag	338.289	<0.09	Mg	285.213	<0.2
Al	396.153	25.9 ± 1.0	Mn	257.610	<0.3
As	228.812	<1	Na	589.592	2.0 ± 0.1
Ba	413.065	<0.2	Nb	309.418	<0.04
Be	234.861	<0.08	Ni	221.648	<0.3
Bi	223.061	<0.4	Pb	220.353	<0.6
Ca	396.366	0.81 ± 0.02	Rb	780.023	<0.5
Cd	228.802	<0.3	Se	196.026	<0.4
Co	228.616	<0.06	Sr	421.552	<0.3
Cr	267.716	<0.2	Ta	240.063	<0.01
Cu	324.752	<0.2	Ti	334.940	3.57 ± 0.07
Fe	238.204	<0.3	Tl	276.787	<0.5
Ga	294.364	<0.2	V	290.880	<0.3
Hf	277.366	<0.13	W	224.876	<0.4
In	325.609	<0.5	Zn	213.876	<0.3
K	766.490	0.79 ± 0.04	Zr	343.823	<0.06
Li	670.784	4.1 ± 0.08			

6.3.2 Purity Determination of the Original SiO₂ Sample Using Gravimetric Analysis

The results of the gravimetric analysis are summarized in Table 6.6. Gravimetric analysis was performed twice: five samples were analyzed in the first group and four samples in the second group, yielding nine gravimetric analysis results. The value accompanying each measurement after ± in Table 6.6 represents the standard uncertainty obtained by combining the uncertainty in weighing using the balance with the measurement uncertainty associated with the ICP-OES measurement of the Si in the filtrate. The mean value of the purity of the raw material as SiO_2 was 99.648%. The standard deviation of the mean value ($RSD/\sqrt{n}$) was considered the standard uncertainty in the gravimetric analysis. Finally, the purity was determined as 99.648% ± 0.074% ($k = 2$).

Table 6.6 The analytical results of the Si in the high-purity SiO₂ using gravimetric analysis

Run no	Sample mass, g	Measured SiO₂, g	Unprecipitated SiO₂, mg	Measured purity values expressed as SiO₂ ± u (k = 1)
1–1	1.16200	1.15629 ± 0.00006	0.519 ± 0.010	99.553% ± 0.0096%
1–2	1.11409	1.11083 ± 0.00006	0.658 ± 0.0036	99.766% ± 0.0084%
1–3	1.10503	1.10112 ± 0.00006	0.501 ± 0.0025	99.691% ± 0.0089%
1–4	1.09772	1.09277 ± 0.00006	0.474 ± 0.0042	99.592% ± 0.0081%
1–5	1.16157	1.15867 ± 0.00006	0.433 ± 0.0020	99.787% ± 0.0060%
2–1	1.12089	1.11809 ± 0.00006	0.0030 ± 0.00002	99.750% ± 0.0052%
2–2	1.14860	1.14497 ± 0.00006	0.0383 ± 0.00098	99.688% ± 0.0050%
2–3	1.09806	1.09287 ± 0.00006	0.0352 ± 0.00026	99.530% ± 0.0053%
2–4	1.16291	1.15678 ± 0.00006	0.0325 ± 0.00045	99.475% ± 0.0050%
			Mean	99.648%
			RSD	0.11%
			RSD/√n	0.037%

6.3.3 Evaluating Preparation Variability of the NMIJ CRM 3645-A Si Standard Solution

Nine Si standard solutions were prepared using several bottles of the purity-determined high-purity SiO₂. Three subsamples were prepared for each standard solution. The molybdenum blue method on CFA was used for measuring the Si in the subsamples to examine the variation in the preparation of the nine Si standard solutions. The procedure for evaluating the preparation variability is shown in Fig. 6.3. The results of the analysis of variance (ANOVA) for the ratio of each measured value to the prepared value are summarized in Table 6.7.

The relative value of the measurement repeatability as standard deviation (s_r) obtained from the square root of the within-bottle variance was 0.093%. The standard deviation between bottles (s_{bb}) and the standard uncertainty between bottles (u_{bb}) were determined based on ISO Guide 35 (JIS Q 0035 2022) using the following equations:

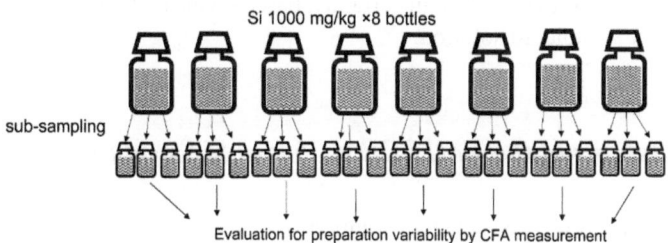

Fig. 6.3 Procedure for evaluating preparation variability of Si standard solution

Table 6.7 Analysis of variance for evaluating the preparation variability of the NMIJ CRM 3645-a Si standard solution

Source of variation	Sum of square	Degree of freedom	Mean square
Between bottle	4.89×10^{-5}	8	6.11×10^{-6}
Within bottle	1.54×10^{-5}	18	8.58×10^{-7}
Total	6.43×10^{-5}	26	

$$s_{bb} = \sqrt{\frac{MS_{among} - MS_{within}}{n}},$$

$$u_{bb} = \sqrt{\frac{MS_{within}}{n}} \sqrt[4]{\frac{2}{\nu_{MS_{within}}}},$$

where, MS_{among} represents the mean square among groups of ANOVA, MS_{within} represents the mean square within a group of ANOVA, and n is the number of observations. The relative values of s_{bb} and u_{bb}, obtained using the equations, were 0.13 and 0.030%, respectively. Therefore, 0.13% was used as the standard uncertainty based on the preparation variability of the NMIJ CRM 3645-a Si standard solution.

6.3.4 Homogeneity of the NMIJ CRM 3645-a Si Standard Solution

The homogeneity of the NMIJ CRM 3645-a was evaluated. The first, fifth, and ninth bottles were extracted from the nine divided bottles during the preparation procedure, and the Si concentration was measured. The procedure for the homogeneity study is shown in Fig. 6.4. Three subsamples were prepared for each standard Si solution. To examine the homogeneity of the NMIJ CRM 3645-a Si standard solution, the Si in the subsamples was precisely measured using ICP-OES. The results of the ANOVA for the ratio of each measured value to the prepared value are summarized in Table 6.8.

The s_{bb} and u_{bb} values were determined based on ISO Guide 35 (JIS Z 0035 2022). The relative values of s_{bb} and u_{bb} were negative and 0.017%, respectively. Therefore, 0.017% was used as the standard uncertainty, based on the homogeneity of the NMIJ CRM 3645-a Si standard solution.

6.3.5 Stability of the NMIJ CRM 3645-a Si Standard Solution

The NMIJ CRM 3645-a Si standard solution was placed in a high-density polyethylene bottle sealed in an aluminum-laminated plastic bag. The Si standard solution was stored in a clean place at 10 °C or less while being kept unfrozen. The results

Fig. 6.4 Procedure of homogeneity study for Si standard solution

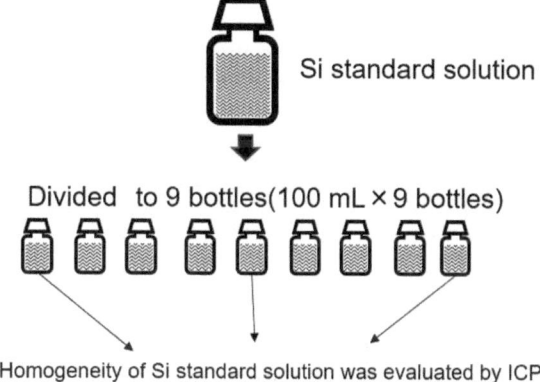

Si standard solution

Divided to 9 bottles(100 mL × 9 bottles)

Homogeneity of Si standard solution was evaluated by ICP-OES measurement

Table 6.8 ANOVA for the homogeneity evaluation of the NMIJ CRM 3645-a Si standard solution

Source of variation	Sum of square	Degree of freedom	Mean square
Between bottle	1.21×10^{-7}	2	6.04×10^{-8}
Within bottle	8.84×10^{-7}	6	1.47×10^{-7}
Total	1.01×10^{-6}	8	

of the stability tests under refrigerated storage conditions show that the NMIJ CRM 3645-a Si standard solution could be stored in a stable state with no change in the mass fraction of Si in the standard solution after 732 days of refrigerated storage. Furthermore, the relative value of mass loss was 0.047% owing to evaporation. The standard uncertainty related to the stability of the Si standard solution was 0.047% owing to evaporation after 732 days.

6.3.6 Calculation of Measurement Uncertainty

The measurement uncertainties are summarized in Table 6.9. The uncertainty in the preparation variability of the Si standard solution represents the main component of the combined standard uncertainty. The relative expanded uncertainty of the certified value of the Si standard solution was evaluated to be 0.28% using coverage factor k = 2, representing a 95% confidence interval.

Table 6.9 Contribution of measurement uncertainty in the NMIJ CRM 3645-a Si standard solution

Uncertainty component	Relative standard uncertainty, %
Purity value of the original SiO_2	0.037
Weighing at the preparation of the Si standard solution	0.0026
Preparation variability	0.13
Stability	0.047
Homogeneity	0.017
Impurities	0.000019
Molecular weights of SiO_2	0.0034
Combined standard uncertainty	0.14
Coverage factor	2
Expanded uncertainty (U), relative (%)	0.28

6.3.7 Validation of the Measured Purity Value of the NMIJ CRM 3645-a Si Standard Solution

The mass fraction of the Si in the NIST SRM 3150 Si standard solution was determined via ICP-OES using a calibration solution prepared from the NMIJ CRM 3645-a. The measured results are summarized in Table 6.10. The measured mass fraction of Si in the NIST SRM3150 is consistent with the certified value within the expanded uncertainty. The calculated E_n value (JIS Z8405 2021) was -0.35, therefore the certified value of NMIJ CRM 3645-a was validated by comparing it with that of NIST SRM 3150 Si standard solution via ICP-OES.

where X_{Lab} represents the value measured in the laboratory, U_{Lab} represents the expanded uncertainty of the measured value in the laboratory, X_{ref} represents the reference value (certified value of the certified reference material), and U_{ref} represents the expanded uncertainty of the reference value (expanded uncertainty of the certified reference material).

Table 6.10 Measured mass fraction of the Si in NIST SRM 3150 for validating the certified value of the NMIJ CRM 3645-a Si standard solution

	Measured mass fraction of Si, mg g^{-1}	Certified value of mass fraction of Si, mg g^{-1}	E_n*
NIST SRM 3150 Si Standard solution	9.888 ± 0.028	9.901 ± 0.023	-0.35

$$^*E_n = \frac{(X_{Lab} - X_{ref})}{\sqrt{U_{Lab}^2 + U_{ref}^2}},$$

6.4 Conclusions

A Si standard solution was developed as a certified reference material in Japan owing to its high demand. A method for assigning a purity value using the gravimetric analysis method of high-purity SiO_2 and a gravimetric preparation method for the Si standard solution was established for its development. The homogeneity and stability of the prepared Si standard solutions were evaluated. Based on the preparation variability and homogeneity of NMIJ CRM 3645-a Si standard solution, the standard uncertainties were 0.13 and 0.017%, respectively. The certified value of the mass fraction of the Si in the Si standard solution had an expanded uncertainty of 0.28%. The certified value of NMIJ CRM 3145-a was validated by comparing it with that of NIST SRM 3150 Si standard solution via ICP-OES.

Competing Interests The authors have no conflicts of interest to declare that are relevant to the content of this chapter.

References

Bièvre, P. D., & Taylor, P. D. P. (1997). Traceability to the SI of amount-of-substance measurements: From ignoring to realizing, a chemist's view. *Metrologia, 34*, 67–75.

Choquette, S. J., Duewer, D. L., & Sharpless, K. E. (2020). NIST reference materials: Utility and future. *Annual Review Analytical Chemistry, 13*, 453–471.

Cheong, C., Nonose, N., Miura, T., & Hioki, A. (2014). Improved accuracy of determination of dissolved silicate in seawater using absorption spectrometry. *Accreditation and Quality Assurance, 19*, 31–40.

Cheong, C., Sakaguchi, A., Sueki, K., & Ohata, M. (2020). Evaluation of the calibration method for accurate analysis of dissolved silica by continuous flow analysis. *Analytical Sciences, 36*, 247–253.

Hioki, A. (2008). A coulometric analysis method and an ion-exclusion chromatographic method for the determination of antimony(V) in large excess of antimony (III). *Analytical Sciences, 24*, 1099–1103.

Japanese Industrial Standards Committee. (2022). *JIS Q 0035:2022 (ISO Guide 35:2017)*. Reference materials—Guidance for characterization and assessment of homogeneity and stability.

Japanese Industrial Standards Committee. (2021). *JIS Z 8405:2021 (ISO 13528:2015)*. Statistical methods for use in proficiency testing by interlaboratory comparison.

Nonose, N., Suzuki, T., Shin, K. C., Miura, T., & Hioki, A. (2017). Characterization of a new candidate isotopic reference material for natural Pb using primary measurement method. *Analytical Chimica Acta, 974*, 27–42.

Milton, M. J. T., & Quinn, T. J. (2001). Primary methods for the measurement of amount of substance. *Metrologia, 38*, 289–396.

Miura, T., & Wada, A. (2022). Precise purity analysis of high-purity lanthanum oxide by gravimetric analysis assisted with trace elemental analysis by inductively coupled plasma mass spectrometry. *Frontiers in Chemistry, 10*, 888636.

Quinn, T. J. (1997). Primary methods of measurement and primary standards. *Metrologia, 34*, 61–65.

Röthke, A., Görlitz, V., Jährling, R., Kipphardt, H., Matschat, R., Richter, S., Rienitz, O., & Schiel, D. (2020). SI-traceable monoelemental solutions on the highest level of accuracy: 25 years from the foundation of CCQM to recent advances in the development of measurement methods. *Metrologia, 57*, 014001.

Suzuki, T., Tiwari, D., & Hioki, A. (2007). Precise chelatometric titrations of zinc, cadmium, and lead with molecular spectroscopy. *Analytical Sciences, 23,* 1215–1220.

Vogl, J. (2018). Roadmap for the purity determination of pure metallic elements. Referred on May 15, 2023, from https://www.bipm.org/wg/CCQM/IAWG/Allow ed/April_2017/ CCQM-IAWG17–28.pdf

Vogl, J., Kipphardt, H., Richter, S., Bremser, W., Torres, M., & d. A., Velina, J., Manzano, L., Buzoianu, M., Hill, S., Petrov, P., Goenaga-Infante, H., Sargent, M., Fisicaro, P., Labarraque, G., Zhou, T., Turk, G. C., Winchester, M., Miura, T., Methven, B., Sturgeon, R., Jährling, R., Rienitz, O., Mariassy, M., Hankova, Z., Sobina, E., Krylov, A. I., Kustikov, Y. A., & Smirnov, V. V. (2018). Establishing comparability and compatibility in the purity assessment of high purity zinc as demonstrated by the CCQM-P149 intercomparison. *Metrologia, 55,* 211–222.

Chapter 7
Development of SI Traceable Si Standard Solution for Nutrient Analysis in Seawater

Chisato Nagaya, Takeshi Fujii, Hitoshi Mitsuda, Naoyuki Tahara, Michio Aoyama, Chikako Cheong, and Tsutomu Miura

Abstract A certified reference material of SI-traceable silicon standard solution (KANSO-Si standard solution) was developed that can be supplied in large quantities to improve the measurement comparability of nutrient analysis in seawater. The high-purity SiO_2 was selected as the starting material of the KANSO-Si. The high-purity SiO_2 was fused by Na_2CO_3 in an electric furnace, followed by dissolution in pure water. The established procedure could be expanded to the production scale of 100 mL $\times$ 1000 bottles per lot at 1000 mg Si kg^{-1} mass fraction of Si. The value assignment of mass fraction of Si in manufactured KANSO-Si was performed by continuous flow analysis calibrated by NMIJ CRM 3645-a Si standard solution. The relative value of expanded uncertainty of the certified value of mass fraction of Si in the KANSO-Si standard solution was 0.4% ($k = 2$) including the standard uncertainties related to the homogeneity, long-term stability, and value assignment of mass fraction of Si.

C. Nagaya · T. Fujii · H. Mitsuda · N. Tahara
KANSO TECHNOS CO., LTD., Katano, Osaka, Japan
e-mail: rminfo@kanso.co.jp

T. Fujii
e-mail: rminfo@kanso.co.jp

H. Mitsuda
e-mail: rminfo@kanso.co.jp

N. Tahara
e-mail: rminfo@kanso.co.jp

M. Aoyama
Japan Agency for Marine-Earth Science and Technology, Yokosuka, Japan

C. Cheong · T. Miura (✉)
National Institute of Advanced Industrial Science and Technology, National Metrology Institute of Japan, Tsukuba, Ibaraki, Japan
e-mail: t.miura@aist.go.jp

C. Cheong
e-mail: c-kato@aist.go.jp

© The Author(s) 2025
M. Aoyama et al. (eds.), *Chemical Reference Materials for Oceanography*, Springer Oceanography, https://doi.org/10.1007/978-981-96-2520-8_7

Keywords Seawater · Nutrients · Continuous flow analysis · SI-traceability ·
Uncertainty · Certified reference material

7.1 Introduction

Different nations have established different standard solutions for nitrates, nitrites, and phosphates in relation to the measurement of nutrient salts in seawater, and in Japan, there are traceable standard solutions that are used as national standards. However, no standard solution for silicate has been satisfactorily established, and the only facilities that provide silicon standard solutions that are traceable based on the *Système International d'unités* (SI) are the US National Institute of Standards and Technology (NIST) and a few overseas manufacturers of reagents. The silicon standard solution provided by these overseas manufacturers is traceable to the NIST standard solution. Because the consistent availability of these silicon standard solutions has been limited, research facilities in various countries sometimes use standard solutions that are prepared independently, and their SI traceability is not guaranteed. Metrological traceability is defined in the ISO/IEC Guide 99: 2007 as "The property of a measurement result that enables that result to be related to a reference via a documented, unbroken chain of calibrations, each contributing to the measurement uncertainty" (International Organization for Standardization, 2007). The use of a standard solution that is SI traceable greatly contributes to ensure the comparability of the measured data. The consequences of the delay in establishing silicon standard solutions were underscored by the results obtained in an international collaborative study in 2018 that was performed by the International Ocean Carbon Coordination Project (IOCCP) and the Japan Agency for Marine-Earth Science and Technology (JAMSTEC) (Aoyama et al., 2018) (Fig. 7.1); the results for silicates exhibited lower comparability than those for nitrates, nitrites, and phosphates, and showed no improvement.

Additionally, a systematic error was confirmed from the standard solution (Aoyama et al., 2018). Moreover, the currently available SI-traceable silicon standard solutions are strongly alkaline, so the required pH adjustment, using acid at the time of use, also reduces the comparability. The colorimetric analysis is used to measure silicates in seawater, and minor differences in pH have major effects on the optical absorbance. It is difficult to align the pH levels of calibration solutions between facilities when using the current standard solutions, and this has led to a lack of progress with respect to the comparability in the measurement of silicate in international collaborative studies. For the Si measurement of reference materials for nutrients in seawater (RMNS), the silicon standard solution produced by an overseas reagent manufacturer was used up to 2019. Figure 7.2 shows the photograph of RMNS developed and supplied by KANSO TECHNOS CO., LTD. (KANSO).

Before the silicon standard development was initiated, the results obtained from the stability tests of RMNS varied within the certified range of ±0.5% and were

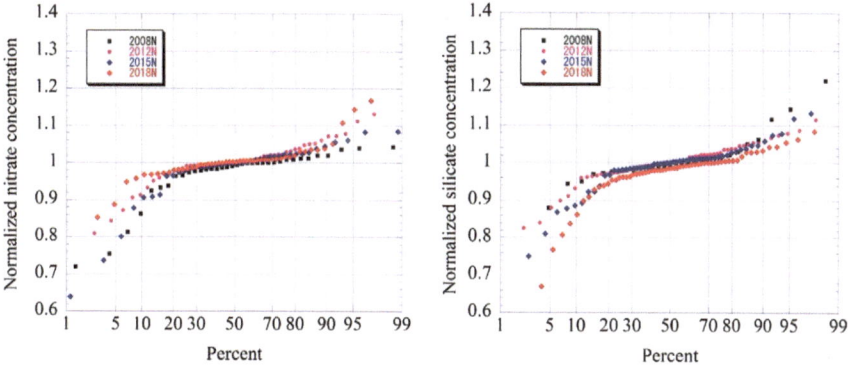

Fig. 7.1 Results of nitrates (left) and silicates (right) in 2018 IOCCP-JAMSTEC international collaborative studies

Fig. 7.2 Reference materials for nutrients in seawater (RMNS)

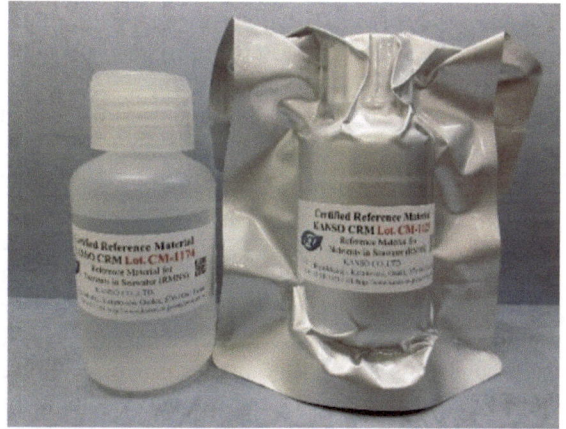

mainly dependent on the purchased lot of the silicon standard solution. These variations were within the range of the expanded uncertainty ($k = 2$) of the silicon standard solutions produced by the overseas manufacturer. However, the uncertainty of the certified value of the silicon standard solution was deemed unacceptable from the perspective of the quality of KANSO's RMNS. The NIST Si standard solution, which is another SI-traceable silicon standard solution, was considered a substitute solution; however, its use was limited by the instability of supply as the solution was sometimes out of stock.

Bearing these limitations in mind, development has begun on producing a silicon standard solution by KANSO (KANSO-Si standard solution) to supply better quality RMNS. Moreover, developed KANSO-Si will aim to supply research facilities in various countries to improve the comparability of silicate measurement on nutrients in seawater.

7.2   Selection of Starting Material

Sodium hexafluorosilicate (Na_2SiF_6), sodium metasilicate (Na_2SiO_3), and silicon dioxide (SiO_2) were considered for use as the raw material in the production of silicon standard solutions. Na_2SiF_6 and Na_2SiO_3 are easy to handle and are soluble in water and alkaline aqueous solutions, respectively. However, Na_2SiF_6 is toxic, thereby hindering import and export, and Na_2SiO_3 has to be made strongly alkaline prior to use. Therefore, the selected preparation method was the alkali fusion of SiO_2. The solution, depending on target concentration, may not need to adjust pH with acid before use.

7.3   Production Method

A KANSO-Si standard solution was prepared by the alkali fusion of SiO_2 (Informative purity; 99.999%, Lot. 5122601, Kojundo Chemical Laboratory Co., Ltd., Tokyo, Japan) with Na_2CO_3, (Informative purity; 99.97%, Lot. AAH200701A, TAKASUGI Pharmaceutical Co., Ltd, Fukuoka, Japan) using a Pt crucible (diameter; 90 mm, height; 40 mm, nominal volume; 250 mL) in an electric furnace HTR-1010 (Lot. 021, AS ONE, Osaka, Japan), followed by dissolution in pure water. In 2019, 30 L of a silicon standard solution with a preparation target value of 1000 mg Si kg^{-1} was produced as a trial product (exp64); the solution was dispensed into 300 polypropylene bottles of 100 mL each. In 2020, the production and operating conditions of the production equipment were investigated, and changes were made to the protocol, developed in 2019, for the preparation of the silicon solution with an aim to expand the production scale to 1000 bottles per lot. By increasing the size of the used Pt crucible, the fine-tuning of the heating conditions in the electric furnace became essential. While the silicon yield was almost 100% in 2019, it decreased to approximately 80% in 2020 under the same conditions, albeit with an expanded production scale. Therefore, the heating conditions for repeated alkali fusion were investigated, and the following sequence of conditions was set: (i) heating to 800 °C over 40 min; (ii) heating to 950 °C over 20 min; (iii) maintaining the temperature at 950 °C for 60 min; (iv) heating to 1150 °C over 30 min; (v) maintaining the temperature at 1150 °C for 60 min; (vi) heating to 1200 °C over 10 min; (vii) maintaining the temperature at 1200 °C for 10 min; (viii) cooling to 1150 °C over 5 min; and (ix) maintaining the temperature at 1150 °C for 30 min, followed by cooling naturally (Fig. 7.3).

The produced Si standard solution, 100 mL aliquots, was dispensed into 1000 bottles at the RMNS production facility described in Chap. 3. The quantity of the Si standard solution (120 L) produced six times performing the alkali fusion was mixed homogeneously in a 200-L tank and was dispensed into a clean room–washed and ultraviolet-sterilized 100-mL polypropylene bottles on a clean bench (Class 100) in a Class 10000 clean room. In addition, to ensure that the Si standard solution did not become more concentrated during bottling, densitometry measurements using

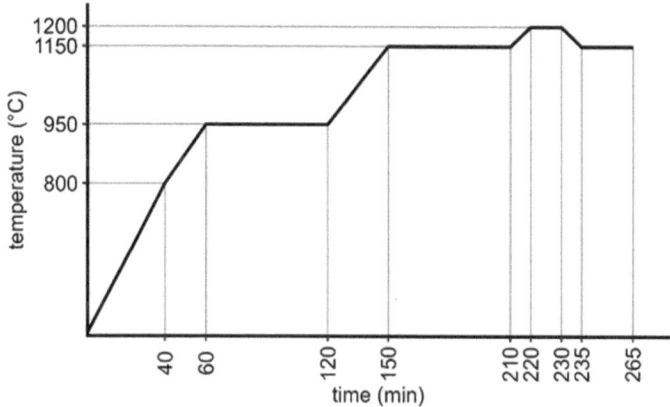

Fig. 7.3 Heating conditions of the electric furnace

DMA™ 5000 M (Anton Paar GmbH) were performed for the bottles selected for homogeneity measurement. After filling, each polypropylene bottle was placed in an aluminum-laminated plastic bag, which was hermetically sealed.

7.4 Value Assignment of KANSO-Si Standard Solution Based on the NMIJ CRM 3645-a Si Standard Solution

The value assignment of KANSO-Si standard solutions produced in 2019 (exp64) and 2020 (exp96) was performed by the National Institute of Advanced Industrial Science and Technology, the National Metrology Institute of Japan (AIST-NMIJ).

The results obtained for the KANSO-Si standard solution as the mass fraction of Si using the continuous-flow analysis method (Cheong et al., 2020) are shown in Table 7.1 and the traceability chain of the CRM is shown in Fig. 7.4. The expanded uncertainty ($k = 2$) of the value assignment by AIST-NMIJ was 0.28% for both exp64 and exp96. The expanded uncertainty of the value assignment was lower than the expanded uncertainty ($k = 2$) of the product from the overseas reagent manufacturer.

Table 7.1 Value assignment of manufactured KANSO-Si standard solutions

Lot No	Prepared mass fraction of Si mg kg^{-1*}	Value assigned mass fraction of Si mg kg^{-1}±expanded uncertainty ($k = 2)^{**}$
exp64	999.3	999.3 ± 2.8
exp96	996.6	997.5± 2.8

*Prepared mass fractions of Si were based on gravimetric preparation
**Value assignments of Si were performed using continuous flow analysis calibrated by NMIJ CRM 3645-a

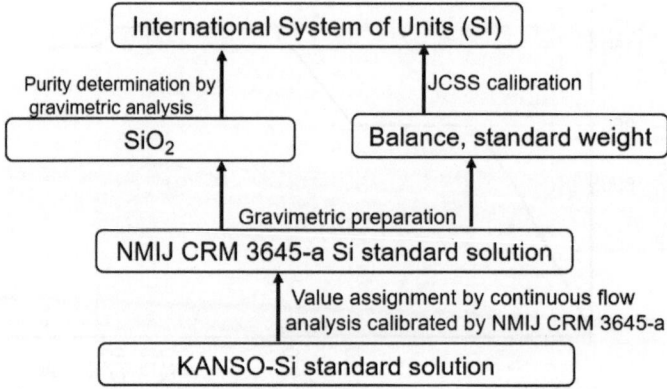

Fig. 7.4 Traceability chain for the value assignment of NMIJ traceable Si standard solution

In Fig. 7.5, the red lines indicate the mass fractions of Si in the prepared materials based on gravimetric preparation. Analysis of the results of the value assignment measurement, for the Si standard solutions produced in 2019 and 2020, revealed that the values estimated from the preparation were within the uncertainty range, confirming that the preparation conditions were appropriate. In addition, measurements were performed on each of the 250 bottles of the trial products from 2020 in the order of dispensing. It was confirmed that dispensing was consistent, with no changes in mass fraction of Si.

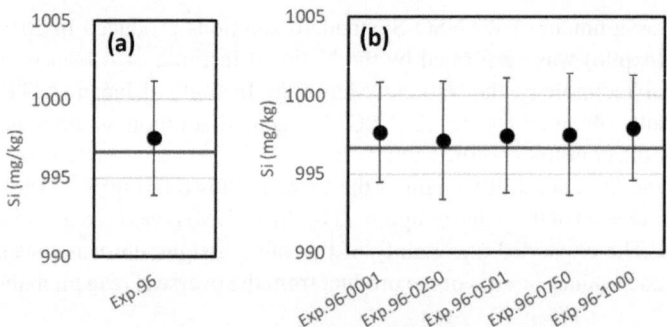

Fig. 7.5 The results of value assignment of Si in KANSO-Si 2020 trial product, exp96. **a** Mean value of the value assigned mass fraction of Si in exp96. **b** Value assignment to each of 250 bottles, in order of dispensing. Red line: Mass fraction of Si in KANSO-Si based on gravimetric preparation

7.5   Homogeneity Study

A homogeneity study for the KANSO-Si standard solution was performed on exp96, which was the trial product produced in 2020. Ten bottles were extracted from the 1000 bottles, and measured variability of mass fraction of Si. The mass fractions of Si in the KANSO-Si standard solutions were measured with a continuous flow analyzer (QuAAtro-2HR; BL-TEC Co., Ltd., Osaka, Japan). A bracketing method was used as calibration on the continuous flow analyzer in the homogeneity study. Solutions were prepared for three points on the calibration solution, namely, low, middle, and high. The middle solution had approximately the same mass fraction of Si as that of exp96 after dilution (175 μmol kg^{-1}); the low and high solutions each had mass fractions of Si differing from the middle solution by 5 μmol kg^{-1}, that is, 170 and 180 μmol kg^{-1}, respectively. A SARTORIUS BP3100S and a SARTORIUS MSA225S chemical balance were used to prepare the calibration solutions and diluted samples using a gravimetric preparation method. The middle solution was measured before and after each sample solution (bracketing method); each sample was measured twice. The advantages of the bracketing method were as follows: (i) the calibration had narrow intervals, which ensured the linearity of the calibration; and (ii) alternate measurement of the sample and a calibration solution of the same mass fraction of Si enabled minor changes in sensitivity to be corrected. In addition, the scatter between measurements was determined by measuring each sample twice. Analysis of the variance of the measurements showed no significant differences between bottles. Furthermore, exp96 was evaluated in accordance with JIS Q 0035:2022 (ISO Guide 35, 2017) using the following equations:

$$s_{bb} = \sqrt{\frac{MS_{among} - MS_{within}}{n}}$$

$$u_{bb} = \sqrt{\frac{MS_{within}}{n} \sqrt[4]{\frac{2}{v_{MS_{within}}}}}$$

where MS_{among} represents the mean square among groups of ANOVA, MS_{within} represents the mean square within a group of ANOVA, and n is the number of observations. The relative values of s_{bb} and u_{bb}, obtained using the equations, were 0.072% and negative, respectively. Therefore, the standard uncertainty of homogeneity of the between bottle homogeneity was 0.072%.

7.6   Stability Evaluation

The mass fraction of Si in the KANSO-Si standard solution measured by a continuous flow analyzer multiple times over a period of 3 years was used to evaluate long-term stability. The evaluated uncertainty related the long-term stability was 0.063%. In addition, the density of the KANSO-Si standard solution, 0 and 4 months after production, was measured using a densitometer. The measured density showed an increase in the density by 0.000125%/month, which was equivalent to the RMNS evaporation rate.

Table 7.2 Contribution of uncertainty in manufactured KANSO-Si standard solution

Uncertainty component	Relative standard uncertainty, %
Value assignment of mass fraction of Si	0.14
Homogeneity	0.072
Stability (Long-term: 3 years)	0.063
Stability (short-term)	0.000
Combined standard uncertainty	0.17
Coverage factor (k)	2
Expanded uncertainty (U), relative (%)	0.4 (Rounded up from 0.34)

7.7   Uncertainty Evaluation

The uncertainty of the KANSO-Si standard solution is listed in Table 7.2. The standard uncertainty related to the value assignment of the mass fraction of Si was 0.14%. Similarly, the evaluated homogeneity and long-term stability are listed in Table 7.2. The short-term stability is not accounted for because no significant variation in mass fraction of Si was detected within the same month. Finally, the relative expanded uncertainty ($k = 2$) was rounded up to 0.4%. The expanded uncertainty ($k = 2$) for the KANSO-Si standard solution was lower than the expanded uncertainty of the product available from other reagent manufacturers.

7.8   Usage of KANSO-Si Standard Solution

At KANSO, the produced SI-traceable silicon standard solution is already used for the certification of silicate in seven lots of RMNS (lot nos.: CL, CM, CN, CO, CP, CQ, and CR). At JAMSTEC, it is used as the standard solution for measuring silicates in seawater on maritime research voyages. However, the international collaborative studies (2020–2021) that IOCCP-JAMSTEC had planned to perform on samples have been postponed because of COVID-19, which impeded the proposed plans.

Competing Interests The authors have no conflicts of interest to declare that are relevant to the content of this chapter.

References

Aoyama, M., Abad, M., Aguilar-Islas, A., Ashraf, P. M., Azetsu-Scott, K., Bakir, A., Becker, S., Benoit-Cattin-Breton, A., Berdalet, E., Björkman, K., Blum, M., De Santis Braga, E., Caradec, F., Cariou, T., Chiozzini, V. G., Collin, K., Coppola, L., Crump, M., Dai, M., Daniel, A., Davis,

C., Solis, M. E., Edelvang, K., Faber, D., Fidel, R., Fonnes, L. L., Frank, J., Frew, P., Funkey, C., Gallia, R., Giani, M., Gkritzalis, T., Grage, A., Greenan, B., Gundersen, K., Hashihama, F., Ibar, V. F. C., Jung, J., Kang, S. H., Karl, D., Kasai, H., Kerrigan, L. A., Kiyomoto, Y., Knockaert, M., Kodama, T., Koo, J. H., Kralj, M., Kramer, R., Kress, N., Lainela, S., Ledesma, J., Lewandowska, J., López, M. d. C. Á., García, P. L., Ludwichowski, K. U., Mahaffey, C., Malien, F., Margiotta, F., Márquez, A., Mawji, E. W., McCormack, T., McGrath, T., Le Merrer, Y., Møgster, J. S., Nagai, N., Naik, H., Normandeau, C., Ogawa, H., Ólafsdóttir, S. R., Ooijen, J. v., Paranhos, R., Park, M. O., Parmentier, K., Passarelli, A., Payne, C., Pierre-Duplessix, O., Povazhnyi, V., Quesnel, S. A., Raimbault, P., Rees, C., Rember, R., Rho, T. K., Ringuette, M., Riquier, E. D., Rodriguez, A., Roman, R. E., Rosero, C., Woodward, E. M. S., Saito, S., Schuller, D., Segal, Y., Silverman, J., Sørensen, D., Stedmon, C. A., Stinchcombe, M., Sun, J., Urbini, L., Wallace, D., Walsham, P., Wang, L., Waniek, J., Yamamoto, H., Yoshimura, T., & Zhang, J. Z. (2018). *IOCCP-JAMSTEC 2018 Inter-laboratory Calibration Exercise of a Certified Reference Material for Nutrients in Seawater*. Retrieved May 15, 2023, from http://www.ioccp.org/ima ges/06Nutrients/IOCCP_JAMSTEC_IC2018report180903protected.pdf.

Cheong, C., Sakaguchi, A., Sueki, K., & Ohata, M. (2020). Evaluation of the calibration method for accurate analysis of dissolved silica by continuous flow analysis. *Analytical Sciences, 36,* 247–253.

International Organization for Standardization. (2007). *ISO/IEC Guide 99:2007 International vocabulary of metrology—Basic and general concepts and associated terms (VIM)*.

JIS Q 0035:2022. (ISO Guide 35: 2017). Reference materials-Guidance for characterization and assessment of homogeneity and stability.

Chapter 8
The History of Standard Seawater for Salinity Measurements

Alan Jenkins and Richard Williams

Abstract The history of Standard Seawater, from its beginnings in 1899 by Martin Knudsen, through to its current preparation at Ocean Scientific International Ltd (OSIL), is described in this chapter. Having been used for over 120 years as a reference material in the determination of salinity, how the standard was developed and maintained to allow for continuity is discussed.

Keywords Standard Seawater Service · Salinity · Martin Knudsen · Primary standard · Practical salinity scale 1978 · International Association for the Physical Sciences of the Oceans (IAPSO)

8.1 Introduction

Salinity is one of the most measured, and arguably one of the most important, parameters of seawater. Not only does it have considerable influence on marine organisms and fish stock, it is vital for measuring global water mass movements. The study of this parameter provides key information that aids researchers studying climate change, as well as allowing for the accurate mapping of the seabed.

It was recognised by Martin Knudsen at the end of the nineteenth century that there needed to be an accurate way for researchers to be able to reliably study and compare salinity data. To that end, he introduced Standard Seawater, a chlorinity standard for seawater that could be used to calculate salinity, and went on to direct what became known as the Standard Seawater Service for the better part of 40 years. Following his passing in 1949, Helge Thomson took over the administrative responsibilities until 1959, when he was succeeded by Knudsen's former assistant, Frede Hermann. A considerable expansion to the Service was overseen by Hermann, driven by increased

A. Jenkins · R. Williams (✉)
Ocean Scientific International Ltd., Havant, Hampshire, United Kingdom
e-mail: Richard.Williams@osil.com

A. Jenkins
e-mail: osil@osil.com

© The Author(s) 2025

M. Aoyama et al. (eds.), *Chemical Reference Materials for Oceanography*, Springer Oceanography, https://doi.org/10.1007/978-981-96-2520-8_8

interest in oceanographic research. From 1974, the Service was directed by Fred Culkin, who, with the aid of many international researchers, helped develop the Practical Salinity Scale 1978 (PSS-78), which allowed for Standard Seawater to be calibrated in terms of conductivity ratio, thus breaking the link between salinity and chlorinity.

When Culkin retired as director in 1989, the Service was transferred to Ocean Scientific International Limited (OSIL), a private company established by Culkin's former colleague Paul Ridout. During the Service's tenure at OSIL, three new Standards with salinity 10, 30 and 38 were introduced to aid researchers operating under different conditions, along with new borosilicate glass bottles to improve Standard longevity and ease of handling. Now, under the directorship of Richard Williams (since 2016), the Standard Seawater Service ships to over 108 countries around the world.

The history of Standard Seawater has been documented in detail previously by Culkin and Smed (1979); this chapter serves as a review of their work and to update developments to the Standard Seawater Service.

8.2 Early Determination of Seawater Salinity

In 1674, British natural philosopher Robert Boyle published his scientific work on ocean salts, "Observations and Experiments about the Saltiness of the Sea". In his work, he shows various experiments that determine the saltiness of the seas around Britain, one such method being:

> Take a clean towel, or any other piece of cloth; dry it well before the sun or before the fire, then weigh it accurately, and note down its weight; dip it in the sea water, and, when taken out, wring it a little till it will not drip when hung up to dry; weigh it in this wet state, then dry it in the sun or at the fire, and when it is perfectly dry, weigh it again: the excess of the weight of the wetted cloth above its original weight, is the weight of the sea water imbibed by the cloth; and the excess of the weight of the cloth after being dried, above its original weight, is the specific gravity of the salt retained by the cloth; and by comparing this weight with the weight of the seawater imbibed by the cloth, we obtain the proportion of salt contained in that species of seawater (Boyle, 1674).

Boyle compiled data pertaining to the saltiness of surface seawater, as well as at depths up to 80 m. In 1693, he demonstrated the use of a solution of silver nitrate to distinguish between fresh and brackish water. He was able to do this owing to the fact that when a silver nitrate solution is added to that containing a chloride, such as seawater, a white precipitate of silver chloride is formed.

From the start of the twentieth century, salinity determination was based on two main facts, namely that the constituents of seawater are constant throughout the major oceans and that chloride, as the major constituent, can be determined with the highest level of accuracy (by titration with silver nitrate).

In 1819, Alexander Marcet (1819) determined that five of the then known major constituents of seawater were present in samples taken from different sources. By also determining the total dissolved salts by evaporation, he concluded that: "All specimens of seawater contain the same ingredients all over the world, these bearing nearly the same proportions to each other so that they differ only as to the total amounts of their saline contents". This was confirmed later by other researchers.

Later in the nineteenth century, ocean studies required the salinity and density to be more accurately determined. At the time, total dissolved salt content was determined through drying samples and weighing the residue, which was both time-consuming and inaccurate. Based on Marcet's statement that seawater has a constant composition, the Danish geologist and mineralogist Johan Georg Forchhammer was able to demonstrate that by measuring a single constituent of seawater, the total dissolved salt content could be calculated (Forchhammer, 1865). Instead of using the evaporation method, he obtained the total dissolved salts by adding together the weights of a number of individual constituents, which he determined separately. By doing this, not only was he able to confirm Marcet's work, but he was able to show that the total dissolved salts could be calculated by multiplying the chloride content by a factor of 1.812. Forchhammer is also credited with introducing the term "salinity", meaning the concentration of total dissolved salts.

In 1884, German-born chemist Wilhelm Dittmar analysed seawater samples collected during the Challenger Expedition (1873–1876) (Dittmar, 1884). He is most commonly credited with establishing the constancy of seawater composition, having shown the constant proportions of seven major components of seawater (other than hydrogen and oxygen), namely, sodium, calcium, magnesium, potassium, chloride, bromide and sulfate. The work performed by Dittmar led to a coefficient of 1.8058 for converting chloride to salinity, a value that was very close to that used in the mid–late 1900s (1.80655).

Towards the end of the nineteenth century, the number of countries interested in salinity data was increasing, and the use of titrating seawater against a solution of silver nitrate and converting the resulting chloride content to salinity using a coefficient was widely accepted. What was lacking was consistency in both the method of analysis and the value of the coefficient used (1.806–1.829), so comparison of data between research groups was poor.

8.3   Chlorinity Determination

For the determination of chlorinity in seawater, only two methods were used to any great extent, namely, the Mohr method (1856) and the Volhard method (1874).

In the Mohr method, the seawater is titrated against a solution of silver nitrate of known concentration with the end point being when all the chloride, and the small amount of bromide, which is also present in seawater, has been precipitated. An indicator is used that changes colour when a slight excess of silver is detected. The Volhard method, developed nearly 20 years later, involves adding an excess of silver nitrate solution to the seawater sample, filtering off the precipitated silver halides, and determining the excess of silver in the filtrate by titration with thiocyanate solution, using ferric alum as an indicator.

Both of these methods make use of silver nitrate to precipitate the halide species present in the seawater sample, but it is important to note that silver nitrate cannot be dried without some decomposition occurring. This makes it a poor primary standard as it is not possible to prepare a silver nitrate solution at a known concentration by dissolving it in water.

At the time, the usual practice has always been to prepare a solution of approximately the desired concentration and then standardise this against a reliable standard of known composition, such as potassium chloride. Whilst this is not an issue in current times, with accurately calibrated analytical balances, highly purified chemicals and ultrapure water, but in the nineteenth century this was a major hurdle. It is also important to note that both of these methods require the use of atomic weights to calculate the concentration of solutions, something that is periodically updated as measurements become more accurate.

8.4 Salinity Research in the Early Twentieth Century

In the late nineteenth century, there was considerable growth in hydrographic investigations which led many European countries to convene a preparatory conference in Stockholm (1899) to establish the International Council for the Exploration of the Sea (ICES). By this time, Martin Knudsen (Fig. 8.1) had improved the Mohr method of chlorinity determination by developing special burettes and pipettes, giving a salinity conversion factor of "1.805Cl + 0.03" (Forch et al., 1902). He also prepared a number of sealed tubes of seawater, the chlorinity of which was determined accurately by the Volhard method, and these had been used successfully to standardise the silver nitrate solutions used in Danish hydrographic work. This had the effect of referring all chlorinity determinations to one standardisation, thus giving greater internal consistency. He therefore submitted to the Stockholm Conference a "Proposal about an international institution for procuring standard water" (Conférence Internationale, 1899a).

In this proposal, Knudsen stressed the importance of the measurement of the salinity of seawater in physical, climatological, and biological investigations. Knudsen pointed out that titration by weighing was at that time a fairly difficult operation but maintained that, whilst not generally achieved by methods then in use, the measurement could be carried out with an accuracy of 0.04‰. Given that it was almost impossible to carry out often enough to obtain the desired accuracy, he thought that the current errors in salinity determinations were usually as high as 0.10–0.15‰. From his own investigations on the "Ingolf" Expeditions of 1895 and 1896 to the waters around the Faroes, Iceland, and Greenland, Knudsen knew that some Atlantic water types differed by only 0.10–0.25‰ in salinity so there was a need for greater accuracy (Knudsen, 1899).

Fig. 8.1 Martin Knudsen,
1871–1949 (Thomsen, 1950)

The proposal also suggested that an institution should be established that would prepare and standardise (in terms of chlorinity) the Standard Seawater, and provide a statement of its physical and chemical properties to interested laboratories. These laboratories could then determine the chlorinity of their seawater samples by comparison with the Standard Seawater, thus eliminating a number of sources of error.

In order to achieve this, Knudsen suggested the institution (consisting of a manager, a physicist, a chemist and two assistants) would first need to obtain a quantity of open Atlantic seawater and then investigate its physical and chemical properties:

> A detailed determination of the total salinity and the quantity of the single salts. Determination of carbonic acid, sulphuric acid, alkalinity, specific gravity, etc., the coefficient of refraction and absorption for light with different wavelengths. Determination of the specific electric resistance, the surface tension and the viscosity. Determinations of freezing point, boiling point, etc.

This seawater would then be sealed in glass ampoules and examined at intervals to study its stability. Alongside this, artificially produced seawater will also be studied. Based on these investigations, which Knudsen expected to take 2 years, it would be possible to determine what should be the future standard for determining the halogen content of seawater.

Persuasive though his arguments were, Knudsen's proposal was not accepted in its entirety by the Conference. Whilst the establishment of an institution devoted to the preparation and study of standard water was not accepted, it could be surmised that the Conference was reluctant to accept the considerable cost that would be involved, some of his other ideas were, though in modified forms. For example, Knudsen was charged with the revision of certain hydrographic tables. (Conférence Internationale, 1899d).

These investigations were carried out at the Technical University in Copenhagen under the direction of a committee consisting of John Murray, Martin Knudsen, Otto Pettersson, Fridtjof Nansen, Otto Krümmel, Henry N. Dickson and Stepan O. Makaroff. Samples of seawater from different parts of the world were collected and the "constants" of seawater were precisely determined, i.e. the relation between the amount of chlorine and the salinity, and between the amount of chlorine and the specific gravity at different temperatures (Thomsen, 1950).

This considerable amount of work was completed and presented at the second ICES at Kristiania (Oslo) in May 1901 (Conférence Internationale, 1901a, 1901b). With these revised hydrographic tables it was possible to determine the chlorine content, salinity and specific gravity of a sample of seawater with great accuracy from a Mohr titration against "Standard Water".

Despite Knudsen's doubts, his arguments for standard water evidently found favour with the members of the Conference:

> The chemical analysis shall be controlled by physical methods and physical determinations by chemical analysis in the following manner. From every collection of samples examined at least three shall be selected and sent to the central bureau. Standard samples shall be sent in return. (Conférence Internationale, 1899b).

The hydrographical programme goes on to say that:

> By Standard water shall be understood samples of filtered seawater, the physical and chemical properties of which are known with all possible accuracy by analysis, and statements of which are sent to the different laboratories, together with samples. In respect to halogen, the ordinary water samples have to be compared with the Standard water by analytical methods.

With that, the Conference established the need for standard water for use in all chlorinity titrations.

8.5 The Beginnings of International Standard Seawater

The Stockholm Conference of 1899 recognised the need for standard water for use in all chlorinity titrations, saying that these samples would be sent out by the central bureau. This relates to a preferred proposal made by one of the Norwegian delegates, Fridtjof Nansen, which said that a Central Laboratory should be established in connection with the Central Bureau of the organisation (International Council for the Exploration of the Sea [ICES]). This Central Laboratory would carry out a number of important investigations of general interest to hydrographic and biological research and would also have the responsibility for supplying standard water. This proposal was incorporated in a recommendation from the Conference (Conférence Internationale, 1899c).

In order to avoid any delay that might result from setting up the Central Laboratory, Knudsen quickly started preparing samples of Standard Water. In describing the standard water used in hydrographical work up to July 1903, Knudsen gave the following reasons for proceeding (Knudsen, 1903):

(1) Though, in my opinion, the members of the Stockholm Conference did not clearly see the advantage of using a standard water for all titrations, I did not hear a single remark against it, and I myself felt convinced that standard water sooner or later would be a great advantage, perhaps a necessity for hydrographical work.

(2) It did not seem probable to me that the whole international cooperation would begin within a short time (it in fact took three years) and in the meantime it would be useful to have at hand reliable standard water.

(3) The researches done for the determination of the constants of different kinds of seawater would offer a convenient occasion for determining the constants of standard water.

(4) In case I should succeed in carrying through the work of the constants-determination and have compiled some Hydrographical Tables that could be generally adopted, I thought it of importance that the standard water used should be investigated with the same means and methods as used in the researches upon which the tables would be found.

Preparation of standard water was nothing new to Knudsen, he had prepared five different batches prior to the Stockholm Conference for use in Danish hydrographical work. His sixth batch, prepared in 1900 in association with the constant determinations, was designated No. VI. The chlorinity of standard water No. VI (19.380‰) was determined by the Danish chemist Soren Sørensen; he prepared a sample of potassium chloride, weighed it, corrected the weight for buoyancy in air, and used this for the titrimetric standardisation of a solution of silver nitrate. This solution was then used for the titration of the chlorinity of the samples of seawater used as Standard Water (Forch et al., 1902; Knudsen, 1903). The chlorinity of Standard Water No. VI, together with all subsequent standards, which are referred to it, are therefore based upon the chlorine content of Sørensen's sample of potassium chloride and consequently depend upon the atomic weights, which he used (Culkin & Smed, 1979).

In 1901, the second preparatory conference was held in Kristiania (now Oslo). The "extreme value" of the Hydrographical Tables that Knudsen compiled was recognised by the Conference along with his provisional report on the determination of the constants of seawater (Conférence Internationale, 1901b). The hydrographic programme went on to say (Conférence Internationale, 1901a):

"Preliminary determinations of salinity may be made on board ship with appropriate instruments; but the exact determinations of the salinity and density of water samples shall take place in a scientific laboratory on shore. The ratios between salinity, density and chlorine given in Dr Martin Knudsen's Hydrographic Tables are to be adopted; and the salinity is to be calculated by the use of these Tables from the determinations of chlorine or from the specific gravity". And that "The same standard seawater shall be employed in all cases for standardising the solutions used for chlorine determinations".

About 80 ampoules of Standard Water No. VI were prepared in April 1900, which were analysed with regards to chlorinity (Volhard titration and a gravimetric method) and specific gravity. They were also subjected to periodic checks of the specific gravity over the following two years which showed an increase of 0.015 kg m^{-3} in σ_0, most likely due to the dissolution of the glass used at the time, though as the glass did not contain any chloride ions the chlorinity was not significantly affected. Unfortunately, this meant that the water in the ampoules could not be used as a density standard, however, the Standard Seawater in modern times is stored using a more resistant glass.

Whilst being called No. VI, this was the first Standard Water for international use and was distributed to Germany, Finland, Norway, Russia and Sweden as well as being used for all Danish titrations. Despite the Central Laboratory not yet being opened, by mid-1902 the stock of Standard Water was running low so Knudsen was asked to prepare a new supply. As before, water was collected from the Atlantic Ocean, which was then diluted with distilled water to give a chlorinity similar to that of standard No. VI before being sealed in ampoules (Knudsen, 1903). 201 Ampoules were prepared, designated VI*a* (Fig. 8.2), with 168 being distributed free of charge to member countries of ICES. Of the remaining ampoules, some were used to determine chlorinity (using standard No. VI as a reference) and density, with the remainder being sent to the Central Laboratory in Kristiania which opened in late 1902.

In July 1903, 100 ampoules of Standard Water (VI*b*) were produced and standardised by comparison with Knudsen's standard VI*a*. It was reported that a new Primary Standard would be produced, which would be compared directly with Knudsen's standard water No. VI, with the intention that this primary standard be used to standardise subsequent batches. However, before this could be prepared, further batches of 250 ampoules of standard water (VI*c* and VI*d*) needed to be produced for general

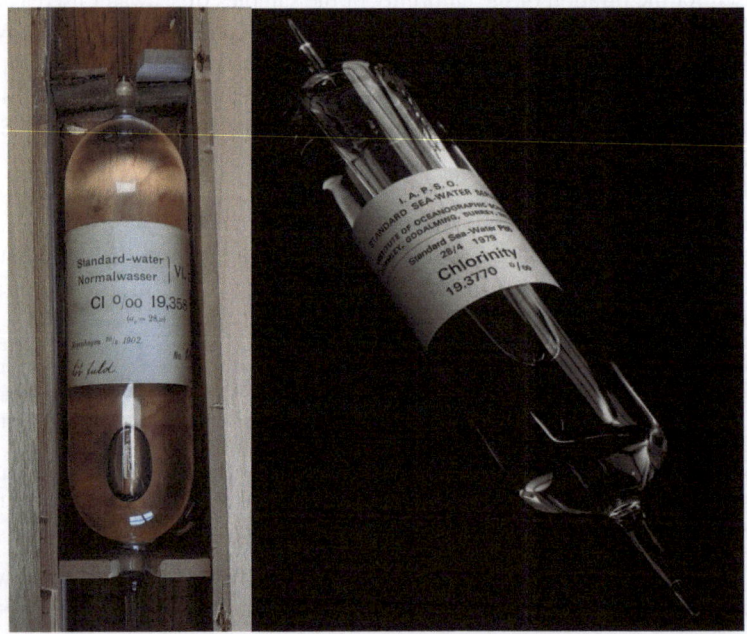

Fig. 8.2 Ampoules of Standard Seawater No. VI*a* (left) and P86 (right)

distribution. In December 1903, 120 ampoules of seawater were reserved as the Primary Standard, though it wasn't until October 1905 that accurate chlorinity analysis could be carried out. By direct comparison with Knudsen's Standard Water No. VI, the chlorinity of the new Primary Standard (designated P) was established as 19.4482‰. Over the following two years, the Central Laboratory produced four batches of Standard Water (P1–P4).

8.6   The Standard Seawater Service

In 1908, the decision was made to close the Central Laboratory, primarily due to Nansen deciding he no longer wanted to continue as director, and assign most of its responsibility's various national laboratories. The preparation of standard water, which was of interest to all hydrographers, would be handed over to Knudsen. The ICES agreed that in the future, in order to cover part of the expenses those who request the standard water would have to financially contribute. The remaining stocks, 83 ampoules of Primary Standard P and 49 ampoules of P4, were transferred to Knudsen at the Danish Hydrographical Laboratory in Copenhagen.

Reports to the ICES over the following years showed that standard water had become widely adopted for measuring salinity. Batches P5–P8 were produced in the years preceding the outbreak of World War I in 1914. To relieve the ICES of the financial responsibility during the war, Knudsen took over the Standard Seawater Service in a personal capacity, an arrangement that continued after the war.

In the years 1920–1937, batches P9–P15 were produced and analysed by Jacob P. Jacobsen, one of Knudsen's collaborators. Each batch was compared against Primary Standard P as well as the previously produced batch using the Volhard method, with agreement being seen between them to show the results were accurate to the third decimal place. Thus, the link with the original Standard Water No. VI was maintained.

With the ever-increasing demand for Standard Water, it became necessary to prepare a new Primary Standard. At the Council meeting in 1936, Knudsen explained that since the Standard Water had become used worldwide, he intended to propose to the International Association for Physical Oceanography (IAPO) that it should direct the preparation of the new Primary Standard as well as cover the associated expenses. Knudsen and Jacobsen had argued that a new definition of salinity was needed that did not depend on ratios of certain atomic weights (of silver nitrate), and that pure silver (Atomgewichtssilber) should be used, which would have the advantage of being independent of atomic weight values (Jacobsen & Knudsen, 1940).

This was accepted by both the ICES and the IAPO, with the new Primary Standard being produced in 1937. The chlorinity of this new standard, called Urnormal or Primary Standard 1937, was found to be 19.3810‰ by direct comparison with Primary Standard P as well as previous standard water samples (Jacobsen & Knudsen, 1940).

Owing to limited demand during World War II, only one batch (P16) of standard seawater was produced in 1943. Hermann (Fig. 8.3), previously an undergraduate

Fig. 8.3 Frede Hermann, 1917–1977 (UK National Oceanography Archives)

who worked on Primary Standard 1937, prepared and analysed P16 with guidance from J.P. Jacobsen.

When Knudsen was approaching his late 70s, he looked to ensure the future continuation of the Standard Seawater Service. Having been responsible for the Service for nearly 40 years, he felt that it should be taken over by a scientific body with interest in the work, so in 1947 he proposed to IAPO that they should undertake future preparations. This was agreed by IAPO at the Oslo meeting in 1948, and under this arrangement batch P17 was produced.

It was the end of an era when, in May 1949, Knudsen passed away. But thanks to his arrangement with IAPO, the Standard Seawater Service was able to continue. At the request of IAPO, Helge Thomsen took over the administrative responsibilities whilst Hermann continued to prepare and analyse the Standard Seawater. Thomsen continued in this role until 1959, overseeing the production of multiple batches (P18–P29), when he made the decision to retire as director of the Service. Given Hermann's long association with the Service, Thomsen proposed that he take over; a decision that was welcomed by IAPO. With Hermann as director from January 1960, the Service saw a large increase in the demand for Standard Seawater, sometimes rising to over 30,000 ampoules per year, in part owing to the considerable expansion in oceanography throughout the world.

For many years, Hermann successfully ran the Seawater Service, producing batches P30–P65, but given his other responsibilities as the head of the Danish Hydrographical Laboratory, he made the decision in 1973 to retire from the Service. He proposed to the International Association for the Physical Sciences of the Oceans (IAPSO, formerly known as IAPO, who updated their name in 1967) that responsibility for running the Service should be passed to Fred Culkin (Fig. 8.4), who had collaborated with him in the chlorinity calibrations for several years. With help from IAPSO, equipment and stock were transferred from Charlottenlund to the Institute of Oceanographic Sciences (IOS) at Wormley (UK) at the end of 1974, with production recommencing in 1975.

Fig. 8.4 Fred Culkin,
1929–2011

Salinometers, instruments that can determine the salinity of seawater using an electrical conductivity method, first began to appear in the 1930s. They worked by measuring the ratio of conductivity between Standard Seawater and a seawater sample, with the ratio being used to calculate the salinity. Over the following decades, salinometers became more readily available and increasingly accurate; In 1961, Bruce Hamon and Neil Brown designed a portable salinometer with a precision of ±0.003‰. Later, in 1975, Tim Dauphinee designed a laboratory salinometer, the Guildline Autosal, which has become synonymous with salinity measurement and is still widely used by oceanographers, reaching accuracies of ±0.002‰.

With the introduction of using conductivity to measure salinity in situ, International Oceanography Tables began to be published that linked conductivity and salinity based on precise determinations of chlorinity (UNESCO, 1965, 1966; Wooster et al., 1969). These tables, and resulting equations, were valid for salinities ranging from 29 to 42‰ at 15 °C, with a correction table for measurements at temperatures from 10 to 31 °C (Brown & Allentoft, 1966; Cox et al., 1967). The equations relied on a conductivity ratio between that of the sample and that of Standard Seawater P31 at 15 °C (chlorinity, 19.374‰).

In 1978, after much research and discussion, the Joint Panel on Oceanographic Tables and Standards (JPOTS) determined that Standard Seawater would be calibrated in terms of its conductivity relative to that of a potassium chloride solution, and that a Practical Salinity Scale would be established (UNESCO, 1978). Intense research from multiple laboratories was used to create the Practical Salinity Scale 1978 (PSS-78) as well as the recommended equation for calculating Practical Salinity from conductivity ratio which is valid at all temperatures and pressures that were of oceanographic interest (UNESCO, 1979).

With the acceptance of the PSS-78, which was valid from salinity 2 to 42 and temperatures from −2 to 35 °C, the link between salinity and chlorinity was broken.

All the batches up to P90 had been calibrated solely in terms of its chlorinity, but following the official adoption of the Practical Salinity Scale by JPOTS in 1980 (UNESCO, 1981a, 1981b, 1981c), batch P91 and all following batches of Standard Seawater were calibrated in conductivity relative to the defined potassium chloride (KCl) standard and labelled with the appropriate K_{15} value (see Sect. 8.8 for definition) (Culkin & Ridout, 1998). For the following 10 years (to 1990, P113), batches were calibrated to both chlorinity and conductivity ratio.

When Culkin retired as director in 1989, the Service was transferred with IAPSO's agreement to Ocean Scientific International Limited (OSIL), a private company established by former colleague Paul Ridout to produce and distribute Standard Seawater. Culkin joined OSIL as a consultant where he remained actively involved. Alongside the Standard Seawater, having a Practical Salinity of approximately 35, three further standards were introduced around 1990. These had practical salinities of 10, 30 and 38 (batch 10L1, 30L2 and 38H1, respectively), and were intended to aid researchers operating under different conditions (Fig. 8.5).

Batches P112–P137 were produced at OSIL in normal glass ampoules then, in 2000, after 8 years of research and development, Standard Seawater began to be stored in borosilicate bottles. This pharmaceutical-grade glass had several benefits, including improved ease of handling and storage, as well as being much easier to fill. Following a move of OSIL from Wormley to Petersfield, it relocated to Havant at the end of 2006 and under the directorship of Richard Williams now ships Standard Seawater to over 108 countries around the world.

It is a remarkable feature of the Standard Seawater Service that the continuity has been maintained from its inception all the way through to the most recent batch, P169

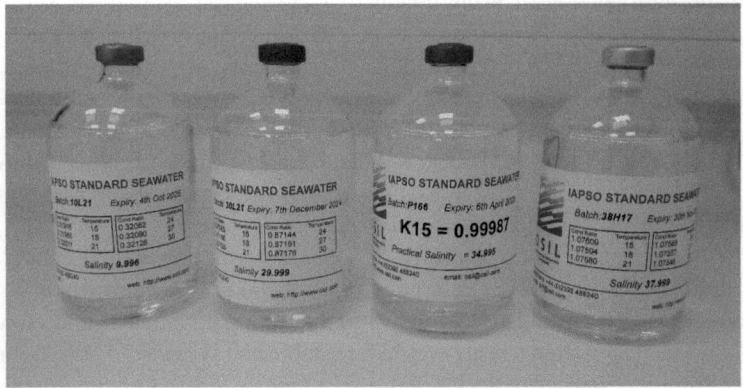

Fig. 8.5 Current range of IAPSO Standard Seawater (left to right: 10L, 30L, P, 38H)

(2024). IAPSO Standard Seawater remains the only standard for Practical Salinity, which is recognised by all major oceanographic bodies around the world.

8.7 Preparation of Standard Seawater

The general preparation method of Standard Seawater and its transfer into ampoules/ bottles has remained largely unchanged since it began, though as demand increased over the years the scale of the production similarly increased.

When Knudsen first began producing Standard Water, he found that a stock of around 50 L was enough to produce enough ampoules for distribution (Fig. 8.6). From 1960, when Hermann oversaw a considerable increase in demand, a stock of 4,000 L was required. With such a large storage tank, and improvements to filtration and circulation, he was able to produce batches of 9,000 ampoules, a far cry from the 80 ampoules of Standard Water No. VI that Knudsen first produced.

Given the large volumes required, surface seawater was collected from a known area in the North Atlantic Ocean well in advance of a new batch being prepared, for which Hermann had an amicable arrangement with a Danish shipping line. The ampoules to be used were washed and dried, and once filled would be labelled, packed and distributed by a part-time helper. The seawater was pumped through filters and stored in a large tank where it was mixed and further filtered. Whilst in the tank, the seawater was slowly diluted with ultra-pure water to the desired salinity, usually just below 35‰, a process that took several weeks. Its temperature was raised to around

Fig. 8.6 Early equipment for preparing Standard Seawater (Jacobsen & Knudsen, 1940)

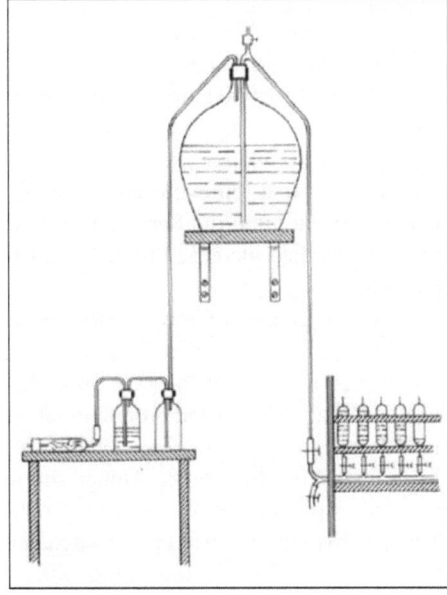

Fig. 8.7 Ampoule filling benches (Culkin & Smed, 1979)

26 °C so that when it was transferred to the glass ampoules (20–24 °C) the seawater was slightly undersaturated with dissolved air.

The seawater was pumped from the storage tank to a horizontal glass tube on the filling table using a powerful pump (whereas Knudsen fed the seawater by gravity). This tube had several outlets to which fitted a clean and unsealed soda-glass ampoule via a rubber tube (Fig. 8.7). The seawater was allowed to fill the ampoule to within 2–3 cm of the top (~275 mL) and then the rubber tubing was closed with a clip and the upper tip of the ampoule was sealed using an acetylene/air flame. After cooling, the ampoule was inverted and similarly sealed, and the seawater in the other tip allowed to drain beforehand. Using this method, and multiple filling benches, it was possible to produce up to 7,000 ampoules in a day.

Over time as technology improved, ultraviolet sterilisation modules were added and filters with smaller pore sizes were introduced, going from the filter papers Knudsen used that remove particulates, to 0.2 μm filters currently used to remove bacteria. By the time the Service moved to the UK at IOS, and later to OSIL, the stock volume had increased to 5,000 L which was stored in a polymer-linked tank (Fig. 8.8).

With the exception of how salinity was measured (chlorinity vs conductivity ratio), this filling process into glass ampoules largely remained the same since Knudsen's first preparations. From the year 2000, having undergone 8 years of research and development, OSIL started producing Standard Seawater batches in pharmaceutical grade Type 1 borosilicate glass bottles as opposed to soda-glass ampoules. Not only did this make the filling process much simpler, but the design made it easier to handle, pack and ship whilst providing a longer shelf-life (3 years) for the Standard. In this case, water is pumped from the storage tank, through a 0.2 μm filter and ultraviolet sterilisation module, to a tube fitted with a number of taps. Bottles are placed on the

Fig. 8.8 5000 L Standard
Seawater tank at OSIL

taps and filled to the desired volume (~210 mL) before being sealed with a secure crimp cap over a halo-butyl bung coated with PTFE to provide a totally impervious barrier to evaporation. Over 8,000 bottles are produced in a filling session; after being stored for a minimum of 2 months, samples are taken throughout the batch to be used for calibration.

8.8   Certification of Standard Seawater—The Primary Standard

At the beginning of the twentieth century, it was understood from the work of Marcet (1819), Forchhammer (1865) and Dittmar (1884) that salinity (i.e. total dissolved salts) could be calculated from chlorinity using a conversion coefficient. When Knudsen first prepared his Standard Water, the accepted definition of chlorinity was:

> By chlorinity is understood the mass of chlorine equivalent to the total mass of halogen contained in the mass of 1 kg of seawater.

In determining the chlorinity of Standard Water No. VI in 1900, Sørensen used both gravimetric and titration techniques. The silver nitrate solution that was used in the titration was standardised against pure potassium chloride using the atomic weights adopted in 1900. The chlorinity of Standard Water No. VI was found to be 19.380‰, with all subsequent Normal or Standard Seawaters up to 1937 being determined either directly or indirectly by comparison to it.

When stocks of the Primary Standard were running low in 1937, a new batch, Primary Standard 1937 (also known as Urnormal), was produced. The chlorinity was determined by Inger Bondorff (formerly Knudsen), Martin Knudsen's daughter, by direct comparison with the previous Primary Standard as well as other standard seawaters. Jacobsen and Knudsen (Jacobsen & Knudsen, 1940) argued that it was unsatisfactory that the definition of chlorinity made it dependent on the ratios of certain atomic weights. This would mean that, if the definition was strictly followed, there would be breaks in the chlorinity determination every time the accepted values of atomic weights changed.

Jacobsen and Knudsen proposed that using a new standard, pure silver (Atomgewichtssilber), would be much better and that a new definition of chlorinity be established based on the following reasoning (Culkin & Smed, 1979):

(1) Let Ag denote the number giving the mass (in grams) of Atomgewichtssilber just necessary to precipitate the halogens of 1 kg of seawater sample of which the chlorinity expressed in ‰ is Cl. It is then natural, by definition, to put Cl proportional to Ag, because Ag is the quantity which can be determined with greater relative accuracy than any other chemical quantity which could be considered in this connection.

$$\text{Thus Cl} = k \times \text{Ag grams Atomgewichtssilber} \qquad (8.1)$$

Where k has the same value for all seawater samples and consequently the same as it has for Urnormal 1937.

(2) A chlorinity of 19.3810‰, as determined by Inger Knudsen by comparison with previous standards, was adopted for Urnormal 1937, thus ensuring continuity between past (as far back as 1902) and future chlorinity determinations.
(3) Investigations carried out by Prof. Otto Hönigschmid in Munchen established that 58.99428 g Atomgewichtssilber were necessary and sufficient to precipitate the halogens in 1 kg of Urnormal 1937.
(4) Substituting these values Eq. (8.1):
19.3810‰ = k × 58.99428 g Atomgewichtssilber
Or Cl = 0.3285234 Ag‰".

Thus the new definition of chlorinity became:

The number giving the chlorinity in per mille of a seawater sample is by definition identical with the number giving the mass with unit gram of Atomgewichtssilber just necessary to precipitate the halogens in 0.3285234 kg of the seawater sample (Jacobsen & Knudsen, 1940).

Whilst this had the advantage of being independent of atomic weight values and providing a continuous link between all chlorinity measurements, the use of pure silver was not feasible for calibrating every batch of standard seawater. Instead, it was more practical to determine the chlorinity of a small number of seawater ampoules against pure silver every few years and then use these ampoules as the practical Primary Standard against which the chlorinity of ordinary standard seawater is determined.

In the early 1970s, the use of in situ electrical conductivity measurements to estimate salinity was increasing. International Oceanographic Tables at the time were
based on a relationship between conductivity, chlorinity and salinity, but did not
go below temperatures of 10 °C, making them unsuited for in situ measurements
(UNESCO, 1966). In 1978–1979, new tables were produced out of intense research
from laboratories in Canada, France and the United Kingdom, and were based on the
conductivity of a seawater sample relative to that of a potassium chloride solution
(UNESCO, 1981a, 1981b, 1981c). From this work, the Joint Panel on Oceanographic
Tables and Standards (JPOTS) (appointed by the United Nations Educational, Scientific and Cultural Organization [UNESCO], ICES, Scientific Committee on Oceanic
Research [SCOR] and IAPSO) formed the Practical Salinity Scale 1978, which
adopted a KCl solution containing 32.4356 g kg^{-1} as the reference standard for
salinity determination (UNESCO, 1981a, 1981b, 1981c; Culkin, 1986).

The new definition, as published by UNESCO in 1981, for Practical Salinity
became:

> The practical salinity, symbol S, of a sample of seawater, is defined in terms of the ratio
> K_{15} of the electrical conductivity of the seawater sample at the temperature of 15 °C and
> the pressure of one standard atmosphere, to that of a potassium chloride (KCl) solution, in
> which the mass fraction of KCl is 32.4356×10^{-3}, at the same temperature and pressure.
> The K_{15} value exactly equal to 1 corresponds, by definition, to a practical salinity exactly
> equal to 35.

So essentially: A seawater sample of Practical Salinity 35 has a conductivity ratio
of unity at 15 °C and 1 atmosphere with a KCl solution containing a mass of 32.4356 g
KCl in a mass of 1 kg of solution.

The Practical Salinity Scale 1978 was accepted by ICES and IAPSO in
1979, SCOR and JPOTS in 1980, and by the Intergovernmental Oceanographic
Commission (IOC) of UNESCO in 1981.

To ensure continuity with the previous scale, the value of Practical Salinity 35 at
unity corresponds to a Standard Seawater whose certified chlorinity was 19.3740‰,
i.e. its salinity was exactly 35‰ on the previous scale.

Potassium chloride was chosen as the reference standard for a number of reasons.
Not only was it the accepted standard used in electrical conductivity measurements,
but Merck "Suprapur" KCl had excellent purity and well-documented chemical analysis. It also had good consistency between batches (Dauphinee et al., 1980). The
major impurity of this material was found to be sodium chloride, but at the level of
interest the molar conductivities of the two salts were sufficiently similar to minimise
the effect of the impurity (Lewis, 1980).

Practical Salinity, S_P, is a dimensionless value, i.e. the algorithms of the Practical
Salinity Scale 1978 were adjusted to eliminate any units. So whilst research papers
may state ppt, ‰ or psu (practical salinity unit), these should not be used. It is
sufficient to state, for example, a "Practical Salinity of 34.995".

The Practical Salinity is defined in terms of the ratio R_t (at temperature, t) by the
following equation:

$$S_p = a_0 + a_1 R_t^{1/2} + a_2 R_t + a_3 R_t^{3/2} + a_4 R_t^2 + a_5 R_t^{5/2}$$

$$+ \frac{(t-15)}{1+k(t-15)}\left\{b_0 + b_1 R_t^{1/2} + b_2 R_t + b_3 R_t^{3/2} + b_4 R_t^2 + b_5 R_t^{5/2}\right\}$$

where:

$a_0 = 0.0080$	$b_0 = 0.0005$
$a_1 = -0.1692$	$b_1 = -0.0056$
$a_2 = 25.3851$	$b_2 = -0.0066$
$a_3 = 14.0941$	$b_3 = -0.0375$
$a_4 = -7.0261$	$b_4 = 0.0636$
$a_5 = 2.7081$	$b_5 = -0.0144$
	$k = 0.0162$

This equation, being based on the new International Oceanographic Tables and utilising electrical conductivity measurements, is valid for salinities ranging from 2 to 42 at temperatures of −2 to 35 °C (UNESCO, 1981a, 1981b, 1981c). As it no longer uses chlorinity, its link with salinity is broken and should now be used as an independent chemical parameter.

It should be noted that the salinities calculated at the time were measured on temperatures according to the International Practical Temperature Scale 1968 (IPTS-68). Temperatures based on the more recent International Temperature Scale 1990 (ITS-90) can be converted to IPTS-68 using the below formula (Saunders, 1990):

$$t_{IPTS68} = 1.00024 \times t_{ITS-90}$$

8.9 Standard Seawater Batch Stability

Standard Seawater was originally intended as a chlorinity standard and was calibrated mainly using the Volhard method which was found to be more accurate than other methods in routine use. From 1969, a combined gravimetric/potentiometric titration method (Hermann & Culkin, 1978) was used, which whilst being operationally simpler, was still time-consuming. When the Seawater Service was due to transfer from Copenhagen to the Institute of Oceanographic Sciences, Wormley, UK, independent checks were made by both sites on batches prepared between 1969 and 1974 (Hermann & Culkin, 1972). The agreement between the two laboratories was better than 1×10^{-4}‰ in chlorinity (std dev 2–3 $\times$ 10^{-4}) which confirmed the reliability of the calibrations but unfortunately revealed little about its stability (Culkin & Ridout, 1998; Culkin & Smed, 1979). The use of glass ampoules for storing the standard has a negligible effect on its stability; given that the glass contains no halogens, the small amount of silica that might dissolve over time will not alter the chlorinity.

Although the standard was only calibrated in terms of chlorinity, from the late 1950s, chlorinity titration was gradually replaced by electrical conductivity measurements. Batch comparisons of Standard Seawaters prepared from 1937 to 1978 were investigated by six laboratories (Mantyla, 1980; Millero et al., 1977; Park, 1964; Poisson et al., 1978) which showed that the conductivity/chlorinity relationship was largely the same but that some variations did occur. Notably, batches P49–P51 were found to have anomalously high conductivities (equivalent to 0.004–0.007‰ in salinity), which were attributed to bacterial contamination, possibly combined with oil pollution (Culkin & Ridout, 1998). Following the adoption of the Practical Salinity Scale in 1978 and better filtration technologies, agreement between batches reportedly improved (Mantyla, 1987; Takatsuki et al., 1991).

In recent years, there have been numerous studies into the stability of Standard Seawater (Aoyama et al., 2002; Bacon et al., 2000, 2007; Culkin & Ridout, 1998; Kawano et al., 2006, 2012; Mantyla, 1980, 1987, 1994; Takatsuki et al., 1991; Uchida et al., 2020), perhaps most notable are those carried out in conjunction with Ocean Scientific International Ltd (OSIL) who currently operate and maintain the Standard Seawater Service. Culkin and Ridout (1998) demonstrated that conductivity changes over time of various batches of standards stored in ampoules could amount to less than 0.001 in salinity (studied over 96 weeks). When this is compared to the accuracy/precision of 0.002 quoted by Guildline Instruments for their widely used Autosal Salinometers, it would suggest that the stability of the standard is not usually the limiting factor. They also showed that there was a variability of less than 0.00001 in K_{15} of a standard (P123) stored in borosilicate bottles over a period of 158 weeks.

An in-depth study by Bacon et al. (2007) examined batch stability and the uncertainty associated with the Standard Seawater manufacturing process, and determined that the expanded uncertainty of the conductivity ratio to be 1×10^{-5}, based on a coverage factor of 2, at the time of manufacture. It found that there was no discernible "within batch" variability. The study also reported on the Standard Seawater "offsets" from the label conductivity ratio of batches P130–P144 over a period of 5 years after manufacture, and showed that no significant change in label conductivity outside the expanded uncertainty was found. This is in contrast to some other studies, and suggestions were made by Bacon as to why this may be the case, but the use of any "correction tables" intended to allow for offsets in modern Standard Seawater should be treated with great caution.

8.10   Final Thoughts

It has never been more important for the health of the oceans, and of the planet, that researchers are able to reliably compare data. Since its inception at the start of the twentieth century, the Standard Seawater Service has provided the means for oceanographic researchers to collect meaningful salinity data that has been used to monitor the oceans. With a well-documented history, and IAPSO's unswerving support, Standard Seawater remains the only Practical Salinity standard recognised by all international oceanographic bodies and will continue to provide researchers with a traceable standard.

Acknowledgements This work was greatly improved by comments from staff at Ocean Scientific International Ltd.

Competing Interests The authors have no conflicts of interest to declare that are relevant to the content of this chapter.

References

Aoyama, M., Joyce, T. M., Kawano, T., & Takatsuki, Y. (2002). Standard seawater comparison up to P129. *Deep-Sea Research, 49*, 1103–1114.

Bacon, S., Snaith, H. M., & Yelland, M. J. (2000). An evaluation of some recent batches of IAPSO standard seawater. *Journal of Atmospheric and Oceanic Technology, 17*, 854–861.

Bacon, S., Culkin, F., Higgs, N., & Ridout, P. (2007). IAPSO standard seawater: Definition of the uncertainty in the calibration procedure and stability of recent batches. *Journal of Atmospheric and Oceanic Technology, 24*(10), 1785–1799.

Boyle, R. (1674). *The saltiness of the sea*. London, Printed by E. Fletcher for R. Davis.

Brown, N., & Allentoft, B. (1966). *Salinity, conductivity, and temperature relationships of seawater over the range of 0 to 50 ppt*. Bissett-Berman Corp San Diego CA.

Conférence Internationale. (1899a). *Conférence Internationale pour l'Exploration de la mer*, réunie à Stockholm 1899, XLII–XLVI.

Conférence Internationale (1899b). *Conférence Internationale pour l'Exploration de la mer*, réunie à Stockholm 1899, 6–7.

Conférence Internationale (1899c). *Conférence Internationale pour l'Exploration de la mer*, réunie à Stockholm 1899, 12–13.

Conférence Internationale. (1899d). *Conférence Internationale pour l'Exploration de la mer*, réunie à Stockholm 1899, 16.

Conférence Internationale. (1901a). *Conférence Internationale pour l'Exploration de la mer*, réunie à Kristiania 1901, 4–5.

Conférence Internationale. (1901b). *Conférence Internationale pour l'Exploration de la mer*, réunie à Kristiania 1901, 26.

Cox, R. A., Culkin, F., & Riley, J. P. (1967). The electrical conductivity/chlorinity relationship in natural seawater. *Deep-Sea Research, 14*(2), 203–220.

Culkin, F. (1986). Calibration of standard seawater in electrical conductivity. *Science of the Total Environment, 49*, 1–7.

Culkin, F., & Ridout, P. S. (1998). Stability of IAPSO standard seawater. *Journal of Atmospheric and Oceanic Technology, 15*, 1072–1075.

Culkin, F., & Smed, J. (1979). The history of standard seawater. *Oceanologica Acta, 2*(3), 355–364.

Dittmar, W. (1884). Report on the scientific results of the exploring voyage of H.M.S. Challenger. *Physics and Chemistry, 1*, HMSO, London.

Dauphinee, T. M., Ancsin J., Klein, H. P., & Phillips, M. J. (1980). The effect of concentration and temperature on the conductivity ratio of potassium chloride solutions to Standard Seawater of salinity 35‰ (Cl 19.3740‰). *IEEE Journal of Oceanic Engineering, OE-5* (1), 17–21.

Forch, C., Knudsen, M., & Sørensen, S. P. L. (1902). Berichte über die Konstantenbestimmungen zur Aufstellungen der hydrografischen Taabellen, K. danske Vidensk. Selsk. Skr., 6, *Raekke, naturvidensk og mathem Afd.*, XII, 1 (p. 22).

Forchhammer, G. (1865). On the composition of seawater, at different depths, and in different latitudes. *Philosophical Transactions of the Royal Society of London, 155*, 203–262.

Hermann, F., & Culkin, F. (1972). A check of the analysis of standard seawater. In *International Council for the Exploration of the sea, Hydrography committee Communication CM 1972/C:33* (5 pp.)

Hermann, F., & Culkin, F. (1978). Preparation and chlorinity calibration of standard seawater. *Deep-Sea Research, 25*(12), 1265–1270.

Jacobsen, J. P., & Knudsen, M. (1940). Urnormal 1937 or primary standard sea-water 1937. *Association D'océanographie Physique, Publication Scientifique, 7*, 1–38.

Kawano, T., Aoyama, M., Joyce, T., Uchida, H., Takatsuki, Y., & Fukasawa, M. (2006). The latest batch-to-batch difference table of standard seawater and its application to the WOCE onetime sections. *Journal of Oceanography, 62*, 777–792.

Kawano, T., Takatsuki, Y., & Aoyama, M. (2012). Comparison of some recent batches of IAPSO standard seawater. *Journal of the Japan Society for Marine Surveys and Technology, 12*(2), 49–55.

Knudsen, M. (1899). The Danish "Ingolf" Expedition. *Hydrography, 1*(2), 37–39.

Knudsen, M. (1903). On the standard-water used in the hydrographical work until July 1903. *Cons. Internat. Pour l'Explor. De la Mer*, Pub. De Circ. No. 2 (9 p.).

Lewis, E. L. (1980). The practical salinity scale 1978 and its antecedents. *IEEE Journal of Oceanic Engineering, OE-5*(1), 3–8.

Mantyla, A. W. (1980). Electrical conductivity comparisons of standard seawater batches P29 to P84. *Deep-Sea Research, 27A*, 837–846.

Mantyla, A. W. (1987). Standard seawater comparisons updated. *Journal of Physical Oceanography, 17*, 543–548.

Mantyla, A. W. (1994). The treatment if inconsistencies in Atlantic deep water salinity data. *Deep-Sea Research, 41*, 1387–1405.

Marcet, A. (1819). On the specific gravity and temperature in different parts of the ocean and in particular seas, with some account of their saline contents. *Philosophical Transactions of the Royal Society of London, 109*, 161–208.

Millero, F. J., Chetirkin, P., & Culkin, F. (1977). The relative conductivity and density of standard seawaters. *Deep-Sea Research, 24*, 315–321.

Mohr, C. F. (1856). Neue massanalytische Bestimmung des Chlors in Verbindungen. *Annalen der Chemie und Pharmacie, 97*, 335–338.

Park, K. (1964). Reliability of standard seawater as a conductivity standard. *Deep-Sea Research, 11*, 85–87.

Poisson, A., Dauphinee, T., Ross, C. K., & Culkin, F. (1978). The reliability of standard seawater as an electrical conductivity standard. *Oceanologica Acta, 1*, 425–433.

Saunders, P. M. (1990). The international temperature scale of 1990, ITS-90. *WOCE Newsletter, 10*, 10.

Takatsuki, Y., Ayoama, M., Nakano, T., Miyagi, H., Ishihara, T., & Tsutsumida, T. (1991). Standard seawater comparisons of some recent batches. *Journal of Atmospheric and Oceanic Technology, 8*, 895–897.

Thomsen, H. (1950). Martin Knudsen 1871–1949. *ICES Journal of Marine Science, 16*(2), 155–159.

Uchida, H., Kawano, T., Nakano, T., Wakita, M., Tanaka, T., & Tanihara, S. (2020). An expanded batch-to-batch correction for IAPSO standard seawater. *Journal of Atmospheric and Oceanic Technology, 37*(8), 1507–1520.

UNESCO. (1965). *First report of the joint panel on oceanographic tables and standards, Copenhagen, October 1964* (UNESCO Technical Paper in Marine Science, No. 1, 1–9).

UNESCO. (1966). *First report of the joint panel on oceanographic tables and standards, Rome, October 1965* (UNESCO Technical Paper in Marine Science, No. 4, 1–9).

UNESCO. (1978). *Eighth report of the joint panel on oceanographic tables and standards, Woods Hole March 1977* (UNESCO Technical Paper in Marine Science, No. 28, 1–35).

UNESCO. (1979). Ninth report of the joint panel on oceanographic tables and standards, Paris, September 1978 (UNESCO Technical Paper in Marine Science, No. 30, 1–32)

UNESCO. (1981a).*Tenth report of the joint panel on oceanographic tables and standards, Sidney, September 1980* (UNESCO Technical Paper in Marine Science, No. 36, 1–25).

UNESCO. (1981b). *Background papers and supporting data on the Practical Salinity Scale 1978* (Sidney UNESCO Technical Paper in Marine Science, No. 37, 1–144).

UNESCO. (1981c). *International oceanography tables* (UNESCO Technical Paper in Marine Science, No. 39, 1–111).

Wooster, W. S., Arthur, A. J., & Dietrich, G. (1969). Redefinition of salinity. *Limnology and Oceanography, 14*(3):437–438.

Volhard, J. (1874). Über eine neue Methode der massanalytischen Bestimmung des Silbers. *Journal Für Praktische Chemie, 117*, 217–224.

Chapter 9
History of Batch-to-Batch Comparative Studies of International Association for the Physical Sciences of the Oceans Standard Seawater

Hiroshi Uchida, Mitsuho Oe, and Masahide Wakita

Abstract Batch-to-batch comparisons of the Practical Salinity (S_P) of International Association for the Physical Sciences of the Oceans standard seawater (SSW) are reviewed. The batch offset values of S_P proposed in previous studies are summarized for SSW batches P29 (1959) to P165 (2021). Comparability of methods is taken into consideration, and blanks in the batch offset values for batches P113 and P117 are filled in. To check the consistency with which long-term batch offset tables are stitched together from batch-to-batch comparisons, the S_P values of historical SSWs (P51–P163) were measured in September 2020. The batch offset table was also evaluated by conducting batch-to-batch comparisons of seawater samples collected during the Japan Meteorological Agency's routine oceanographic observation cruises. Because the cause of batch-to-batch differences larger than the expanded uncertainty of the SSW label values is still unclear, a reference seawater that is more robust and more stable than SSW might be required to establish a high level of international comparability of S_P measurements.

Keywords Comparability of practical salinity · Chlorinity · Electrical conductivity · Composition of standard seawater · Batch offset table

H. Uchida (✉)
Research Institute for Global Change, Japan Agency for Marine-Earth Science and Technology, Yokosuka, Japan
e-mail: huchida@jamstec.go.jp

M. Oe
Japan Meteorological Agency, Minato-ku, Tokyo, Japan
e-mail: mitsuho.oe@met.kishou.go.jp

M. Wakita
Mutsu Institute for Oceanography, Japan Agency for Marine-Earth Science and Technology, Mutsu, Japan
e-mail: mwakita@jamstec.go.jp

9.1 Introduction

In 2010, the Thermodynamic Equation of Seawater 2010 (TEOS-10) was adopted as the standard description for the thermodynamic properties of seawater in oceanography. TEOS-10 replaced the previous standard, the equation of state of seawater 1980 (EOS-80). Millero (2010) has reviewed the history of the equation of the state of seawater, and Pawlowicz et al. (2016) have reviewed the history of oceanic salinity measurements. Since 1902, and through a variety of changes in the definitions of standards and salinity, a certified reference material known as standard seawater (SSW) has played an important role in ensuring the comparability of observed salinity data (Bacon et al., 2007; Culkin & Ridout, 1998; Culkin & Smed, 1979). SSW is created in batches and the most recent batch is numbered P165. Each batch of SSW is created by sampling seawater from a particular region of the North Atlantic, filtering it to remove organic material that could cause problems in long-term storage, carefully measuring a particular characteristic of the seawater, and then bottling and labeling it. Before 1978, the characteristic was chlorinity; since then, the characteristic has been the conductance ratio relative to a specified KCl solution (Chap. 8).

Because of changes to the definition of salinity and to the container used for distributing SSW, the consistency among definitions and the comparability of certified values of SSW between different batches have been examined in a number of studies (e.g., Aoyama et al., 2002; Bacon et al., 2000, 2007; Culkin & Ridout, 1998; Kawano et al., 2000, 2001, 2006; Mantyla, 1980, 1987, 1994; Millero et al., 1977; Park, 1964; Poisson, 1975; Poisson et al., 1978; Takatsuki et al., 1991; Uchida et al., 2020). Although the consistency between batches of SSW has improved in recent decades (Bacon et al., 2007; Kawano et al., 2006; Takatsuki et al., 1991), an inconsistency of about ± 0.001 in Practical Salinity (S_P) still exists (Uchida et al., 2020). The magnitude of this inconsistency is significantly greater than the expanded uncertainty of ± 0.0004 for the certified values (in S_P) estimated for batches of SSW prepared shortly before 2007 by Bacon et al. (2007). Careful corrections for this inconsistency have made it possible to detect small changes in S_P (-0.0006 ± 0.0001 dec^{-1}) in the deep ocean that might be related to global climate change (Uchida et al., 2020). This means that salinity measurements conducted in the deep ocean for climate studies must be of the highest possible accuracy, as close to the resolution of the salinometers (0.0002 in S_P) as possible.

In this study, we review the history of those studies of batch-to-batch comparisons and present new results of batch-to-batch comparisons of old and recent SSWs. We also fill in the blanks in the long-term batch offset table for batches P113 and P117. We then describe our current effort to establish a high level of comparability between salinity measurements.

9.2   History of Batch-to-Batch Comparative Studies

Since the 1950s, electrical conductivity meters (salinometers) have been widely used to measure the salinity of seawater (e.g., Park & Burt, 1965a, 1965b). A salinometer is usually calibrated with SSW by adjusting the salinometer to read the conductivity corresponding to the chlorinity of the SSW on the assumption that the relationship between chlorinity and conductivity is the same for all batches of the SSW.

Park (1964) first examined the reliability of SSW as a conductivity standard. Computed salinities from certified chlorinities were compared with salinities estimated from conductance measurements by using an inductive salinometer for batches P15 (prepared in 1937), P18, P24–P26, P29–P32, P35, and P36. Poisson (1975) also conducted measurements of electric conductivity relative to a KCl solution for batches P37, P49, P50, P53, P56, P62, and P64 by using a Jones-type bridge and a cell with bright platinum electrodes. These two studies revealed that the salinity estimated from the conductivity of some batches was higher than that calculated from the certified chlorinity. The Joint Panel on Oceanographic Tables and Standards (JPOTS) recommended that the measurements be repeated by other laboratories, and 26 batches of SSW prepared during the period 1962–1975 were distributed to four laboratories. The results of one laboratory were published independently (Millero et al., 1977), and Poisson et al. (1978) summarized the results of the inter-laboratory, batch-to-batch comparisons using salinometers with the results from the previous studies of Park (1964) and Poisson (1975) for batches P37–P41, P44, P46–P56, P59–P62, P64, and P66–P71. Chlorinity-based salinities (S_{Cl}) were computed from certified chlorinities (Cl) using the equation $S_{Cl} = 1.80655$ Cl. The discrepancy between the S_{Cl} values and salinities estimated from conductance measurements (mean $\pm$ SD was $0.0018 \pm 0.0020‰$) was considered significant, and it was suggested that variability in the relationship between the chlorinity and conductivity of SSW was the cause of this discrepancy. Although the density, pH, and concentrations of silicate and dissolved organic carbon were also measured for those batches, those measurements did not suggest a clear explanation for the discrepancy. Poisson et al. (1978) concluded that the electrical conductivity as well as chlorinity of each batch of SSW should be certified as soon as possible.

After considering a new equation of state of seawater, the JPOTS recommended the Practical Salinity Scale 1978 (PSS-78) (UNESCO, 1981), and the relationship between chlorinity and S_{Cl} was subsequently abandoned in favor of a conductivity–S_P relationship (Millero, 2010). Batches of SSW since P91, labeled with the results of conductance measurements, have all been made in basically the same way by Ocean Scientific International Ltd. (UK).

Because SSW has been considered to have a lifetime of a few years, long-term monitoring of SSW characteristics has been carried out by performing batch-to-batch comparisons during periods of overlapping lifetimes. These comparisons have then been cumulatively "stitched together" to form a table of long-term batch offsets. Mantyla (1980) used a salinometer to conduct batch-to-batch comparisons of batches P35–P37, P39–P47, P49, P51, P52, P54, P56–P59, P61, P63, P65, P66 P69, P70, and P72–P84. Mantyla (1980) then combined these results with the results of 10 other published and unpublished batch-to-batch comparison experiments, including the results from Park (1964), Poisson (1975), Millero et al. (1977), and Poisson et al. (1978), to make a batch offset table for batches P29–P84. Because all of these comparisons were relative to an arbitrarily selected reference batch of SSW, adjustments to a common batch of SSW were usually made based on overlaps of the same batches of SSW in the various comparisons. Because batch P79 was the standard used to determine the KCl concentration for PSS-78, the batch offset table for batches P29–P84 was proposed relative to batch P79.

Mantyla (1987) used a salinometer to continue batch-to-batch comparisons for batches P36, P40–P47, P52, P59, P61, P63, P70, P73, and P76–P102 and updated the batch offset table for batches P29–P102 by combining those results with the results of Mantyla (1980). The offsets were calculated relative to a new reference based on the assumption that the average of the batch offsets for P91–P102 was zero. The rationale was that SSW was certified by the K_{15} value defined by the PSS-78 rather than by the chlorinity from batch P91, where K_{15} is the electrical conductivity ratio relative to a KCl solution (32.4356 g kg^{-1}) at a temperature of 15 °C (International Practical Temperature Scale of 1968 [IPTS-68]) and a pressure of one standard atmosphere.

After Mantyla (1987), Takatsuki et al. (1991) used a salinometer to conduct batch-to-batch comparisons for batches P90, P100, P104, P106, P108, P111, and P112. Mantyla (1994) also used a salinometer to make batch-to-batch comparisons for batches P103–P110. Aoyama et al. (2002) used salinometers to conduct interlaboratory batch-to-batch comparisons for batches P110, P112, P114, P116, P118–P124, and P127–P129 and expanded the batch offset table proposed by Mantyla (1987) up to P129 by combining those results with the results from Takatsuki et al. (1991) and Mantyla (1994). Because the results of Aoyama et al. (1998) (non-peer-reviewed letter, International WOCE Newsletter, 32, 5–7) were later published as a peer-reviewed paper (Aoyama et al., 2002), those results are referred to as Aoyama et al. (2002) in this manuscript.

Culkin and Ridout (1998) recalibrated SSW batches P120–P129 relative to newly prepared solutions of KCl and examined the variation with age of their K_{15} values for a maximum of two years. They also examined the shelf life of SSW stored in a borosilicate glass bottle for batch P123 and confirmed no significant change in the K_{15} value for three years. Bacon et al. (2000) also examined aging variations of batches P115–P132 by using salinometers to conduct batch-to-batch comparisons. The results were compared with the batch-to-batch offsets proposed by Aoyama et al. (2002).

Following Aoyama et al. (2002), Kawano et al. (2000) conducted batch-to-batch comparisons of batches P70, P88, P94, P114, P116, P119, P121, P123, P124, P127–P129, and P132-P135, and Kawano et al. (2001) used salinometers to conduct batch-to-batch comparisons of batches P123 and P132–140. Kawano et al. (2006) used salinometers to conduct interlaboratory batch-to-batch comparisons of batches P133 and P135–P145 and expanded the batch offset table of Aoyama et al. (2002) up to P145 by including the results of Kawano et al. (2000, 2001). Kawano et al. (2006) also filled in the blanks of the batch offset table for batches P115, P125, P126,

P130, and P131 by using results from Culkin and Ridout (1998) and Bacon et al. (2000). Kawano et al. (2006) reanalyzed those results (Table 5 in Kawano et al., 2006) because they were regarded as batch-to-batch comparison experiments and were consistent with the other results of Kawano et al. (2006). Finally, Kawano et al. (2006) proposed a new batch offset table relative to the new reference. The new batch offset table assumed the average of the batch offsets for P130–P145 to be zero because consistency among those batches was much better than the consistency of the older batches. Kawano et al. (2006) demonstrated a reduction of inconsistency at 105 crossover points of the World Ocean Circulation Experiment (WOCE) sections by adding 14 crossover points in the Indian Ocean to the 91 crossover points in the Pacific Ocean and Atlantic Ocean reported by Aoyama et al. (2002). The standard deviation of duplicate measurements of S_P was reduced from 0.0020 to 0.0017 by applying the SSW batch offset correction. Kawano et al. (2006) also demonstrated a reduction of unrealistic S_P changes with time in the deep ocean by applying batch offset corrections to WOCE sections and their repeat hydrographic sections 10 year apart.

Bacon et al. (2007) estimated for the first time the uncertainty of the label value of manufactured SSW. Following Culkin and Ridout (1998), they recalibrated SSW batches P130–P144 with reference to newly prepared solutions of KCl. No significant change was found in the label value outside the expanded uncertainty with a coverage factor of 2 (S_P difference of 0.0004) for as long as 5 year after the SSW batch manufacture.

The container used for SSW was changed from a soda-glass ampoule (up to P139) to a borosilicate glass bottle (from P140) in 2000. The only exceptions were P138 (both ampoule and bottle) and P142 (ampoule). For batch P138, Kawano et al. (2006) evaluated the bottle-type SSW, and Bacon et al. (2007) evaluated the ampoule-type SSW.

Following Kawano et al. (2006), Uchida et al. (2020) used salinometers to conduct batch-to-batch comparisons of batches P138, P141, P142, and P144–P163 and expanded the batch offset table of Kawano et al. (2006) up to P163 with a revision of the offset for P145. Because the Bacon et al. (2007) results suggested little change in the characteristics of SSW between batches, in apparent contradiction to the batch-offset tables being determined by other workers, Uchida et al. (2020) compared batch-to-batch offsets for batches P130–P144 with the corresponding batch-to-batch differences estimated from Bacon et al. (2007) and found no significant change in label S_P outside the expanded uncertainty (S_P difference of 0.0004) in both cases. Thus, it appears that the batches considered by Bacon et al. (2007) were of unusually consistent quality.

In practical terms, the use of batch corrections does have consequences for ocean climate research. Uchida et al. (2020) detected recent freshening (-0.0006 ± 0.0001 g kg^{-1} dec^{-1}) in the deep North Pacific by applying the batch-to-batch corrections to conductivity–temperature–depth (CTD) data (36 CTD profiles during two recent decades) accumulated at time series station K2 (47°N, 160°E; water depth 5215 m). It was difficult to detect such a slight trend in salinity without applying the batch corrections. Uchida (2019) expanded the batch offset correction table up to

P165 following Uchida et al. (2020) (the expanded batch offsets were −0.0005 and −0.0001 in S_P for batches P164 and P165, respectively).

The batch offset values proposed by those previous studies were summarized in three papers (Kawano et al., 2006; Mantyla, 1987; Uchida et al., 2020). Figure 9.1 shows the batch offsets of S_P for all of these studies relative to the new reference proposed by Kawano et al. (2006) and fills in the blanks in the batch offset values for batches P113 and P117 (see Sect. 9.6). The batch-to-batch variation was quite large for batches before P90 (1980), which were calibrated based on chlorinity, but it has been significantly smaller for batches after P91 (1980) which were calibrated based on KCl reference. However, there has been a long-term trend in the offsets, which have decreased by about 0.0025 over the last 40 year. Uchida (2019) has provided the batch offsets for batches P29–P165, and that online dataset will be expanded for the most recent batches.

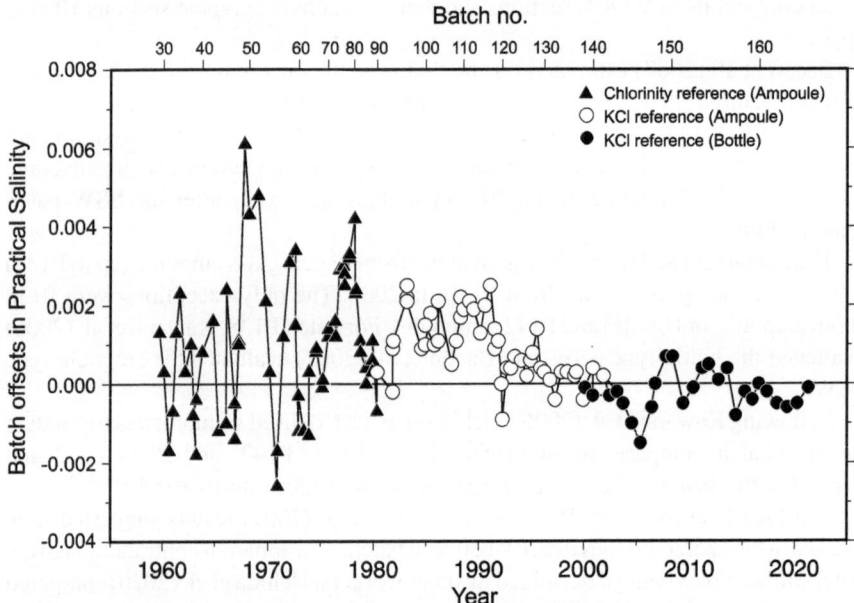

Fig. 9.1 Batch offsets of S_P for IAPSO SSW batches P29–P165. The offset values reported in Mantyla (1987) (P29–P90), Kawano et al. (2006) (P91–P144, except for P113 and P117), Uchida et al. (2020) (P145–P163), Uchida (2019) (P164 and P165), and this study (P113 and P117) are plotted relative to the average of the batch-to-batch offsets for batches P130–P145 reported by Kawano et al. (2006) and also available online (Uchida, 2019). Batch offsets are to be added to S_P measured using a particular batch for standardization of the salinometer

9.3   Evaluation of the Batch Offset Table for Historical SSWs

Although SSW has a finite lifetime, unused vials of older batches can sometimes be found in the storage areas of marine institutions, leftover as new batches are received and integrated into standard measurement protocols. In an attempt to check the consistency of the long-term batch correction table (Fig. 9.1), which was created by stitching together batch-to-batch comparisons made over many years, a collection of historical SSW stored from as far back as 1969 was created, and as salinometer (Autosal model 8400B; Guildline Instruments, Ltd., Canada) was used to conduct batch-to-batch comparisons of this collection at the Japan Agency for Marine-Earth Science and Technology (JAMSTEC) at the time of the batch offset determination for P164 in September 2020 (Uchida, 2019). The dates of measurement were 7 September (for P155–P164), 8 September (for P148–P154), 9 September (for P138, P141, P143–P147), 17 September (for P130–P137, P139, P142), and 18 September (P51, P82, P84, P85, P88, P93, P103, P112, P114, P119, P121–P124, P127, P128). At least five bottles of P161 were measured on each day of the measurements to correct for temporal drift of the salinometer. Measured S_P was calibrated against the batch-offset-corrected values by adjusting the mean value for the reference batches P160–P163.

Figure 9.2 shows the differences in S_P from the batch-offset-corrected values. The differences for older SSWs that had been sealed in soda-glass ampoules and stored for more than 20 year were highly variable (maximum S_P difference of 0.08), very much greater than the proposed batch offsets, and they tended to be positive (Fig. 9.2a). Although evaporation of water was unlikely to have affected the salinity of the SSW in the ampoules, we observed suspended matter resembling glass flakes and crystals on the inner wall of the glass. The seawater inside the ampoules as well as the glass container may have undergone chemical changes. Therefore, these results were probably unreliable. However, the oldest SSW, P51 (produced in 1969), seemed to have been kept in a refrigerator for a long time (but recently stored at room temperature for several years), and the average of the differences of the S_P for five ampoules was close to zero (−0.0023), although the variation was large (SD of 0.0167).

However, differences for more recent SSWs stored in borosilicate glass bottles were close to zero (maximum S_P difference of 0.0012) (Fig. 9.2b). With the exception of the oldest batch (P138) among the borosilicate glass bottles and batches P146–P150 (12–15 year-old), the differences of S_P were within ±0.0003, smaller than the expanded uncertainty of SSW S_P of 0.0004 according to Bacon et al. (2007). However, for P146–P150, there was a negative bias (Fig. 9.2b). The history of the batch-to-batch comparisons suggested that the S_P of these batches had drifted with time at a rate of about −0.0005 per decade, albeit with considerable uncertainty (Fig. 9.3) (Uchida, 2019).

9.4 Changes to the Composition of SSW

Changes in conductivity/density/salinity relationships in real seawater have been linked to changes in the concentrations of seawater constituents that are involved in biogeochemical cycling (Pawlowicz et al., 2011). To determine whether similar factors were related to batch offsets, total alkalinity (TA) and concentrations of dissolved inorganic carbon (DIC), nitrate, silicate, ammonia, and dissolved organic carbon (DOC) were measured for batches P138, P141, P143–P151, P155, and P164 in 2020–2021 (Table 9.1). The focus of the study was the temporal drift of the S_P of the SSW (Fig. 9.3).

Uchida et al. (Chap. 10) have examined changes to the composition of SSW since 2010, including the results reported here. They found that (i) silicate increased with time (5.47 μmol kg^{-1} year^{-1}), (ii) DIC decreased with time (-13.5 μmol kg^{-1} year^{-1}), (iii) TA, nitrate was constant, and (iv) DOC and ammonia were constant for batches after P148, although they were highly variable for batches before P147. For batch P145, TA and DIC were much lower than usual (~420 μmol kg^{-1} and ~300 μmol kg^{-1}, respectively). As Poisson et al. (1978) also found, there was no clear relationship between these compositional changes and the batch offsets of S_P (ΔS_P values in Table 9.1). However, the decrease of S_P (e.g. $-$0.0013 per 15 year for P147) might have been partly related to the rate of decrease of DIC (-203 μmol kg^{-1} per 15 year) reported by Uchida et al. (Chap. 10), because a DIC change of 203 μmol kg^{-1} will change S_P by 0.0014 (Pawlowicz, 2010).

9.5 Evaluation of the Batch Offset Table for Recent SSWs by the Japan Meteorological Agency

The batch offset table, mostly based on measurements in shore laboratories, was also evaluated by conducting batch-to-batch comparisons during routine oceanographic observation cruises by the Japan Meteorological Agency (JMA). When a new batch of SSW is introduced, batch-to-batch comparisons are performed with the previously used one or two batches to check the inter-cruise consistency of S_P measurements (Table 9.2). Because of the small number of batches in each batch-to-batch comparison, the measured batch-to-batch offset difference between two batches was compared with the corresponding batch-to-batch offset difference calculated from the offset table (Fig. 9.4). The measured differences agreed well with the calculated differences from the offset table, except for the comparison in 2017. The offset difference between P159 and P160 in 2017 differed significantly (S_P difference of -0.0011) from the offset difference calculated from the offset table. Because the results for P157 and P159 in 2016 and for P160 and P161 in 2018 were consistent with the offset table, either the P159 or P160 measurements in 2017 could be the source of the discrepancy.

Table 9.1 Concentrations of dissolved inorganic carbon (DIC), total alkalinity (TA), nitrate, silicate, ammonia, and dissolved organic carbon (DOC) in SSW measured in 2020–2021. Batch ages at the time of measurement and differences in S_P (ΔS_P) from batch-offset-corrected values (mean ± SD) shown in Fig. 9.2 are also shown. The number of bottles measured for S_P is shown in parentheses. Values from the standard seawater composition model (SSW76) by Pawlowicz (2010) are also shown. TA from SSW76 is the same as TA from the reference composition (Millero et al., 2008), but DIC from SSW76 is slightly larger than DIC from the reference composition (see Pawlowicz, 2010)

Batch no.	Age of batch (year)	DIC (μmol kg^{-1})	TA (μmol kg^{-1})	Nitrate (μmol kg^{-1})	Silicate (μmol kg^{-1})	Ammonia (μmol kg^{-1})	DOC (μmol kg^{-1})	ΔS_P (PSS-78)
P138	20.7	1767.3	2288.6	0.08	124.31	0.89	56.3	0.0005 ± 0.0011 (5)
P141	18.3	1793.1	2292.4	0.19	161.70	1.97	83.4	0.0002 ± 0.0005 (5)
P143	17.6		2253.9	0.08	146.52	2.28		0.0000 (2)
P144	17.0	1793.7	2319.3	0.06	158.41	1.69	66.7	0.0003 ± 0.0008 (5)
P145	16.2	1550.5	1883.5	0.04	182.02	1.66	93.8	−0.0001 ± 0.0004 (5)
P146	15.4	1794.3	2285.6	0.45	166.66	1.87	70.8	−0.0007 ± 0.0007 (5)
P147	14.8	1835.7	2304.9	1.31	186.53	1.12	60.8	−0.0012 ± 0.0004 (5)
P148	14.0	1961.7	2296.8	0.96	96.63	0.04	53.8	−0.0005 ± 0.0002 (5)
P149	13.0	1976.1	2308.3	0.03	67.47	0.18	51.3	−0.0006 ± 0.0001 (5)
P150	12.4	1973.0	2305.5	0.31	78.50	0.03	49.9	−0.0007 ± 0.0002 (5)
P151	11.4	1985.8	2283.8	0.00	63.18	0.07	48.7	−0.0002 ± 0.0003 (5)
P155	8.0	1992.6	2308.8	1.04	61.68	0.01		0.0001 ± 0.0003 (5)
P164	1.0	2081.5	2303.1	0.06	27.02	0.17	51.7	±0.0001 (5)
SSW76		2080	2300	0	0	0	0	

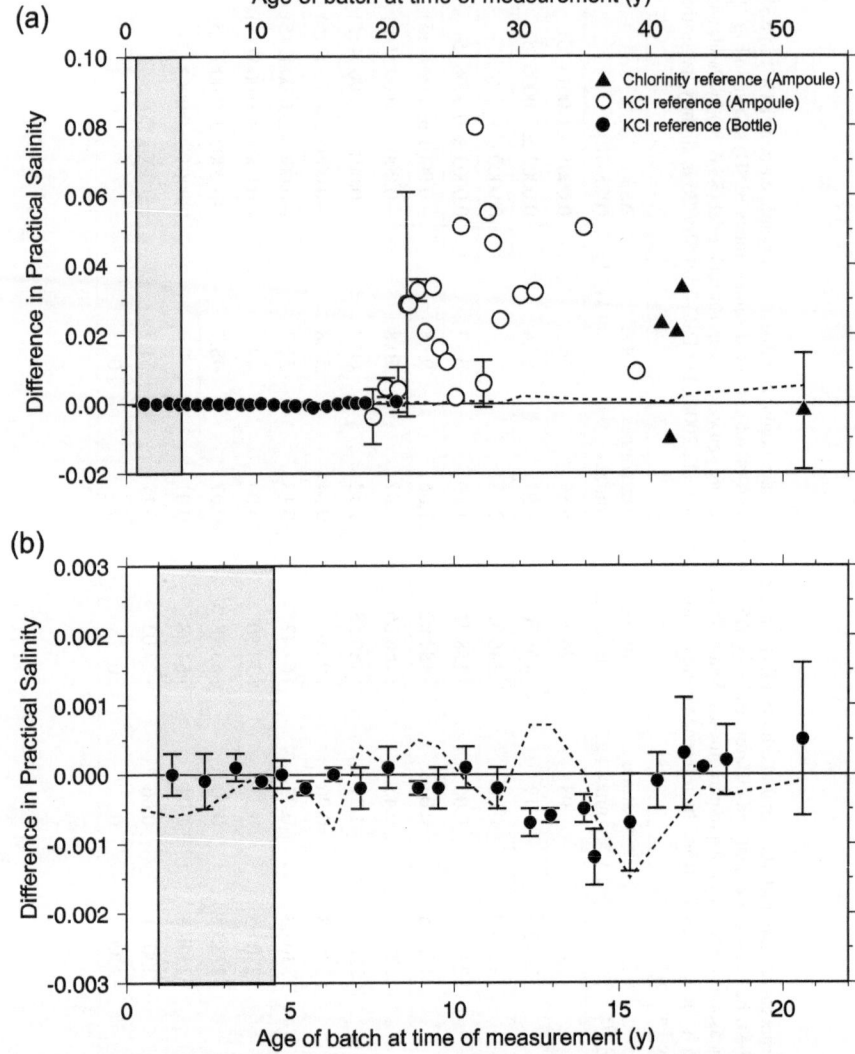

Fig. 9.2 Differences in S_P from the batch offset-corrected values for measurements in 2020 of historical batches stored since as far back as 1969 for **a** all batches and **b** recent batches in borosilicate glass bottles. Measurements were conducted in September 2020, and details of the data are available online (Uchida, 2019). Differences for batches (P160–P163) that were used as the reference are shaded. Vertical bars show their SDs for the batches measured for more than two bottles. Dashed lines show the batch offset values (Fig. 9.1) for comparison

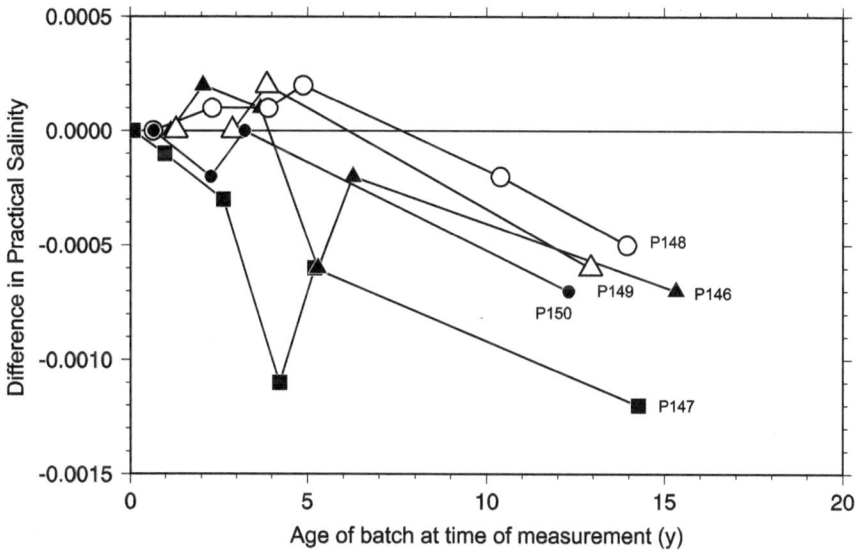

Fig. 9.3 Temporal drift of differences in S_P from the batch offset-corrected values for P146–P150. Details of the data are available online (Uchida, 2019)

Following the methodology of Saunders (1986), who used the deep ocean as a natural calibration tank, S_P of North Pacific bottom water at the same location (40°N, 165°E) measured on each cruise in 2014, 2016, 2017, 2020, and 2021 were compared to determine the source of the discrepancy (Table 9.3). The standard deviation of the batch-offset-corrected S_P of the bottom water was small (0.0002) and was equal to the resolution of the salinometers. Because the bottom water S_P in 2017 was calibrated with P159, the offset for P160 measured in 2017 must have deviated by about −0.0011 from the batch offset value of Uchida et al. (2020). Because the offset difference between P160 and P161 in 2018 was consistent with the offset difference of Uchida et al. (2020), only the S_P of the bottles for P160 that were obtained on the KS1704 cruise in 2017 may have been low, but for reasons that are unknown.

9.6   Estimation of Batch Offsets for SSW Batches P113 and P117

There are blanks (batches P113 and P117) in the batch offset table for SSW batches P29–P145 reported by Kawano et al. (2006). Batch offsets for batches P113 and P117 were therefore estimated to fill the blanks by using results of batch-to-batch comparisons reported in previous publications.

Joyce et al. (1992) measured conductivity ratios with a salinometer for several ampoules of batches P113 and P114 during an oxygen/salinity comparison cruise

Table 9.2 Offset of S_P ($\times 10^{-3}$) from the label value of the IAPSO SSW based on routine measurements by the JMA. The date of measurements, cruise number, batch number of the SSW used in standardization (STD) of the salinometers, and batch numbers examined are listed. SD is the standard deviation of the S_P ($\times 10^{-3}$)

Date of measurements	Cruise no.	Batch no./offset (no. of bottles measured, SD)	
		Used in STD	Examined batches
17 Apr 2012	RF1203_1	P153/0.0 (29, 0.2)	P152/−0.3 (4, 0.1), P154/−0.4 (27, 0.3)
9 May 2012	RF1203_2	P153/0.0 (24, 0.2)	P152/0.1 (8, 0.2), P154/−0.4 (21, 0.2)
15 June 2013	RF1305	P154/0.0 (26, 0.2)	P153/−0.1 (9, 0.2), P155/−0.3 (26, 0.2)
13 May 2014	RF1404	P155/0.1 (22, 0.3)	P156/0.1 (20, 0.3)
28 May 2014	KS1404	P155/0.0 (39, 0.2)	P156/−0.2 (34, 0.2)
23 May 2015	KS1505	P156/0.0 (32, 0.3)	P157/−1.3 (30, 0.2)
22 May 2016	KS1605	P157/0.3 (33, 0.3)	P159/0.5 (33, 0.4)
13 May 2017	KS1704	P159/0.1 (24, 0.2)	P160/−0.6 (25, 0.1)
18 June 2018	KS1805	P160/0.3 (23, 0.2)	P161/0.4 (23, 0.2)
9 May 2019	KS1904	P161/−0.1 (17, 0.1)	P162/−0.7 (16, 0.1)
11 July 2020	RF2005	P162/0.1 (10, 0.3)	P163/0.1 (10, 0.1)
8 May 2021	RF2104	P163/0.0 (10, 0.1)	P162/−0.1 (10, 0.1), P164/0.1 (10, 0.1)
4 May 2022	RF2203_1	P164/−0.1 (14, 0.2)	P163/−0.3 (10, 0.2), P165/0.2 (10, 0.2)
26 May 2022	RF2203_2	P164/0.1 (10, 0.1)	P163/−0.2 (10, 0.2), P165/0.3 (10, 0.1)

in 1991. Measured S_P were 34.9947 (18:25, July 1st), 34.9937 (18:15, July 2nd), and 34.9970 (18:00 July 5th) for batch P113; they were 34.9963 (18:45, June 30th), 34.9967 (20:30, July 1st), 34.9967 (18:00, July 3rd), 34.9967 (19:30, July 3rd), and 34.9967 (18:00, July 6th) for batch P114. Averages of the S_P were 34.9951 and 34.9966 for batches P113 and P114, respectively. The label values were 34.9937 and 34.9945 for the S_P of batches P113 and P114, respectively, and the batch offset of the S_P was 0.0020 for batch P114. The batch offset of the S_P of batch P113 could therefore be estimated to be 0.0013.

Parrilla (2007) measured 8 ampoules of batch P120 after the salinometer was standardized with batch P117 and reported that the label value (S_P of 34.994) was within 0.0002 of the S_P. Therefore, the batch offset for batch P117 can be estimated to be the same as the batch offset for P120 (−0.0009 in S_P).

The estimated batch offset for batch P117 was evaluated in the deep ocean by applying the batch offset correction to the WOCE hydrographic data (available from CCHDO https://cchdo.ucsd.edu/). Bottle-sampled S_P data calibrated with batches P117, P123, and P126 obtained during cruises 18HU93019_1 (June 1993),

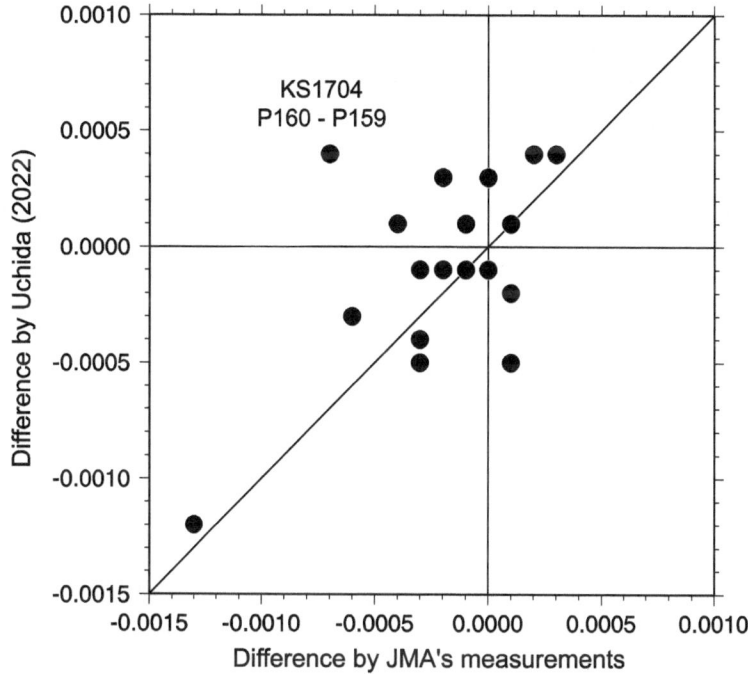

Fig. 9.4 Difference of the offsets of S_P from the label values between the batch of SSW examined and the batch used in standardization listed in Table 9.2. The differences from the JMA's routine measurements are compared with the corresponding differences of the batch offsets calculated by Uchida (2019)

Table 9.3 Comparison of mean S_P for bottom water in the western North Pacific (40°N, 165°E) derived from measurements of water samples obtained at 3 depths (5000, 5250, and ~5460 m [10 m above the bottom]). The water samples were obtained during the same cruise for each year as the cruise shown in Table 9.2. The mean S_P after applying the batch offset correction by Uchida et al. (2020) is also listed

Year	Cruise no.	Station no.	Reference batch no.	S_P Mean ± SD	Batch-offset-corrected S_P
2014	KS1404	KS4174	P155	34.6886 ± 0.0000	34.6887
2016	KS1605	KS4785	P157	34.6896 ± 0.0003	34.6888
2017	KS1704	KS5125	P159	34.6894 ± 0.0002	34.6890
2020	RF2005	RF6736	P162	34.6897 ± 0.0003	34.6892
2021	RF2104	RF6852	P163	34.6897 ± 0.0005	34.6891
Average (SD)				34.6894 (0.0005)	34.6890 (0.0002)

18HU19940524 (May 1994), and 18HU95011_1 (June 1995), respectively, along the WOCE AR7W line across the Labrador Sea were compared. Data obtained at three stations between latitudes of 57.36°N and 58.23°N for each cruise were used. The $T–S$ relationship for waters deeper than 3000 dbar (maximum pressure of 3657 dbar) was estimated by fitting a line for each cruise, and the S_P at a potential temperature of 1.7 °C (average pressure of 3446 dbar) was extracted from the regression line. The extracted S_P were 34.8906, 34.8867, and 34.8869 for 1993, 1994, and 1995, respectively. The agreement between them was improved by application of the batch offset correction (S_P values of 34.8897, 34.8874, and 34.8875 for 1993, 1994, and 1995, respectively).

9.7 Discussion

Although we believe that the batch offsets and long-term drifts summarized here are reliable estimates of the variability in SSW, these batch offsets are not the only possible source of systematic errors in salinity measurements. Gouretski and Jancke (2001) have reported inter-cruise offsets of salinity for a global hydrographic dataset. They have estimated the offsets for historical cruises to be on average 3–6 times the modern cruise offsets. Such large offsets could not be explained by SSW batch offsets. Purkey and Johnson (2013) have also reported inter-cruise offsets of S_P for WOCE cruises and their reoccupation cruises. Although the application of SSW batch offset corrections improved the agreement among occupations of each section, it did not eliminate the need for ad-hoc offsets to reduce inter-cruise S_P biases. Purkey et al. (2019) have closely examined inter-cruise offsets in S_P for the WOCE cruises and their reoccupation cruises in the South Pacific Ocean and have found that the difference in S_P between cruises occupied after 2000 has been less than 0.0001. They therefore applied an additional ad hoc S_P offset, ranging from −0.0037 to 0.0001, to most of the WOCE occupations in the 1990s. This purely empirical result is consistent with our analysis in suggesting that the change of SSW from soda glass ampoules to borosilicate glass bottles in 2000 improved the stability of the S_P of SSW.

Mantyla (1994) accounted for salinity offsets due to (i) different types of sampling bottles (epoxy-lined Nansen bottles or PVC plastic Niskin bottles), (ii) time lags between water sampling and S_P determination, (iii) batch offsets of SSW, (iv) salinometer response shifts, and (v) differences between up (water sampling) and down (CTD profiling) cast values because of hysteresis in pressure, temperature, or conductivity sensors. Gouretski and Jancke (2001) have suggested an additional possible source of systematic errors in salinity: (vi) the order of samples for salinity drawn from sampling bottles. Specifically, condensation of water may occur on the bottle interior of the headspace, according to the ambient air conditions, if salinity sampling is conducted later (head space becomes large) and is delayed for a long time.

Although there are several possible sources of systematic errors in salinity as described above, only the following possible sources related to SSW are discussed here: (i) within-batch differences, (ii) variations of the chlorinity–conductivity rela-tionship, (iii) initial offsets of label values, and (iv) variation of the conductivity of SSW with age.

Aoyama et al. (2002) have examined within-batch differences in terms of the stan-dard deviation of the S_P of several ampoules from the same batch of SSW (12 batches among P64–P128). They found that the standard deviations of S_P were 0.0001–0.0004 during the first four years after production. For recent SSWs in borosilicate glass bottles, the standard deviations of S_P measured within the 3 year shelf life were also 0.0001–0.0004 for 13 batches from P152 to P165 (Table 9.2). The effect of the within-batch differences on an estimation of the batch-to-batch offsets is therefore expected to be smaller than the typical batch offsets if SSWs are measured within the 3 year shelf life.

The main cause of the batch-to-batch differences among batches up to P90 might be the variation in the relationship between chlorinity and conductivity because the old batches were calibrated based on chlorinity (Poisson et al., 1978). Although the definition of salinity was changed to the definition in PSS-78 for batches after P91, the SSW labels showed both the conductivity ratio and chlorinity until batch P113. Millero and Huang (2009) have compared the S_P calculated from the conductivity ratio with the salinity calculated from chlorinity (S_{Cl}) by using the values shown on the labels. In this study, the batch-offset-corrected S_P was compared with the S_{Cl} (Fig. 9.5). The systematic bias (-0.0017 in terms of S_P) of the differences was greatly reduced to -0.0003 by applying the batch offset corrections. However, the variability of the comparison was not greatly reduced. This remaining variability (SD of 0.0019) may then be explained by the variation of the chlorinity–conductivity relationship, because the magnitude of the variation is the same as the magnitude of the batch offset variations for batches earlier than P90 using the chlorinity reference (SD of 0.0018) (Fig. 9.1).

For the initial offsets of label values of SSW, Kawano et al. (2005) examined supplier and lot dependencies of the electrical conductivity of KCl solutions due to impurities in the KCl reagents. They found that this effect on the uncertainty of SSW label S_P values was about 0.0012. However, Bacon et al. (2007) have examined the effect of impurities in the KCl and have found that it is smaller than a S_P difference of 0.0001, although the batch offsets for the period examined by Bacon et al. (2007) were small (S_P SD of 0.0003) (Uchida et al., 2020). Bacon et al. (2007) have estimated the overall expanded S_P uncertainty for the label value of SSW to be 0.0004 and have suggested that the largest uncertainties contributing to the overall uncertainty are (i) measurement of the KCl solution conductivity ratio, (ii) solvent conductivity in the KCl solutions (effectively CO_2 saturation uncertainty), and (iii) measurement of the new SSW conductivity ratio. In fact, Bacon et al. (2007) have reported that the recalibrated S_P values of SSW in reference to KCl solutions vary by ± 0.0004 for batches P141 and P144 that were measured within a month. Because the K_{15} value of the SSW is rounded to five significant digits, resolution of S_P calculated from the certified value is 0.0004, although the resolution of the salinometer model Autosal 8400B corresponds to an S_P difference of 0.0002. The expanded uncertainty of the SSW estimated by Bacon et al. (2007) should be increased slightly to a S_P difference

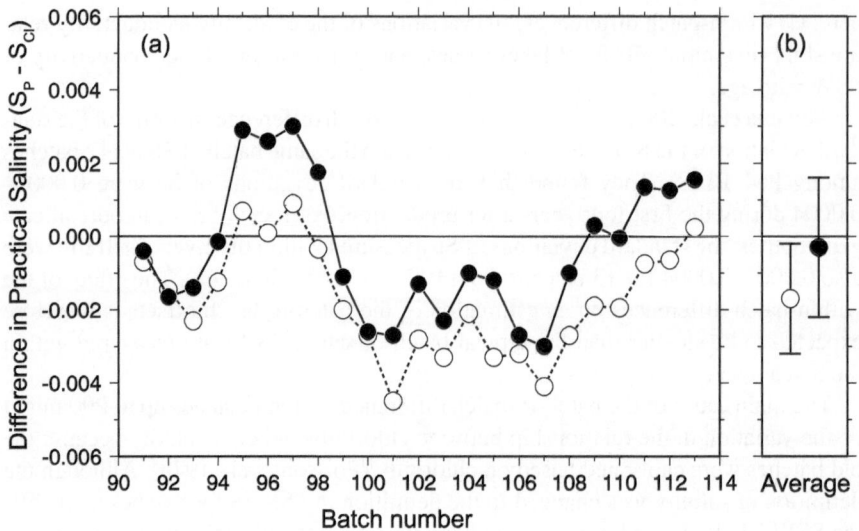

Fig. 9.5 Difference between the S_P calculated from the conductivity ratio and the salinity calculated from chlorinity based on the values shown on the SSW labels (open circles) for **a** batches P91–P113 and **b** their average (the vertical bar shows SD). The results for the batch offset-corrected S_P calculated from the conductivity ratio are also shown (closed circles)

of 0.0005 to take into consideration the uncertainty due to the resolution (0.0002/ $\sqrt{3}$). An initial offset equal to an S_P uncertainty of ±0.0005 may therefore exist for the label value.

The values recalibrated with reference to KCl solutions for batches P120–P129 have been reported in Table 1 of Culkin and Ridout (1998). The corresponding values for batches P130–P144 that were reported in Table 5 of Bacon et al. (2007) were reanalyzed (Tables 9.4 and 9.5) using the batch offset estimation by Uchida et al. (2020) by assuming that the results obtained on the same measurement day used the same KCl solution. Offsets for the KCl solutions were adjusted so that overlapping data between measurement days were consistent with each other and so that the mean of the average offsets for P120–P144 matched the mean of the corresponding batch offsets (Uchida, 2019). Note that the relative variability of the average offsets was independent of the batch offset table. The adjustment values for each measurement day varied from −0.0003 to 0.0007, and the offset values for each batch also varied from −0.0003 to 0.0006. The average offsets agreed well with the batch offsets (Fig. 9.6). The standard deviation of the S_P differences between the estimated offsets and the batch offsets (Uchida, 2019) was 0.0003. These results suggested that the batch offsets were mostly explained by initial offsets of the label values for batches after P91.

The average of the estimated initial offset of S_P values calibrated in reference to KCl solutions for batches P130–P144 (Bacon et al., 2007) was close to zero (mean ± SD is 0.0001 ± 0.0002), although the S_P offset for batches P120–P129

Table 9.4 Offsets in S_P ($\times 10^{-3}$) from the label values estimated from the results of Culkin and Ridout (1998) by reanalyzing their initial (displayed in italics) and recalibrated data by using the KCl standard solutions. Values from the batch offset table (Fig. 9.1) are also shown. Offsets with an asterisk were used assuming that the same KCl solution was used as that used on the measurement day because its measurement was conducted within 7 days. Offsets with double asterisks for the batch offset table were estimated by combining the results of Culkin and Ridout (1998) and Bacon et al. (2000) (see Kawano et al., 2006). Data for P130 were derived from Bacon et al. (2007), and the average offset for P130 was determined in Table 9.5

Batch no.	1992 Sep. 8th	1993 Jan. 19th	1993 May 8th	1994 Jan. 13th	1994 July 27th	1994 Nov. 22nd	1995 Feb. 7th	1995 July 18th	1995 Nov. 21st	1996 Mar. 19th	Average offset	Batch offset table
Adjustment value	0.5	0.5	0.5	0.4	0.1	0.4	0.7	0.5	0.2	−0.1		
P120	0.1	0.1	0.1	0.0							0.1	−0.9
P121	*0.5*	0.5	0.5	0.8							0.6	0.4
P122		*0.5**	0.5	0.4	0.5	0.4					0.5	0.4
P123			0.5*	0.4	0.5*	0.4	0.7				0.5	0.7
P124				*0.4**	0.5	0.4	0.7				0.5	0.6
P125					*0.1**	0.4	−0.1	0.1			0.1	0.2**
P126						0.4*	0.3	0.5	0.2	0.7	0.4	0.6**
P127							*0.7**	0.5	0.6	0.7	0.6	0.8
P128								*0.5*	0.6	0.3	0.5	1.4
P129									*0.2**	0.3	0.3	0.4
P130										*−0.1*		0.3

Note: In this table, the offset of the KCL standard solution used on each measurement date is estimated based on the measurement values of the same batch measured on different dates. The data marked with an asterisk are not the results of the same measurement date, but are assumed to have been measured using the same KCL standard solution because the measurement dates are close (within 7 days)

(Culkin & Ridout, 1998) was slightly biased (mean $\pm$ SD is 0.0004 $\pm$ 0.0002). The average of the recent batch S_P offsets for P130–P165 was also consistently close to zero (mean $\pm$ SD was −0.0001 $\pm$ 0.0005) (Fig. 9.1). Therefore, considering the fact that the systematic bias between the S_P calculated from the conductivity ratio and the salinity calculated from chlorinity by using the values shown on the labels for batches P91–P113 was greatly reduced by applying the batch offset corrections (Fig. 9.5), the arbitrarily selected reference for the batch offset table (the batch offsets for P130–P145 was assumed to be zero) would seem appropriate.

In addition to an initial offset, there may also be an aging component (as found for batches P146–P150, Fig. 9.3). Changes in properties of SSW over time, which can occur during storage due to reaction with the glass of the container and microbial activity, may contribute to conductivity changes (Poisson et al., 1978; Mantyla, 1980; also see Sect. 9.4). Changes in properties may have been large for the SSWs sealed in soda-glass ampoules because the quality of the soda glass of the ampoules was lower than that of the borosilicate glass of the bottles used for more recent SSWs, and the air space was larger in the ampoules than in the bottles. The larger air space

Table 9.5 Same as Table 9.4, but for the results of Bacon et al. (2007). Offsets shown in parentheses were not used to estimate the adjustment values, and the average offsets since the age of the batch at measurement was older than the 3 year shelf life. The measurement day for offsets marked # were modified because the measurement day listed in Bacon et al. (2007) was a misprint. The adjustment value for the measurement day of 19 March 1996 was determined in Table 9.4

Batch no.	1996 Mar. 19th	1996 Oct. 10th	1997 Apr. 9th	1997 Nov. 11th	1998 June 4th	1999 Feb. 9th	1999 Apr. 16th	1999 June 9th	1999 Dec. 9th	2000 June 2nd	2000 Nov. 10th	2001 Nov. 11th	2002 May 9th	2002 May 15th	2003 Feb. 25th	2003 Oct. 23rd	2004 July 9th	Average offset	Batch offset table
Adjustment value	−0.1	−0.1	−0.3	0.1	0.3	−0.3	0.2	0.3	−0.3	0.5	0.2	0.1	0.0	0.0	−0.1	0.5	0.1		
P130	*−0.1*	−0.1*	−0.3	−0.7													0.1	−0.3	0.3
P131		*−0.1*	0.1	0.1														0.1	0.1
P132			*−0.3*	0.1	0.3	−0.3	−0.2			(0.2)								−0.2	−0.4
P133				*0.1*	−0.1	0.1	−0.6											−0.1	0.3
P134					*0.3*	0.5	0.6	0.3	0.5				(0.4)					0.4	0.3
P135						*−0.3*	0.2	−0.1*	0.1				(0.4)				(0.5)	0.0	0.2
P136							*0.2*	0.3	0.1		0.2		0.4					0.2	0.3
P137									−0.3	−0.3#	−0.2	−0.3	−0.4					−0.3	−0.4
P138								*0.3*				0.1	0.0					0.1	−0.1
P139											*0.0#*	0.5		0.4	0.3			0.3	0.4
P140											0.2	0.1		0.0			(0.1)	0.1	−0.3
P141																	0.5*	0.5	−0.3
P142												*0.1*		0.4		0.1	0.5	0.3	0.2
P143															*−0.1*	0.1	−0.3*	−0.1	−0.2
P144																	−0.3/0.5*	0.1	−0.5

Note: In this table, the offset of the KCL standard solution used on each measurement date is estimated based on the measurement values of the same batch measured on different dates. The data marked with an asterisk are not the results of the same measurement date, but are assumed to have been measured using the same KCL standard solution because the measurement dates are close (within 7 days)

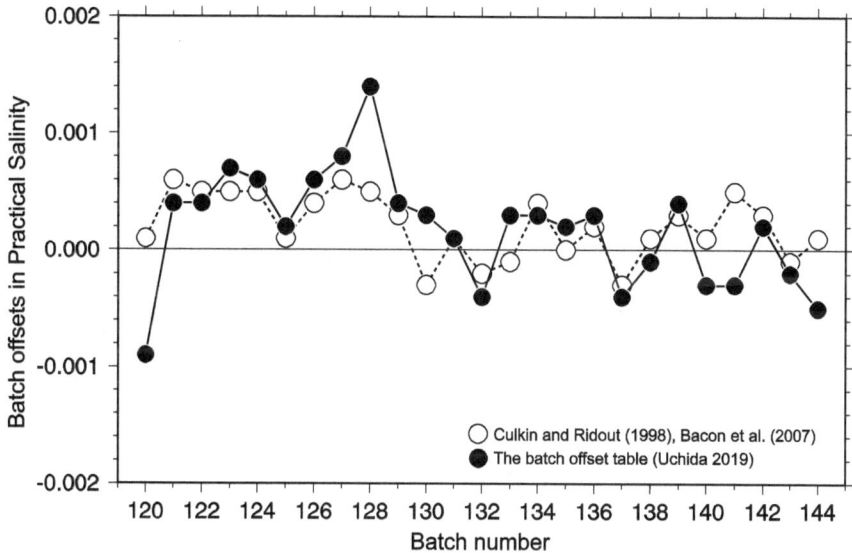

Fig. 9.6 Comparison of the batch offsets estimated from the results of Culkin and Ridout (1998) and Bacon et al. (2007) with the batch offset table (Uchida, 2019)

allowed more "sloshing" inside the ampoules with any motion (Bacon et al., 2007). In fact, the S_P values of SSWs stored in ampoules increased greatly during long-term storage (Fig. 9.2a). A temporal drift and shift of S_P sometimes occurred after the initial calibration, even for SSWs stored in bottles, although the S_P values were relatively stable over time compared to the S_P values of SSWs stored in ampoules.

The S_P values for batches P146–P150 gradually decreased with time, as shown in Sect. 9.3. Batch P145, for example, was calibrated in July 2004 and was measured at JAMSTEC, Japan in March 2005 and at the Woods Hole Oceanographic Institution (WHOI), U.S.A. in April 2005 (see Kawano et al., 2006). The S_P offsets from the label value for P145 were estimated to be 0.0002 and 0.0001 based on the JAMSTEC and WHOI results, respectively. However, the S_P offset of about −0.001 that occurred within the 3 year shelf life after these measurements persisted for 15 year: −0.0010 in 2006, −0.0011 in 2007, −0.0009 in 2009, −0.0015 in 2010, and −0.0011 in 2020 for the measurements at JAMSTEC and −0.0011 in 2011 for the measurement at JMA (see Uchida, 2019; Uchida et al., 2020). In the case of batch P160, only the S_P values of bottles used on the KS1704 cruise in 2017 may have decreased by about −0.001, as described in Sect. 9.5. These results suggest that the S_P values of SSWs are usually stable within their 3 year shelf life but may tend to increase if they are stored in ampoules (batches before P139) and decrease with time for at least some SSWs stored in bottles (batches after P140).

A lack of traceability of the results of S_P measurement in the International System of Units (SI) is a fundamental problem with the current definition and technology of measuring S_P in seawater (Pawlowicz et al., 2016; Seitz et al., 2011). In contrast to fundamental physical phenomena, measurement standards such as SSW that are prepared by humans are inevitably subject to variations over time (Seitz et al., 2011). At present, it is possible to certify the electrical conductivity of SSW with traceability to the SI within an uncertainty of about 0.02% at the highest level (Seitz et al., 2010, 2019) and the corresponding S_P uncertainty of 0.008. Unfortunately, this level of uncertainty is too large to use in climate studies in the deep ocean (Uchida et al., 2020). Although the batch corrections for SSW are efficient in practice for establishing comparability for S_P measurements, errors in the offset estimation by chaining the batch-to-batch comparison experiments can accumulate. Pawlowicz et al. (2016) have therefore recommended the development of a parallel system of measuring the density of SSW to establish traceability of Absolute Salinity because density measurement results are traceable to the SI at nearly the precision required, and the density uncertainty will not increase with time. An expanded uncertainty with a coverage factor of 2 in seawater density measurement is 1.4 ppm at the highest level (Kayukawa & Uchida, 2021), which is equivalent to 0.002 g kg^{-1} in Absolute Salinity.

A practical way to establish a high level of international comparability of S_P measurements (e.g., <0.001) would be to have a large amount of secondary reference seawater that was more robust and stable than SSW to estimate batch-to-batch differences of the SSWs due to initial offsets and changes in SSW conductivity ratios over time. Such a secondary reference seawater would require long-term stability in terms of not only S_P (based on conductivity) but also Absolute Salinity (based on density). Howell et al. (2010) have detected inconsistency (0.0013) of the S_P values of SSWs based on their in-house standard seawater when they began using SSW P151 instead of using P150 for the calibration of their salinometer during a period of a few months. Such in-house standard seawater is inappropriate for long-term use, although these inconsistencies could be completely explained by batch offsets (Uchida et al., 2020). The Chinese Primary Standard Seawater for S_P measurements in China is distributed in borosilicate glass bottles (Li et al., 2016) similar to the way IAPSO SSW is distributed and therefore will likely have the same stability problems as SSW.

Multiparametric Standard Seawater (MSSW) (KANSO TECHNOS Co., Ltd., Japan) might be an alternative candidate for such reference seawater, because MSSW is expected to be more stable than SSW with respect to both S_P and density (or Absolute Salinity). MSSW is created from North Pacific Intermediate Water collected from a depth of 400 m in Suruga Bay, south of Shizuoka, Japan, and stored in 500 mL aluminum beverage bottles with plastic inner caps with high gas and water vapor impermeability and is produced by a method similar to that used for the Reference Material for Nutrients in Seawater (RMNS) (KANSO TECHNOS Co., Ltd.), with no air space in the bottle. The use of aluminum bottles and plastic inner caps is meant to avoid the long-term increase in silicate seen in borosilicate glass bottles (Budéus, 2018; Uchida et al. 2011; Chap. 10) as well as other chemical changes (for more details, see Chap. 12).

Uchida et al. (2020) have reported the stability of the S_P of MSSW (lot Pre16) over seven years. They used 8 batches (SSW P153–P162 except for P158 and P160) of IAPSO SSW for the calibration of their salinometers. The standard deviation of the repeated S_P measurements for MSSW was reduced from 0.00036 to 0.00026 by applying the batch offset correction for IAPSO SSW. An ultra-high-resolution (~0.00013 g kg^{-1} in Absolute Salinity) density sensor based on measurements of refractive indexes by the interference method (Uchida et al., 2019) will contribute substantially toward establishing the traceability of Absolute Salinity measurements to the SI, and it will be suitably calibrated with MSSW. We have started the evaluation of Practical and Absolute Salinities for IAPSO SSWs using MSSW (330 bottles of lot Pre20 produced in January 2022), hopefully for use over the next two decades in laboratories and on research vessels beginning in 2022.

Acknowledgements H. Uchida would like to thank the late Michio Aoyama for encouraging him to carry out this research. We thank the marine technicians of Marine Works Japan, Ltd., who analyzed the composition of the SSWs. Comments by Rich Pawlowicz of the University of British Columbia helped to improve the paper. We also thank Richard Williams of Ocean Scientific International, Ltd., for providing information for the SSW batches P113 and P117. This paper is a contribution to the tasks of the Joint SCOR/IAPWS/IAPSO Committee on the Properties of Seawater (JCS).

Competing Interests Part of this work was supported by the Japan Society for the Promotion of Science KAKENHI Grants 18K03752, 21H01165, and 20H04349.

References

Aoyama, M., Joyce, T. M., Kawano, T., & Takatsuki, Y. (2002). Standard seawater comparison up to P129. *Deep-Sea Research I, 49*, 1103–1114. https://doi.org/10.1016/S0967-0637(02)00018-3

Bacon, S., Culkin, F., Higgs, N., & Ridout, P. (2007). IAPSO standard seawater: Definition of the uncertainty in the calibration procedure, and stability of recent batches. *Journal of Atmospheric and Oceanic Technology, 24*, 1785–1799. https://doi.org/10.1175/JTECH2081.1

Bacon, S., Snaith, H. M., & Yelland, M. J. (2000). An evaluation of some recent batches of IAPSO standard seawater. *Journal of Atmospheric and Oceanic Technology, 17*, 854–861. https://doi.org/10.1175/1520-0426(2000)017%3c0854:AEOSRB%3e2.0.CO;2

Budéus, G. T. (2018). Potential bias in TEOS10 density of sea water samples. *Deep-Sea Research, I, 134*, 41–47. https://doi.org/10.1016/j.dsr.2018.02.005

Culkin, F., & Ridout, P. S. (1998). Stability of IAPSO standard seawater. *Journal of Atmospheric and Oceanic Technology, 15*, 1072–1075. https://doi.org/10.1175/1520-0426(1998)015%3c1072:SOISS%3e2.0.CO;2

Culkin, F., & Smed, J. (1979). The history of standard seawater. *Oceanologica Acta, 2*, 355–364.

Gouretski, V. V., & Jancke, K. (2001). Systematic errors as the cause for an apparent deep water property variability: Global analysis of the WOCE and historical hydrographic data. *Progress in Oceanography, 48*, 337–402. https://doi.org/10.1016/S0079-6611(00)00049-5

Howell, G. H., Guenther, R, Janzen, C. D., & Larson, N. G. (2010). On the use of a secondary standard to improve Autosal calibration. In *Ocean sciences meeting* (Abstract 751348). American Geophysical Union, Portland, Oregon. https://www.seabird.com/cms-portals/seabird_com/cms/documents/SecStandardToImproveAutosalCalHandout-3Pages.pdf

Joyce, T., Bacon, S., Kalashnikov, P., Romanov, A., Stalcup, M., & Zaburdaev, V. (1992). Results of an oxygen/salinity comparison cruise on the R/V Vernadsky. In *WHP office report WHPO 92-3* (WOCE Report 93/92). https://doi.org/10.1575/1912/244

Kawano, T., Aoyama, M., & Takatsuki, Y. (2005). Inconsistency in the conductivity of standard potassium chloride solutions made from different high-quality reagents. *Deep-Sea Research I, 52*, 389–396. https://doi.org/10.1016/j.dsr.2004.11.002

Kawano, T., Aoyama, M., Joyce, T., Uchida, H., Takatsuki, Y., & Fukasawa, M. (2006). The latest batch-to-batch difference table of standard seawater and its application to the WOCE onetime sections. *Journal of Oceanography, 62*, 777–792. https://doi.org/10.1007/s10872-006-0097-8

Kawano, T., Takatsuki, Y., & Aoyama, M. (2000). Comparison of some recent batches of IAPSO Standard Seawater. *Journal of Japan Society for Marine Surveys and Technology, 12*, 49–55 (in Japanese with English abstract). https://doi.org/10.11306/jsmst.12.2_49

Kawano, T., Takatsuki, Y., Imai, J., & Aoyama, M. (2001). A comparison of recent standard seawater and quality evaluation of the standard seawater supplied in a bottle. *Journal of Japan Society for Marine Surveys and Technology, 13*, 11–18 (in Japanese with English abstract). https://doi.org/10.11306/jsmst.13.2_11

Kayukawa, Y., & Uchida, H. (2021). Absolute density measurements for standard sea-water by hydrostatic weighing of silicon sinker. *Measurement: Sensors, 18*. https://doi.org/10.1016/j.measen.2021.100200

Li, Y., Luo, Y., Kang, Y., Yu, T., Wang, A., & Zhang, C. (2016). Chinese primary standard seawater: Stability checks and comparisons with IAPSO standard seawater. *Deep-Sea Research I, 113*, 101–106. https://doi.org/10.1016/j.dsr.2016.04.005

Mantyla, A. W. (1980). Electrical conductivity comparisons of standard seawater batches P29 to P84. *Deep-Sea Research, 27A*, 837–846. https://doi.org/10.1016/0198-0149(80)90047-3

Mantyla, A. W. (1987). Standard seawater comparisons updated. *Journal of Physical Oceanography, 17*, 543–548. https://doi.org/10.1175/1520-0485(1987)017%3c0543:SSCU%3e2.0.CO;2

Mantyla, A. W. (1994). The treatment of inconsistencies in Atlantic deep water salinity data. *Deep-Sea Research I, 41*, 1387–1405. https://doi.org/10.1016/0967-0637(94)90104-X

Millero, F. J. (2010). History of the equation of state of seawater. *Oceanography, 23*(3), 18–33. https://doi.org/10.5670/oceanog.2010.21

Millero, F. J., Chetirkin, P., & Culkin, F. (1977). The relative conductivity and density of standard seawaters. *Deep-Sea Research, 24*, 315–321. https://doi.org/10.1016/0146-6291(77)91000-1

Millero, F. J., Feistel, R., Wright, D. G., & McDougall, T. J. (2008). The composition of standard seawater and the definition of the reference-composition salinity scale. *Deep-Sea Research I, 55*, 50–72. https://doi.org/10.1016/j.dsr.2007.10.001

Millero, F. J., & Huang, F. (2009). The density of seawater as a function of salinity (5 to 70 g kg^{-1}) and temperature (273.15 to 363.15 K). *Ocean Science, 5*, 91–100. https://doi.org/10.5194/os-5-91-2009

Park, K. (1964). Reliability of standard sea water as a conductivity standard. *Deep-Sea Research, 11*, 85–87. https://doi.org/10.1016/0011-7471(64)91084-8

Park, K., & Burt, W. V. (1965a). Electrolytic conductance of sea water and the salinometer, part 1. *Journal of Oceanographic Society of Japan, 21*, 25–36. https://doi.org/10.5928/kaiyou1942.21.69

Park, K., & Burt, W. V. (1965b). Electrolytic conductance of sea water and the salinometer, part 2. *Journal of Oceanographic Society of Japan, 21*, 124–132. https://doi.org/10.5928/kaiyou1942.21.124

Parrilla, G. (2007). Documentation from Cruise 29HE06_1, PDF version. CCHDO. Retrieved July 27, 2022 from https://cchdo.ucsd.edu/cruise/29HE06_1

Pawlowicz, R. (2010). A model for predicting changes in the electrical conductivity, practical salinity, and absolute salinity of seawater due to variations in relative chemical composition. *Ocean Science, 6*, 361–378. https://doi.org/10.5194/os-6-361-2010

Pawlowicz, R., Feistel, R., McDougall, T. J., Ridout, P., Seitz, S., & Wolf, H. (2016). Metrological challenges for measurements of key climatological observables part 2: Oceanic salinity. *Metrologia, 53*, R12–R25. https://doi.org/10.1088/0026-1394/53/1/R12

Pawlowicz, R., Wright, D. G., & Millero, F. J. (2011). The effects of biogeochemical processes on oceanic conductivity/salinity/density relationships and the characterization of real seawater. *Ocean Science, 7*, 363–387. https://doi.org/10.5194/os-7-363-2011

Poisson, A. (1975). Measurement of absolute electrical conductivity of standard seawater on the basis of KCl as standard. In *UNESCO technical papers in marine science* (Vol. 24, p. 61).

Poisson, A., Dauphinee, T., Ross, C. K., & Culkin, F. (1978). The reliability of standard seawater as an electrical conductivity standard. *Oceanologica Acta, 1*, 425–433.

Purkey, S., & Johnson, G. C. (2013). Antarctic Bottom Water warming and freshening: Contributions to sea level rise, ocean freshwater budgets, and global heat gain. *Journal of Climate, 26*, 6105–6122. https://doi.org/10.1175/JCLI-D-12-00834.1

Purkey, S., Johnson, G. C., Talley, L. D., Sloyan, B. M., Wijffels, S. E., Smethie, W., Mecking, S., & Katsumata, K. (2019). Unabated bottom water warming and freshening in the South Pacific Ocean. *Journal Geophysical Research: Oceans, 124*, 1778–1794. https://doi.org/10.1029/201 8JC014775

Saunders, P. M. (1986). The accuracy of measurement of salinity, oxygen and temperature in the deep ocean. *Journal of Physical Oceanography, 16*, 189–195. https://doi.org/10.1175/1520-048 5(1986)016%3c3C0189:TAOMOS%3e3E2.0.CO;2

Seitz, S., Feistel, R., Wright, D. G., Weinreben, S., Spitzer, P., & De Bievre, P. (2011). Metrological traceability of oceanographic salinity measurement results. *Ocean Science, 7*, 45–62. https:// doi.org/10.5194/os-7-45-2011

Seitz, S., Jakobsen, P. T., & Mariassy, M. (2019). Metrological advances in reference measurement procedures of electrolytic conductivity. *Metrologia, 56*, 034003. https://doi.org/10.1088/1681-7575/ab1527

Seitz, S., Spizer, P., & Brown, R. J. C. (2010). CCQM-P111 study on traceable determination of practical salinity and mass fraction of major seawater components. *Accreditation and Quality Assurance, 15*, 9–17. https://doi.org/10.1007/s00769-009-0578-8

Takatsuki, Y., Aoyama, M., Nakano, T., Miyagi, H., Ishihara, T., & Tsutsumida, T. (1991). Standard seawater comparison of some recent batches. *Journal of Atmospheric and Oceanic Technology, 8*, 895–897. https://doi.org/10.1175/1520-0426(1991)008%3c0895:SSCOSR%3e2.0.CO;2

Uchida, H., Kawano, T., Aoyama, M., & Murata, A. (2011). Absolute salinity measurements of standard seawaters for conductivity and nutrients. *La Mer, 49*, 119–126. http://www.sfjo-lamer.org/la_mer/49-3_4/49-3-4-5.pdf

Uchida, H. (2019). The latest batch-to-batch correction table for IAPSO standard seawater. https:// doi.org/10.17596/0001983. Accessed 2022-07-27.

Uchida, H., Kawano, T., Nakano, T., Wakita, M., Tanaka, T., & Tanihara, S. (2020). An expanded batch-to-batch correction for IAPSO standard seawater. *Journal of Atmospheric and Oceanic Technology, 37*, 1507–1520. https://doi.org/10.1175/JTECH-D-19-0184.1

Uchida, H., Kayukawa, Y., & Maeda, Y. (2019). Ultra high-resolution seawater density sensor based on a refractive index measurement using the spectroscopic interference method. *Scientific Report, 9*, 15482. https://doi.org/10.1038/s41598-019-52020-z

UNESCO. (1981). The practical salinity scale 1978 and the international equation of state of seawater 1980. In *UNESCO technical papers in marine science* (Vol. 36, p. 25).

Chapter 10
Changes in the Composition of International Association for the Physical Sciences of the Oceans Standard Seawater

Hiroshi Uchida, Masahide Wakita, Akiko Makabe, Akihiko Murata, and Alexander Petrovic

Abstract Changes in the composition of International Association for the Physical Sciences of the Oceans Standard Seawater (SSW) for salinity measurements were examined for borosilicate glass bottle-type SSW batches P138 (2000) to P165 (2021) to understand their impact on salinity and density stability of SSW. Results showed that (i) silicate increased with time (5.47 µmol kg^{-1} year^{-1}); (ii) dissolved inorganic carbon (DIC) decreased with time (-13.5 µmol kg^{-1} year^{-1}); (iii) total alkalinity, nitrate, nitrite, and phosphate were constant; and (iv) ammonium, dissolved organic carbon (DOC), and total dissolved nitrogen were constant for batches after P148, although they were highly variable for batches before P147. The rate of increase of silicate was similar to that for low-salinity standard seawater (10L-Series) and for the old batches sealed in soda-glass ampoules. A glucose addition experiment was used to examine the effect of DOC on density. The results enable estimation of the Absolute Salinity anomaly (δS_A) of SSW as a function of time (t in year) from calibration of SSW with changes of silicate and DIC: δS_A (g kg^{-1}) $= 0.0024 + 0.00021$ t (for t > 0.3 year). For batch P149 (2007), mass fractions of major constituents (Na$^+$, Mg^{2+}, Ca^{2+}, K$^+$, Sr^{2+}, Cl$^-$, SO$_4{}^{2-}$, and Br$^-$) were measured in 2020, and the mass fractions were generally consistent with results from previous studies in 2010 and the mass fractions of the Reference Composition, which is currently the best estimate of the

H. Uchida (✉) · A. Murata
Research Institute for Global Change, Japan Agency for Marine-Earth Science and Technology, Yokosuka, Japan
e-mail: huchida@jamstec.go.jp

M. Wakita
Mutsu Institute for Oceanography, Japan Agency for Marine-Earth Science and Technology, Mutsu, Japan

A. Makabe
Institute for Extra-cutting-edge Science and Technology Avant-garde Research, Japan Agency for Marine-Earth Science and Technology, Yokosuka, Japan

A. Petrovic
King Abdullah University of Science and Technology, Thuwal, Saudi Arabia

© The Author(s) 2025
M. Aoyama et al. (eds.), *Chemical Reference Materials for Oceanography*, Springer Oceanography, https://doi.org/10.1007/978-981-96-2520-8_10

chemical composition of SSW, although it is necessary to improve the accuracy of the measurements to detect temporal variations of mass fractions of major constituents of SSW.

Keywords Compositional changes of standard seawater · Practical salinity · Absolute salinity · Leaching of silicate

10.1 Introduction

An internationally recognized standard seawater for salinity measurements have been used worldwide for more than 120 year and has played an important role in ensuring the intercomparability of observed salinity data since 1902 (e.g., Culkin & Smed, 1979; Pawlowicz et al., 2016; Ridout, 2011). This standard seawater is produced in batches and is endorsed by the International Association for the Physical Sciences of the Oceans (IAPSO). Since batch P91, each batch has been labeled with a conductance measurement meant for standardization of salinity determinations according to the procedures outlined in the Practical Salinity Scale 1978 (UNESCO, 1981) (Bacon et al., 2007; Chap. 8). Earlier batches were labeled using a titration measurement that was meant for standardization of titration-based salinity determinations (Culkin & Smed, 1979).

Batch-to-batch comparative studies of the Practical Salinity (S_P) of IAPSO standard seawater (SSW) have been reviewed by Uchida et al. (Chap. 9). Although the batch offset corrections for certified values of SSW are efficient in practice for establishing intercomparability of S_P measurements (Uchida et al., 2020), errors in the offset estimation can accumulate because of the lack of traceability of the results of S_P measurements in the International System of Units (SI) at the required level of precision. Pawlowicz et al. (2016) have therefore recommended development of a parallel system that involves measuring the density of SSW to establish traceability of the salinity defined by the Thermodynamic Equation of Seawater 2010 (TEOS-10) (IOC et al., 2010), the Absolute Salinity (S_A). This recommendation follows from the fact that the S_A can be determined based on density measurements, and density measurement results, unlike conductivity measurements, are traceable to the SI at the necessary precision.

A hydrostatic weighing apparatus can be used to measure the density of seawater, and such measurements are among the primary methods used in metrology (e.g., Kayukawa & Uchida, 2021). However, hydrostatic weighing is time-consuming and difficult. Therefore, an oscillation-type density meter (Picker et al., 1974) is normally used to measure seawater density in routine analyses.

Oscillation-type density meters are usually calibrated with air and pure water (e.g., Wolf, 2008) because the density of pure water can be accurately calculated at any temperature from the equation of state of water (International Association for the Properties of Water and Steam [IAPWS] R6-95, IAPWS, 2018), which is itself based on hydrostatic weighing measurements. However, measurement of the density of SSW using an oscillation-type density meter requires an extrapolation of the standard curve because seawater densities are typically about 2–4% higher than the densities of pure water. Furthermore, the measured density tends to be lower than the density calculated using the TEOS-10, and the difference varies widely (-0.016 to -0.001 kg m^{-3}) from study to study (Feistel et al., 2010; Romeo et al., 2019; Schmidt et al., 2018; Uchida et al., 2011a; Woosley et al., 2014). Most of these inconsistent discrepancies may be due to individual differences in the nonlinearity of the oscillation-type density meters, because the density of SSW recently measured with a hydrostatic weighing apparatus agrees well with the calculated density (Kayukawa & Uchida, 2021). It may therefore be that oscillation-type density meters should be calibrated with pure water and an appropriate standard seawater, such as IAPSO SSW or Multiparametric Standard Seawater (Uchida et al., 2020), to establish comparability of seawater density measurements.

The salinity variable in the TEOS-10 standard is called the Absolute Salinity (S_A). The S_A (also called the "density salinity") of seawater can be directly estimated from the measured density of the seawater by back-calculating from the TEOS-10 equations using the temperature and pressure at which the density was measured. However, the S_A introduced in TEOS-10 can also be linked to a conductance determination using $S_A = S_R + \delta S_A$, where S_R is the Reference-Composition Salinity and is calculated from the Practical Salinity ($S_R = 35.16504$ g kg^{-1}/35 $\times$ S_P), and δS_A is the Absolute Salinity anomaly, which accounts for the way that changes in the relative composition of dissolved sea salts affect the conductivity and density of seawater.

In the open ocean, δS_A can be estimated using additional measurements of silicate, nitrate, total alkalinity (TA), and dissolved inorganic carbon (DIC) (Pawlowicz et al., 2011), because the changes of the concentrations of these chemical constituents of seawater from place to place in the ocean are thought to most affect the conductivity and density. If the changes of the concentrations of these constituents are known for SSW (and are sufficient to adequately describe the complete composition of stored SSW as it ages), evaluation of the S_R (or S_P) of SSW can be made using the S_A estimated from density measurements and can provide an alternative path for traceability to the SI.

SSW is derived from water obtained near the surface of the northeast Atlantic Ocean (Bacon et al., 2007), which after filtration and UV treatment to remove organic matter, is placed into ampoules (until batch P139) or sealed bottles (batches P140 onwards). An increase of silicate in stored samples for both ampoule-type and bottle-type SSWs has been reported in previous studies (Poisson et al., 1978; Higgs & Ridout, 2009 [unpublished manuscript, see Fig. 3 in Budéus, 2018]; Uchida et al., 2011a). Because silicate is predominantly a nonionic constituent of SSW, conductivity (and therefore S_P) does not change if the silicate concentration changes. There might hence be no problem in using SSW as a standard for S_P. However, the density (or S_A) of SSW would change because of an increase of silicate over time. Moreover, the observation of decreases of S_P over time for some batches of bottle-type SSW (Chap. 9) suggests that changes of composition may arise from changes of chemical constituents other than silicate.

Although previous studies have measured the composition of a small number of batches of the currently used bottle-type SSW (e.g., Woosley et al., 2014) or the mass fractions of major constituents of a single batch (Feistel et al., 2010; Seitz et al., 2010), no previous studies have examined the temporal variation of any constituent of SSW other than silicate.

In this study, we examined changes of the composition (nutrient concentrations and carbonate system parameters) of bottle-type SSW for batches P138 (2000) to P165 (2021) by expanding the results of Uchida et al. (2011a) to complement the results for the S_P of SSW (Chap. 9). We also measured mass fractions of major constituents for batch P149 (2007) and compared our measurements with results from the inter-laboratory comparisons published in 2010 and the TEOS-10 Reference Composition (Millero et al., 2008), which is currently the best estimate of the chemical composition of SSW. We then evaluated the equation that uses density measurements to estimate the δS_A of SSW.

10.2   Materials and Methods

We used the P-series (S_P of close to 35) of borosilicate glass bottle-type SSW from batch P138 to P165 (Ocean Scientific International, Ltd. [OSIL]), except for P139, P140, and P142. The mean ± SD of the S_P for these batches was 34.994 ± 0.002. We also used the low-salinity 10L series of SSW with an S_P of about 10 (batch 10L15 calibrated on 17 June 2015) and Atlantic Seawater, which is the source of filtered seawater for SSW (S_P of 35) and was calibrated on 15 May 2019 (OSIL) and distributed in a 5 L, high-density polyethylene (HDPE) bottle.

The S_P was measured with a salinometer (Autosal model 8400B; Guildline Instruments, Ltd.). We used P-series SSW to calibrate the salinometer. The S_P was measured at 24 °C following the method of Kawano (2010) (for more details, see Uchida et al., 2020 and Chap. 9).

The density was measured with an oscillation-type density meter (model DMA 5000 M with an Xsample 122 sample changer; Anton-Paar GmbH). The density meter was calibrated by using ultrapure water (Milli-Q water from Millipore Sigma) and P-series SSW, and the density was measured at 20 °C following the method of Uchida et al. (2011a).

Most of the nutrients (nitrate, nitrite, silicate, phosphate, and ammonium), TA, and DIC were measured on the Global Ocean Ship-based Hydrographic Investigations Program (GO-SHIP) cruises (e.g., Uchida et al., 2011b, 2021) or cruises equivalent to GO-SHIP by the R/V *Mirai*. Some of the data were measured in a laboratory of the Japan Agency for Marine-Earth Science and Technology (JAMSTEC) by using systems identical to or similar to those on the R/V *Mirai*.

The nutrients were measured with a continuous flow analyzer (model TRAACS 800 system [in 2010]; BRAN + LUEBBE, or model QuAAtro 2-HR system [since 2011]; BL TEC K.K.). These values were calibrated by using reference material for nutrients in seawater (RMNS) and certified reference material (CRM) for nutrients (KANSO TECHNOS Co., Ltd.).

The TA was measured based on spectrophotometry with a custom-made system (Nippon ANS, Inc.), and the DIC was measured with a coulometric system that measured total CO_2 (model 3000; Nippon ANS, Inc.). The TA and DIC values were calibrated by using certified reference material (Dickson-RM) provided by Prof. A. G. Dickson (Scripps Institution of Oceanography, University of California San Diego) or CO_2 in seawater reference material (KANSO-RM), which was prepared by KANSO TECHNOS using a method similar to the method used to produce the Dickson-RM, and the values of which were determined by measuring TA and DIC relative to the Dickson-RM.

The dissolved organic carbon (DOC) and total dissolved nitrogen (TDN) concentrations were measured with a total organic carbon analyzer (model TOC-V or TOC-L; Shimadzu Co.). The DOC and TDN values were calibrated following the method of Ogawa et al. (2003) (see also Wakita et al., 2016) using consensus reference material provided by Prof. D. A. Hansell (Miami University).

The mass fractions of major constituents (Na^+, Mg^{2+}, Ca^{2+}, K^+, Sr^{2+}, Cl^-, SO_4^{2-}, and Br^-) were measured with an inductively coupled plasma optical emission spectrometer (ICP-OES) (model SPS5510; Hitachi High-Tech Science Co.) or an ion chromatograph (IC) (model Dionex ICS-2100; Thermo Fisher Scientific) after appropriate sample dilution using ultrapure water (Milli-Q water) following the method of Miyazaki et al. (2017). The mass fractions values were calibrated using a standard solution for ICP-OES (Certipur Multi-Element Standards IV for Inductively Coupled Plasma Spectroscopy [product code 1.11355.0100]; Merck KGaA) and standard solutions for IC (Multication Standard Solution III [product code 137-14611]; Anion Mixture Standard Solution I [product code 019-24011]; Sodium Standard Solution [Na 1000] [product code 199-10831]; Chloride Ion Standard Solution [Cl-1000] [product code 038-16153]; FUJIFILM Wako Pure Chemical Co.).

10.3   Measured Changes of Nutrient Concentrations and Carbonate System Parameters

Since 2010, concentrations of nutrients (nitrate, nitrite, silicate phosphate, and ammonium) (Table 10.1), DIC, TA (Table 10.2), DOC, and TDN (Table 10.3) in bottle-type SSWs have been measured. Those measurements include the results reported by Uchida et al. (2011a). Nitrate, nitrite, phosphate, and total alkalinity have been constant. Concentrations of silicate and DIC have changed linearly with time. Concentrations of ammonium, DOC, and TDN have been constant in batches after P148, although they were highly variable for batches before P147. Note that we do not normalize these results by salinity even though the salinity in different P-numbered batches is not identical because the effects of such normalization would be much smaller than the uncertainties in the nutrient and carbonate system parameters measurements themselves. These patterns are described in detail below.

Silicate concentrations increased linearly with time (Fig. 10.1). The characteristics of the increase in silicate concentrations were similar for ampoule-type and bottle-type SSWs: silicate increased at a constant rate ($5.47 \, \mu mol \, kg^{-1} \, year^{-1}$) until approximately 5 year after calibration of the SSW. After that time, the variability increased, especially for the ampoule-type SSW. These characteristics were consistent with the results reported by Higgs and Ridout (2009) (unpublished manuscript, see Fig. 3 in Budéus (2018)). This rate of increase of silicate was also similar to that of the low-salinity 10L-series SSW, which is distributed in the same borosilicate glass bottles used for the P-series SSW (see Table 10.1). The regression line (Fig. 10.1) indicates that the silicate concentration was $25.16 \, \mu mol \, kg^{-1}$ at the time the SSW was calibrated, but this value is surprisingly large and is atypical of the source seawater (northeast Atlantic surface water). In fact, the silicate concentration measured in the Atlantic Seawater distributed by the OSIL in HDPE bottles from the same location was only $0.84 \, \mu mol \, kg^{-1}$ (Table 10.1). The implication is that the silicate concentration in the SSW probably increased rapidly immediately after bottling and subsequently increased at a slower rate. This possibility is further discussed in Sect. 10.5.

In contrast, the DIC concentration decreased linearly with time (Fig. 10.2). The DIC concentration changed at a constant rate ($-13.5 \, \mu mol \, kg^{-1} \, year^{-1}$) until approximately 5 year after calibration. After that time, the variability increased slightly. The DIC of a 16.2 year-old sample of P145 was much lower than usual, and the TA of the same bottle of P145 was also much lower than usual (see caption of Fig. 10.3).

The TA otherwise remained constant with time, although the variability increased slightly 5 year after calibration (Fig. 10.3). The exception was the one datum for P145 mentioned above. Millero et al. (2011) measured the TA of P146 within 5 year of calibration, and our value ($2302.6 \, \mu mol \, kg^{-1}$) for P146 measured 5.3 year after calibration agreed with theirs ($2306 \pm 3 \, \mu mol \, kg^{-1}$). Woosley et al. (2014) have measured the TA for six batches from P149 to P155. Their TA was also constant with time (mean $\pm$ SD of $2293 \pm 6.5 \, \mu mol \, kg^{-1}$) and consistent with our data. However, our data differed systematically from the TA of nine batches of ampoule-type SSW (mean $\pm$ SD of $2351 \pm 20 \, \mu mol \, kg^{-1}$) reported by Goyet et al. (1985).

Like TA, concentrations of DOC (Fig. 10.4), ammonium (Fig. 10.5), and TDN (not shown in a figure) were constant with time for batches after P148. However, they were highly variable for batches before P147. For P146, the DOC concentration measured within 5 year of calibration was $57 \pm 2 \, \mu mol \, kg^{-1}$ (Millero et al., 2011), and it increased to $70.8 \, \mu mol \, kg^{-1}$ after 15.7 year of calibration (Table 10.3). The mean $\pm$ SD of the DOC concentration for batches before P147 was $72.0 \pm 14.2 \, \mu mol \, kg^{-1}$ and was consistent with DOC values reported by Poisson et al. (1978) for 13 batches of ampoule-type SSW (mean $\pm$ SD of $72.7 \pm 19.1 \, \mu mol \, kg^{-1}$).

If particulate organic matter (POM) is present in SSW, microbial activity could degrade POM and increase the concentration of dissolved organic matter (DOC and TDN). Ammonification of the TDN could increase the concentration of ammonium. However, because the method of SSW production, including the filtering process,

Table 10.1 Concentrations of nitrate, nitrite, silicate, phosphate, and ammonium (in μmol kg^{-1}) in SSW. Date of measurement and age of batch at the time of measurement (in year) are also listed. Batch no. 10L15 is the 10L-series (Sp of 9.9958), and AS stands for Atlantic Seawater (Sp of about 35). "n/a" indicates that the datum is not available

Batch no.	Date of measurement	Age of batch	Nitrate	Nitrite	Silicate	Phosphate	Ammonium
P144	14 July 2010	6.8	0.08	0.01	90.52	0.071	n/a
P145	14 July 2010	6.0	0.06	0.01	69.77	0.079	n/a
P146	14 July 2010	5.2	0.47	0.07	78.48	0.087	n/a
P147	14 July 2010	4.1	1.36	0.05	69.40	0.123	n/a
P148	14 July 2010	3.8	0.96	0.30	45.16	0.085	n/a
P149	14 July 2010	2.8	0.04	0.00	40.95	0.044	n/a
P150	14 July 2010	2.1	0.39	0.09	38.91	0.064	n/a
P151	14 July 2010	1.2	0.03	0.00	32.39	0.040	n/a
P152	14 July 2010	0.2	0.04	0.00	22.16	0.040	n/a
P153	17 Apr 2011	0.1	0.46	0.04	15.67	0.064	n/a
P152	28 Apr 2011	1.0	0.04	0.01	35.15	0.045	n/a
P149	1 May 2011	3.6	0.11	0.00	49.25	0.072	n/a
P150	1 May 2011	2.9	0.73	0.08	45.84	0.123	n/a
P151	1 May 2011	1.9	0.09	0.00	41.61	0.059	n/a
P152	1 May 2011	1.0	0.08	0.00	34.85	0.055	n/a
P151	24 Apr 2012	2.9	0.01	0.00	45.34	0.039	n/a
P152	24 Apr 2012	2.0	0.01	0.00	41.07	0.042	n/a
P153	24 Apr 2012	1.1	0.38	0.03	30.33	0.050	n/a
P154	24 Apr 2012	0.5	1.06	0.03	24.72	0.087	n/a
P144	2 Feb 2013	9.4	0.04	0.03	102.81	0.086	n/a
P152	8 Jun 2014	4.1	0.01	0.01	46.53	0.029	−0.01
P153	8 Jun 2014	3.3	0.32	0.03	39.06	0.049	0.18
P154	8 Jun 2014	2.6	1.09	0.03	38.37	0.093	0.23
P155	8 Jun 2014	1.7	0.60	0.21	33.97	0.052	−0.02
P156	8 Jun 2014	0.9	0.51	0.09	29.60	0.028	0.05
P157	5 Dec 2014	0.6	0.52	0.08	31.11	0.268	−0.07
P153	27 Dec 2015	4.8	0.40	0.03	46.63	0.044	0.02
P154	27 Dec 2015	4.2	1.10	0.11	45.47	0.087	−0.01
P155	27 Dec 2015	3.3	0.56	0.26	44.46	0.051	0.00
P156	27 Dec 2015	2.4	0.49	0.09	39.52	0.030	0.05
P157	27 Dec 2015	1.6	0.57	0.12	36.85	0.047	0.01
P158	27 Dec 2015	0.8	0.45	0.10	30.96	0.034	−0.01
10L15	27 Dec 2015	0.5	0.42	−0.01	23.40	0.064	0.11
P141	17 Feb 2017	15.3	0.26	0.10	97.34	0.054	1.02

(continued)

Table 10.1 (continued)

Batch no.	Date of measurement	Age of batch	Nitrate	Nitrite	Silicate	Phosphate	Ammonium
P144	17 Feb 2017	13.4	0.02	0.02	126.87	0.065	1.73
P148	17 Feb 2017	10.4	0.94	0.27	85.22	0.071	0.03
P159	17 Feb 2017	1.2	0.32	0.00	30.30	0.032	0.01
10L15	17 Feb 2017	1.7	0.36	−0.01	32.76	0.080	0.12
P157	27 Jul 2017	3.2	0.55	0.07	47.45	0.046	0.09
P160	27 Jul 2017	1.0	0.01	0.06	28.06	0.019	0.03
P157	24 Jul 2018	4.2	0.46	0.26	47.93	0.048	0.02
P158	24 Jul 2018	3.3	0.55	0.01	42.04	0.033	0.02
P159	24 Jul 2018	2.6	0.40	0.01	36.69	0.034	0.00
P160	24 Jul 2018	2.0	−0.01	0.08	34.60	0.019	−0.01
P161	24 Jul 2018	1.2	1.01	0.07	30.80	0.069	0.00
P162	1 Jun 2019	1.1	0.04	0.01	34.69	0.039	0.03
P159	25 Jan 2020	4.1	0.31	0.00	44.19	0.032	0.03
P160	25 Jan 2020	3.5	0.01	0.13	40.94	0.021	0.02
P161	25 Jan 2020	2.7	1.02	0.08	39.37	0.069	0.02
P162	25 Jan 2020	1.8	0.01	0.09	37.58	0.033	0.02
P163	25 Jan 2020	0.8	0.05	0.06	27.17	0.066	0.01
P138	4 Oct 2020	20.7	0.08	0.05	124.31	0.027	0.89
P141	4 Oct 2020	18.3	0.19	0.09	161.70	0.053	1.97
P143	4 Oct 2020	17.6	0.08	0.03	146.52	0.083	2.28
P144	4 Oct 2020	17.0	0.06	0.02	158.41	0.063	1.69
P145	4 Oct 2020	16.2	0.04	0.01	182.02	0.060	1.66
P146	4 Oct 2020	15.4	0.45	0.07	166.66	0.101	1.87
P148	4 Oct 2020	14.0	0.96	0.32	96.63	0.066	0.04
P149	4 Oct 2020	13.0	0.03	0.00	67.47	0.020	0.18
P150	4 Oct 2020	12.4	0.31	0.22	78.50	0.041	0.03
P151	4 Oct 2020	11.4	0.00	0.00	63.18	0.017	0.07
P155	4 Oct 2020	8.0	1.04	0.18	61.68	0.047	0.01
P164	4 Oct 2020	0.5	0.06	0.02	27.02	0.164	0.17
P147	16 Mar 2021	14.8	1.31	1.13	186.53	0.167	1.12
P164	16 Mar 2021	1.0	0.07	0.01	30.82	0.171	0.18
P165	2 May 2022	1.0	0.06	0.00	29.33	0.106	0.15
AS	24 Oct 2022	3.4	0.01	0.00	0.84	0.044	0.02

Table 10.2 Same as Table 10.1, but for dissolved inorganic carbon (DIC) and total alkalinity (TA) (in μmol kg^{-1})

Batch no.	Date of measurement	Age of batch	DIC	TA
P144	12 Sep 2010	7.0	n/a	2318.6
P145	12 Sep 2010	6.2	n/a	2317.3
P146	12 Sep 2010	5.3	n/a	2302.6
P147	12 Sep 2010	4.3	n/a	2309.5
P148	12 Sep 2010	4.0	n/a	2307.6
P149	12 Sep 2010	2.9	n/a	2307.4
P150	12 Sep 2010	2.3	n/a	2306.2
P151	12 Sep 2010	1.3	n/a	2303.7
P152	12 Sep 2010	0.4	n/a	2298.9
P152	7 Nov 2010	0.5	2055.9	n/a
P153	27 Apr 2011	0.1	2058.1	2288.7
P144	4 Dec 2012	9.2	1834.2	2319.1
P144	3 Feb 2013	9.4	1830.0	2314.1
P152	15 Oct 2014	4.4	2019.8	2307.7
P153	15 Oct 2014	3.6	2012.7	2302.3
P154	15 Oct 2014	3.0	2030.4	2307.3
P155	15 Oct 2014	2.1	2030.9	2303.5
P156	15 Oct 2014	1.2	2010.3	2290.3
P157	9 Jan 2015	0.7	2048.7	2305.3
10L15	31 Dec 2015	0.5	643.2	n/a
P153	31 Dec 2015	4.8	2011.1	2305.0
P154	31 Dec 2015	4.2	2029.2	2311.5
P155	31 Dec 2015	3.3	2026.9	2308.0
P156	31 Dec 2015	2.4	2013.6	2293.9
P157	31 Dec 2015	1.6	2039.3	2308.8
P158	31 Dec 2015	0.8	2064.6	2303.8
P159	23 Feb 2017	1.2	2066.0	2304.3
P157	15 Jul 2017	3.2	2021.6	2312.1
P160	15 Jul 2017	1.0	2063.0	2304.1
P161	26 Jul 2018	1.2	2067.3	2301.2
P162	27 May 2019	1.1	2080.9	2306.6
P163	12 Jan 2020	0.8	2091.5	2298.5
P138	4 Oct 2020	20.7	1767.3	2288.6
P141	4 Oct 2020	18.3	1793.1	2292.4
P143	4 Oct 2020	17.6	n/a	2253.9
P144	4 Oct 2020	17.0	1793.7	2319.3

(continued)

Table 10.2 (continued)

Batch no.	Date of measurement	Age of batch	DIC	TA
P145	4 Oct 2020	16.2	1550.5	1883.5
P146	4 Oct 2020	15.4	1794.3	2285.6
P148	4 Oct 2020	14.0	1961.7	2296.8
P149	4 Oct 2020	13.0	1976.1	2308.3
P150	4 Oct 2020	12.4	1973.0	2305.5
P151	4 Oct 2020	11.4	1985.8	2283.8
P155	4 Oct 2020	8.0	1992.6	2308.8
P147	12 Mar 2021	14.8	1835.7	2304.9
P164	12 Mar 2021	1.0	2081.5	2303.1
P165	1 May 2022	1.0	2066.5	2302.0

did not change between batches P147 and P148 (Richard Williams of OSIL, 2022, personal communication), the cause of the large variation before P147 is unclear.

The characteristic values of SSW are summarized in Table 10.4 based on data within 5 year of calibration. The data are summarized as regression lines as a function of time for silicate and DIC and mean values for the other parameters. The mean nitrate, nitrite, phosphate, and ammonium values are almost the same as the values of Atlantic Seawater (AS of Table 10.1). The mean TA (2303.5 ± 5.7 μmol kg^{-1}) and DOC (54.3 ± 3.8 μmol kg^{-1}) concentrations and the DIC concentration (2071.5 ± 17.1 μmol kg^{-1}) at the time of calibration estimated from the regression line are consistent with values of real seawater collected from northeast Atlantic surface water. The mean $\pm$ SD concentrations at a depth of 10 m for 6 stations between 47 and 56°N along 20°W (WOCE A16N meridional section in 2003, Hansell et al., 2021) are 2305.1 ± 3.7 μmol kg^{-1}, 56.9 ± 3.2 μmol kg^{-1}, and 2056 ± 17 μmol kg^{-1} for TA, DOC, and DIC, respectively. The mean TA concentration is also consistent with the concentrations (2300 μmol kg^{-1}) of the Reference Composition (Millero et al., 2008) and the standard seawater composition model (SSW76) used by Pawlowicz (2010) to investigate effects of compositional changes on conductivity and density. Our estimated DIC value at the time of calibration was quite similar to the value of 2080 μmol kg^{-1} in the SSW76 composition, although it was about 100 μmol kg^{-1} larger than that implied by the Reference Composition (about 1963 μmol kg^{-1}) which was itself defined on the basis of the assumption of atmospheric saturation in the 1970s (Pawlowicz, 2010).

Table 10.3 Same as Table 10.1, but for dissolved organic carbon (DOC) and total dissolved nitrogen (TDN) (in μmol kg^{-1})

Batch no.	Date of measurement	Age of batch	DOC	TDN
P152	15 Oct 2014	4.4	54.1	3.7
P153	15 Oct 2014	3.6	52.7	3.7
P154	15 Oct 2014	3.0	53.0	4.6
P155	15 Oct 2014	2.1	51.7	3.8
P156	15 Oct 2014	1.2	53.0	4.1
P157	8 Jan 2015	0.7	50.8	3.6
P153	15 Jan 2017	5.9	56.5	3.9
P154	15 Jan 2017	5.2	55.8	4.8
P155	15 Jan 2017	4.3	53.3	4.5
P156	15 Jan 2017	3.5	54.9	4.2
P157	15 Jan 2017	2.7	56.9	4.0
P158	15 Jan 2017	1.8	56.9	4.5
P159	15 Jan 2017	1.1	58.1	4.1
P160	22 May 2019	2.8	55.6	4.0
P161	22 May 2019	2.1	52.6	4.7
P162	22 May 2019	1.1	50.6	3.5
P138	26 Jan 2021	21.0	56.3	4.5
P141	26 Jan 2021	18.6	83.4	6.3
P144	26 Jan 2021	17.4	66.7	5.5
P145	26 Jan 2021	16.5	93.8	5.6
P146	26 Jan 2021	15.7	70.8	6.2
P147	26 Jan 2021	14.7	60.8	6.6
P148	26 Jan 2021	14.3	53.8	4.6
P149	26 Jan 2021	13.3	51.3	3.2
P150	26 Jan 2021	12.7	49.9	3.5
P151	26 Jan 2021	11.7	48.7	3.3
P156	26 Jan 2021	7.5	50.4	4.1
P163	26 Jan 2021	1.8	50.8	3.6
P164	26 Jan 2021	0.8	51.7	3.9
P165	1 Sep 2022	1.4	66.0	3.9

10.4　Comparison of Major Constituents

In November 2020, mass fractions of major constituents (Na^+, Mg^{2+}, Ca^{2+}, K^+, Sr^{2+}, Cl^-, SO_4^{2-}, and Br^-) were measured for SSW batch P149, which was calibrated on 5 October 2007. Five samples of P149 diluted 2000 or 200 times with ultrapure water were prepared from one bottle of P149 and were measured by IC or ICP-OES

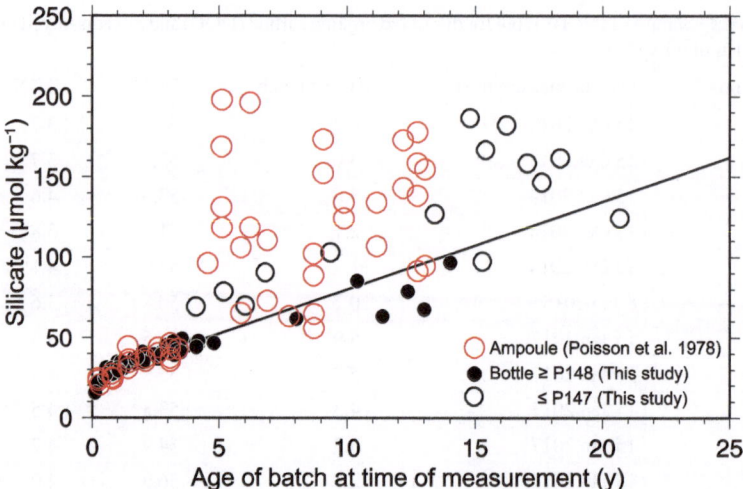

Fig. 10.1 Silicate plotted against age of batch at time of measurement for bottle-type SSW (closed black circles for batches after P148 and open black circles for batches before P147). The regression line for the bottle-type SSW estimated using the data within 5 year from calibration is also shown. Data for ampoule-type SSW by Poisson et al. (1978) are also shown (red circles)

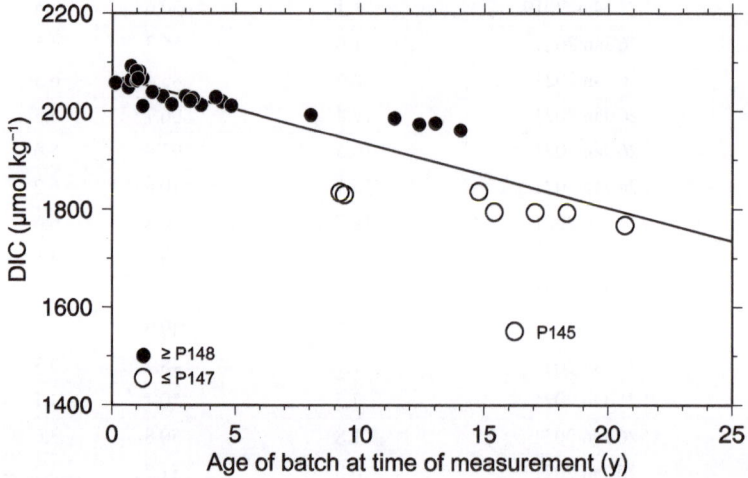

Fig. 10.2 Dissolved inorganic carbon (DIC) plotted against age of batch at time of measurement for bottle-type SSW (closed circles for batches after P148 and open circles for batches before P147). The regression line estimated using the data within 5 year of calibration is also shown

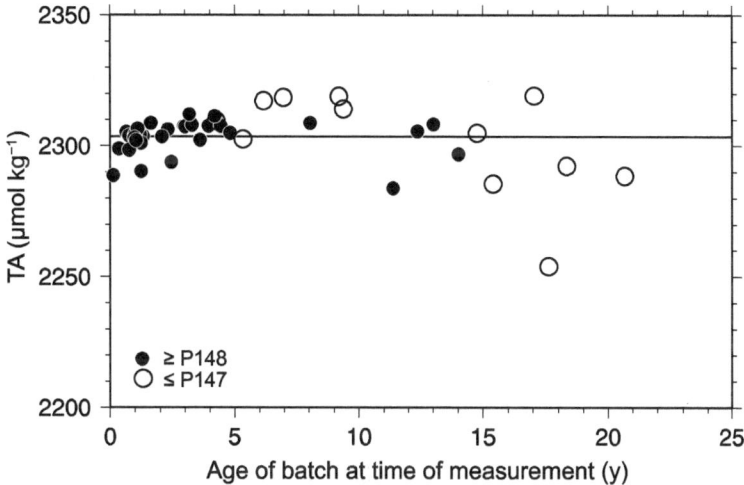

Fig. 10.3 Total alkalinity (TA) plotted against age of batch at time of measurement for bottle-type SSW (closed circles for batches after P148 and open circles for batches before P147). The horizontal line shows the mean value calculated using the data within 5 year of calibration. One point for P145 (year = 16.2, TA = 1883.5 μmol kg^{-1}) is outside the range of the figure

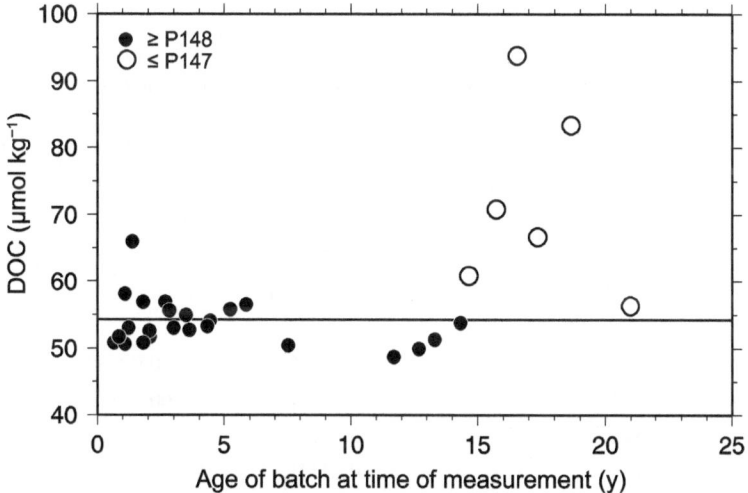

Fig. 10.4 Same as Fig. 10.3, but for dissolved organic carbon (DOC)

(Table 10.5). Minimum, lower quartile, median, upper quartile, and maximum values were determined from these data and are shown as box-and-whisker plots (Fig. 10.6). The median values with uncertainties are listed in Table 10.6.

Mass fractions of major constituents for SSW batch P149 were also measured by Seitz et al. (2010) in an inter-laboratory comparison study for Na$^+$, Mg^{2+}, Sr^{2+},

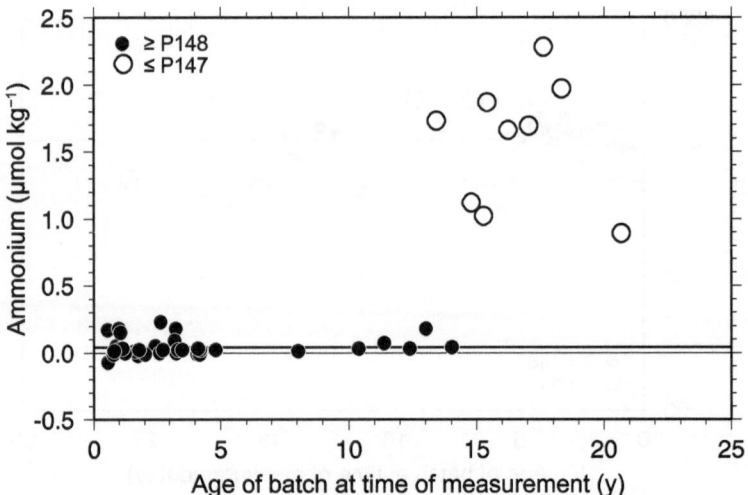

Fig. 10.5 Same as Fig. 10.3, but for ammonium

Table 10.4 Characteristic values (in µmol kg^{-1}) of concentrations of seawater constituents of SSW batches after P148. Regression lines determined by least squares fitting as a function of age of batch (t in year) or averages with standard deviations (SD) calculated for the data within 5 year of calibration are listed. For silicate and DIC, the SDs from the regression lines are shown. The number of data used is shown in brackets. Values from the reference composition (Millero et al., 2008) are also shown, except for DIC which is from the standard seawater composition model (SSW76) by Pawlowicz (2010) (see text for details)

Property	Regression line or average	SD [number]	Reference composition
Silicate	25.16 + 5.47 t	3.04 [32]	0
Nitrate	0.33	0.35 [32]	0
Nitrite	0.05	0.05 [32]	0
Phosphate	0.066	0.052 [32]	0
Ammonium	0.04	0.08 [19]	0
DIC	2071.5–13.5 t	17.1 [22]	2080 (SSW76)
TA	2303.5	5.7 [26]	2300
DOC	54.3	3.8 [17]	0
TDN	4.0	0.4 [17]	0

Cl^{-}, and SO$_4^{2-}$ and by Feistel et al. (2010) for Cl^{-} and SO$_4^{2-}$ within 2 year of the calibration of P149. We calculated the statistical properties of these data based on reports from eight or nine laboratories and compared them with the measurements made in this study 10 year after the earlier measurements (Fig. 10.6 and Table 10.6).

The variabilities of the results from the eight or nine laboratories reported by Seitz et al. (2010) and Feistel et al. (2010) tended to be larger than those of this study, but median values were generally consistent with the results of this study. The results of

Table 10.5 Mass fractions of major constituents of SSW batch P149 (calibrated in October 2007) measured by ion chromatography (IC) or inductively coupled plasma optical emission spectrometry (ICP-OES) in November 2020. The seawater was diluted 2000 or 200 times with ultrapure water (Milli-Q water). Two measurements were averaged for each ICP-OES diluted sample. Dilution ratio was precisely determined by Practical Salinity measurements using a salinometer and weighing method using an electric balance for the first three values and the other two values, respectively

Seawater component	Mass fractions (g kg^{-1})					Measurement procedure (dilution ratio)
Na$^+$	10.830	10.880	10.977	11.229	11.114	IC (2000×)
Mg^{2+}	1.265	1.266	1.272	1.280	1.265	ICP-OES (2000×)
	1.252	1.255	1.281	1.289	1.304	ICP-OES (200×)
Ca^{2+}	0.424	0.396	0.394	0.396	0.393	ICP-OES (2000×)
	0.391	0.385	0.397	0.403	0.407	ICP-OES (200×)
K$^+$	0.445	0.417	0.424	0.435	0.434	IC (2000×)
Sr^{2+}	0.00773	0.00789	0.00789	0.00794	0.00790	ICP-OES (2000×)
	0.00780	0.00778	0.00802	0.00807	0.00797	ICP-OES (200×)
Cl$^-$	19.276	19.320	19.409	19.586	19.342	IC (2000×)
SO$_4{}^{2-}$	2.683	2.692	2.704	2.728	2.696	IC (2000×)
Br$^-$	0.0608	0.0612	0.0633	0.0640	0.0631	IC (200×)

this study were also consistent with the mass fractions of the Reference Composition (Millero et al., 2008), except for Br$^-$, Na$^+$, and K$^+$.

Although it is necessary to improve the accuracy of the measurements to detect temporal variations of mass fractions of major constituents of SSW, it might be possible to detect batch-to-batch variations by measuring many batches simultaneously.

10.5   Discussion of Compositional Changes

The composition of SSW appeared to be remarkably stable, with two exceptions. First, silicate values were significantly higher than expected, even when regressed back to the time of calibration, and they continued to slowly increase over time. Second, DIC values slowly decreased with time.

The increase of silicate concentrations in SSW is thought to be due to leaching from the glass container because the concentrations found were higher than would be expected in the low-nutrient, North Atlantic surface water from which SSW is prepared (Poisson et al., 1978). As described in Sect. 10.3, the silicate concentration in Atlantic Seawater stored in HDPE bottles was reasonably small (0.84 µmol kg^{-1}) (Table 10.1). However, the value measured for Atlantic Seawater and the value at the

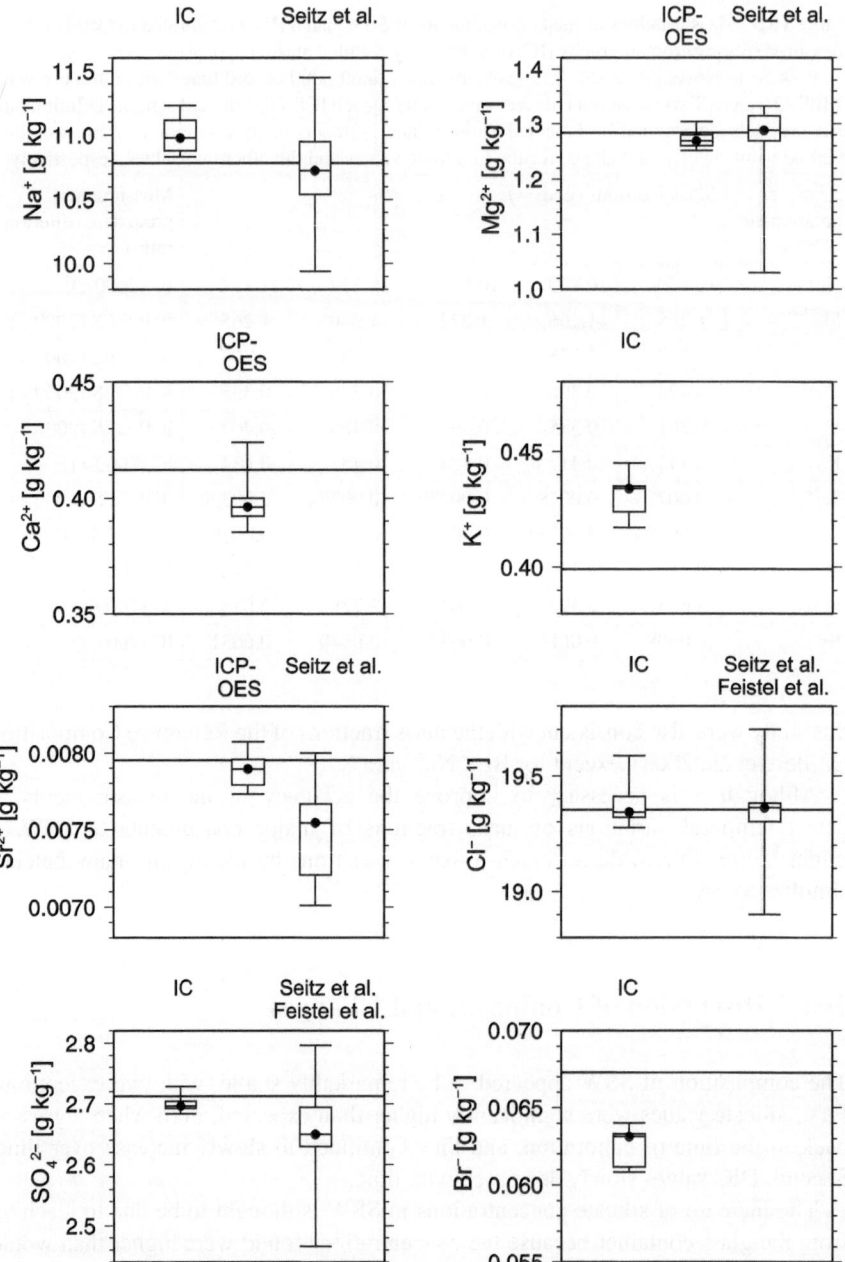

Fig. 10.6 Box-and-whisker plots of the mass fractions of major constituents of SSW batch P149 measured by ion chromatography (IC) or inductively coupled plasma optical emission spectrometry (ICP-OES) in November 2020. Dots, boxes, and vertical bars indicate the median, interquartile range (IQR), and minimum/maximum of the data, respectively. Results from Seitz et al. (2010) and Feistel et al. (2010) in Table 10.6 are also shown. Horizontal line indicates the value of the Reference Composition (Millero et al., 2008)

Table 10.6 Comparison of the mass fractions of SSW batch P149. Median values of measurements by IC or ICP-OES are compared with median values of results of the inter-laboratory comparison by Seitz et al. (2010) combined with results of Cl^- and SO_4^{2-} from Feistel et al. (2010). Uncertainty (95% confidence intervals) of the median can be given by 1.58 IQR/$\sqrt{N}$, where IQR is interquartile range and N is number of measurements and is shown with the median value. Values of the Reference Composition (Millero et al., 2008) are also shown. "n/a" indicates that the data is not available

Seawater constituent	This study (g kg^{-1})	Seitz et al. (2010) and Feistel et al. (2010) (g kg^{-1})	Reference composition (g kg^{-1})
Na^+	10.977 ± 0.165	10.725 ± 0.229	10.78145
Mg^{2+}	1.269 ± 0.010	1.288 ± 0.026	1.28372
Ca^{2+}	0.396 ± 0.004	n/a	0.41208
K^+	0.434 ± 0.008	n/a	0.39910
Sr^{2+}	0.00790 ± 0.00008	0.00755 ± 0.00025	0.00795
Cl^-	19.342 ± 0.063	19.360 ± 0.042	19.35271
SO_4^{2-}	2.696 ± 0.008	2.650 ± 0.034	2.71235
Br^-	0.0631 ± 0.0015	n/a	0.06728

time of calibration (25.16 µmol kg^{-1}) estimated from the regression line (Fig. 10.1) were very different.

This difference arising from storage in borosilicate glass was also tested in samples of real seawater collected in the Red Sea. Silicate, nitrate, and phosphate concentrations were measured for 59 pairs of samples in both glass bottles for DIC measurements and in polyacrylate vials for measurement of nutrient concentrations. The glass bottles were previously used SSW bottles, and the seawater samples were poisoned with mercuric chloride. The seawater samples stored in polyacrylate vials were refrigerated until measurement. Measurements were conducted 0.63 year after sample collection. Results for the glass bottles and polyacrylate vials were in good agreement for nitrate and phosphate (mean ± SD [in µmol kg^{-1}] of 5.86 ± 8.04 [glass bottles] and 5.83 ± 7.91 [polyacrylate vials] for nitrate and 0.37 ± 0.44 [glass bottles] and 0.34 ± 0.43 [polyacrylate vials] for phosphate), but silicate concentrations were high in the glass bottles (mean ± SD [in µmol kg^{-1}] of 29.60 ± 4.03 [glass bottles] versus 4.70 ± 5.45 [polyacrylate vials]). The increase of the silicate concentration (24.91 µmol kg^{-1}) was similar to the concentration at the time of calibration estimated from the regression line in Fig. 10.1.

However, such a rapid rise from the time of sampling was inconsistent with the long-term regression itself. This rapid increase of the silicate concentration was then investigated in a leaching experiment using Atlantic Seawater stored in two borosilicate glass bottles that had previously contained SSW. Silicate concentrations were 1.06, 1.77, and 11.60 µmol kg^{-1} after 0.25, 3.23, and 64.04 days, respectively, from bottling one bottle and 2.36, 3.63, and 17.13 µmol kg^{-1} after 2.88, 5.85, and 66.67 days, respectively, from bottling the other bottle, although the concentrations of nitrate, nitrite, phosphate, and ammonium were almost the same as the initial concentrations (AS of Table 10.1). Because SSW is typically left in bottles

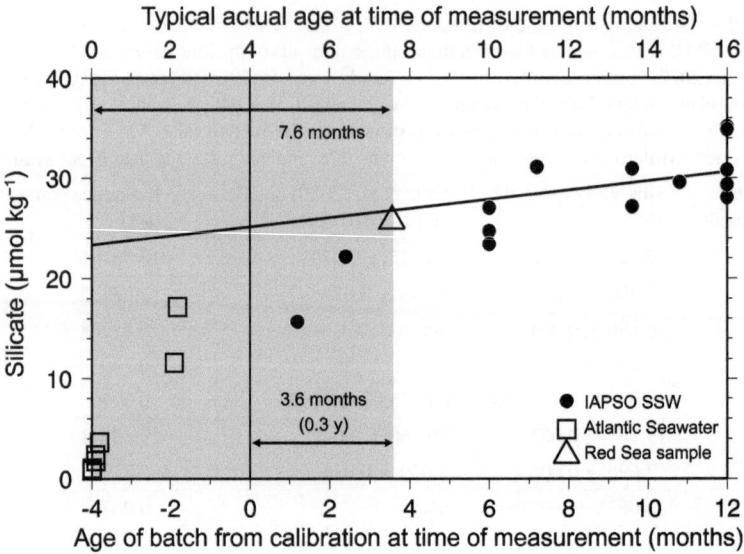

Fig. 10.7 Silicate concentration plotted against age of batch at the time of measurement for bottle-type SSW enlarged for the first year of Fig. 10.1. The increase of silicate concentrations for the Red Sea samples and results of the leaching experiment by using the Atlantic Seawater is also plotted against typical age at time of measurement (see text for details). An approximate period of the rapid initial increase is shown by the shading

for several months prior to calibrating (Richard Williams of OSIL, 2022, personal communication), the experimental data were compared with the SSW results on the assumption that the actual age from bottling the SSW was four months older than the age when the batch was calibrated (Fig. 10.7).

Although the silicate system is apparently quite complicated in water (e.g., Duedall et al., 1976), the accepted equations for the leaching reaction that occurs in a glass container of SSW can be described as follows (e.g., Gin et al., 2021; Hahn & van Duin, 2019; Li et al., 2021; Yamashita, 2011). The first step is the exchange reaction between soluble ions such as alkali metal ions like Na^+ and hydronium ions in water as follows

$$\equiv Si-O-Na + H_3O^+ \text{ (aq)} \rightarrow \equiv Si-OH + Na^+ \text{ (aq)} + H_2O. \quad (10.1)$$

This reaction forms a hydrated alteration layer with water and soluble ions on the glass surface. On the bulk side of the glass, the gradient due to the decreased concentration of soluble ions causes diffusion of soluble ions from the bulk side to the hydrated alteration layer and formation of a diffusion layer. Within the hydrated alteration layer, hydrolysis breaks the silica network

$$\equiv Si-O-Si \equiv +H_2O \rightarrow 2 \equiv Si-OH \quad (10.2)$$

and the reverse reaction causes condensation of silanol groups. A porous gel layer with a loose network of siloxane bonds is then generated

$$2 \equiv Si-OH \rightarrow \equiv Si-O-Si \equiv +H_2O. \tag{10.3}$$

Further hydrolysis leads to desorption of silicate, and dissolution proceeds as follows

$$\equiv Si-O-Si(OH)_3 + H_2O \rightarrow \equiv Si-OH + Si(OH)_4 \text{ (aq).} \tag{10.4}$$

The initial stage of dissolution is controlled by diffusion because of the growth of diffusion layers, and the amount of dissolution is proportional to the square root of time. As dissolution progresses, the growth rate of the diffusion layer on the bulk side decreases and eventually equals the dissolution rate of silicate at the glass surface. The amount of dissolution then becomes proportional to time (Yamashita, 2011).

For SSW, an approximate period of the initial stage of dissolution can be estimated to be about 7.6 months (or 3.6 months [0.3 year] from calibration) (Fig. 10.7).

There was concern that the dissolution behavior of silicate might differ between new and used bottles. In other words, one could assume that the dissolution rate of silica at the glass surface of the used bottles would be slow and that the silicate concentration would be proportional to time as shown in Fig. 10.1 immediately after bottling. In reality, however, the silicate concentration in the bottles used for the leaching experiment with the Red Sea samples increased rapidly immediately after bottling as would be expected for new bottles. The lack of a difference between new and used bottles may indicate that diffusion of Na^+ ions in the diffusion layer proceeds without dissolution of Na^+ ions from the surface of the bottle (i.e., Eq. (10.1)), and that the gradient of Na^+ ions in the glass matrix disappeared when the bottle was emptied after use with SSW.

The leaching reaction causes the Na^+ concentration and pH to increase. Although we did not measure pH directly here, increases of pH for ampoule-type SSW have been reported in previous studies (Goyet et al., 1985; Poisson et al., 1978). If TA remains constant, an increase in pH would be associated with a decrease of DIC in the carbonate system. To quantify the expected decrease of DIC, we calculated the DIC from the measured pH and TA in the dataset of Goyet et al. (1985) by using the CO2SYS program developed by Lewis and Wallace (1998). The silicate concentration of the SSW was estimated from the regression line proposed in this study and used in the CO2SYS program. For the dataset of Poisson et al. (1978), the DIC was estimated from the measured pH and silicate by using the CO2SYS program assuming that TA was constant (2300 $\mu mol\ kg^{-1}$). The calculated DIC concentrations trended downward, similar to the result of the bottle-type SSW (Fig. 10.2), although there was a large amount of variation and a slight systematic bias (Fig. 10.8).

An increase in the concentration of Na^+ ions causes an increase of TA (e.g., Wolf-Gladrow et al., 2007). However, because the TAs were constant (Fig. 10.3), some cations associated with the decrease of DIC must have been removed from the SSW. Because the surface water in the eastern North Atlantic is currently supersaturated

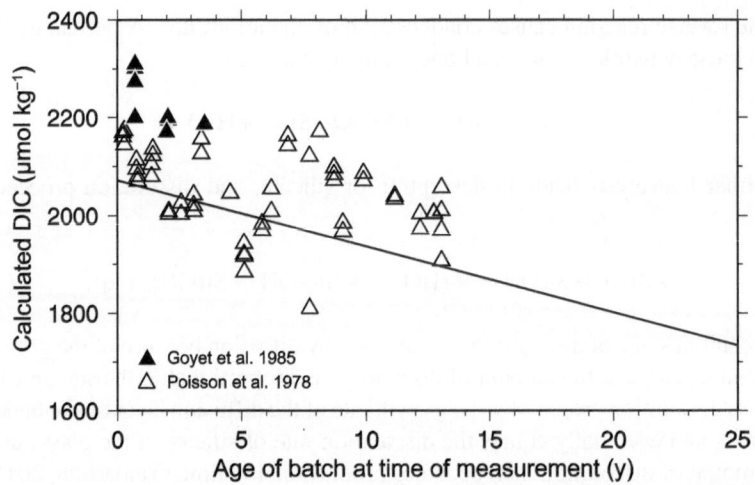

Fig. 10.8 Calculated DIC plotted against the age of batch at the time of measurement for ampoule-type SSW. Closed triangles are values estimated from total alkalinity and pH by Goyet et al. (1985), and open triangles are estimates from pH data by Poisson et al. (1978) assuming TA to be constant. The line is the regression line for bottle-type SSW shown in Fig. 10.2

with respect to calcium carbonate ($CaCO_3$), the decrease in DIC might have been due to $CaCO_3$ precipitation (Higano, 1977). Equations (10.1)–(10.4) lead to the conclusion that a one-mole increase of $Si(OH)_4$ would correspond to a two-mole increase of Na^+. To keep the TA constant, Ca^{2+} must decrease by one mole for every two-mole increase of Na^+. Because the CO2SYS program estimates that HCO_3^- is greatly reduced by the observed compositional changes listed in Table 10.4, the precipitation of $CaCO_3$ can be described by

$$Ca^{2+} + 2HCO_3^- \rightarrow CaCO_3 + CO_2 + H_2O, \tag{10.5}$$

and the increase of CO_2 may lead to outgassing into the head space of the glass bottle (Wolf-Gladrow et al., 2007). This reaction corresponds to a two-mole decrease of DIC for every one-mole increase of $Si(OH)_4$ and may therefore largely explain the observed changes of silicate (5.47 μmol kg^{-1} year^{-1}) and DIC (−13.5 μmol kg^{-1} year^{-1}). In reality, it is possible that the reaction is not complete after Eq. (10.1). The amount of desorbed silicate might therefore be smaller than the expected amount (Masaru Yamashita, 2022, personal communication).

In that case, the Na^+ and Ca^{2+} concentration in batch P149 at the time of measurement in November 2020 would have been 0.003 g kg^{-1} (2 × 71 × 10^{-6} mol kg^{-1} × 22.98976928 g mol^{-1}) larger than the initial concentration because of the increase of the silicate (71 μmol kg^{-1} per 13 year) and 0.004 g kg^{-1} (0.5 × 176 × 10^{-6} mol kg^{-1} × 40.078 g mol^{-1}) smaller than the initial concentration because of the decrease of the DIC (−176 μmol kg^{-1} per 13 year). Although the estimated increase of the Na^+ concentration is small in magnitude, the estimated decrease of Ca^{2+} is comparable in magnitude to the difference between the measured mass fraction and the Reference Composition value (0.20 g kg^{-1} for Na$^+$ and -0.016 g kg^{-1} for Ca^{2+}, Table 10.6).

Steiner et al. (2021) have reported the Ca^{2+} concentrations in batch P157. The mass fraction value (0.4099 g kg^{-1}) at the time of measurement in August 2016 (2.3 year of age, Zvi Steiner of GEOMAR Helmholtz Centre for Ocean Research Kiel, 2022, personal communication) would have been 0.0006 g kg^{-1} less than the initial value because of the decrease of the DIC. The estimated initial value (0.4105 g kg^{-1}) was also consistent with the mass fractions of the Reference Composition (0.41208 g kg^{-1}).

10.6 Effect of DOC on Density

SSW is used as the international standard for measuring S_P. SSW is also sometimes used as a reference in measuring seawater density to ensure comparability of seawater density measurements (e.g., Uchida et al., 2011a). Because changes of conductivity/density/salinity relationships in real seawater have been linked to changes in the concentration of seawater constituents that are involved in biogeochemical cycling (Pawlowicz et al., 2011), here we discuss changes of S_A in SSW and take these factors into account.

First of all, the effect of DOC on seawater density was examined because that effect was not considered as a future task in the conductivity/density/salinity relationships examined by Pawlowicz et al. (2011). Although the concentration of DOC in the source seawater for SSW is 54.3 ± 3.8 µmol kg^{-1} (Table 10.4) as described in Sect. 10.3, DOC is not included in the Reference Composition, which is the best estimate of the chemical composition of SSW. We investigated the relationship between DOC and density using glucose, which is often used to create calibration curves for DOC measurements (Ogawa et al., 2003). The increase of the density caused by adding glucose was plotted against DOC for seawater and ultrapure water (Milli-Q water) (Fig. 10.9). The seawater was collected at depths deeper than 3000 m from the North Pacific (47°N, 160°E). The seawater was filled in a 20 L polyethylene tank without filtration and was stored in a refrigerator. The concentration of DOC in the deep seawater was 37 µmol kg^{-1}. Density values without DOC were estimated from the intercepts of the regression lines, and differences from the estimated densities at zero DOC were plotted. Although densities calculated from the equation of state (TEOS-10) with S_R measurements were constant, the densities of ultrapure water and seawater increased in a linear manner with increasing DOC. The slopes of the regression lines were 1.22 × 10^{-5} and 1.34 × 10^{-5} (kg m^{-3}) (µmol kg^{-1})$^{-1}$ for ultrapure water and seawater, respectively, and the corresponding change of S_A was estimated to be 1.76 × 10^{-5} (g kg^{-1}) (µmol kg^{-1})$^{-1}$ from the equation of state (TEOS-10).

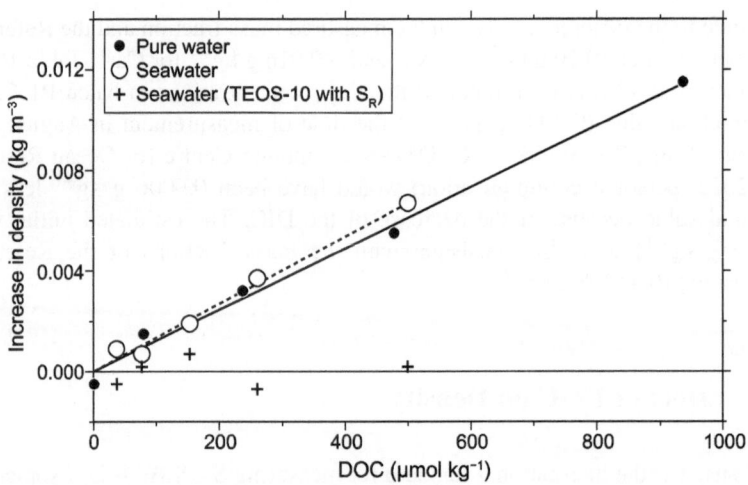

Fig. 10.9 Increase of densities at 20 °C by adding glucose plotted against DOC for seawater and ultrapure water. Densities at zero DOC are estimated from the regression lines (solid line is for ultrapure water, and dotted line is for seawater), and differences from the estimated densities at zero DOC are plotted. For seawater, densities are also calculated from the equation of state (TEOS-10) with S_R measurements, and differences from mean values are also plotted

10.7 A Model for the Salinity Anomaly of Aging SSW

By combining the rate of increase in S_A obtained from the glucose addition experiment and the δS_A model of Pawlowicz et al. (2011), we modified the δS_A model as follows:

$$\delta S_A \left[g\ kg^{-1} \right] = (5.07Si(OH)_4 + 3.89NO_3 + 5.56\Delta NTA$$
$$+ 0.47\Delta NDIC + 1.76\Delta NDOC) \times 10^{-5}, \qquad (10.6)$$

where $Si(OH)_4$ is the silicate concentration; NO_3 is the nitrate concentration; ΔNTA is the TA anomaly (TA $- 2300 \times S_P/35$); $\Delta NDIC$ is the DIC anomaly (DIC $- 2080 \times S_P/35$); and $\Delta NDOC$ is the DOC anomaly (DOC $- 54.3 \times S_P/35$), respectively, in $\mu mol\ kg^{-1}$. For SSW, if there is no temporal variation in S_R, δS_A can be estimated as a function of time (t in year) from Eq. (10.6) and the characteristic values (Table 10.4).

$$\delta S_A \left[g\ kg^{-1} \right] = 0.0014 + 0.00021\ t \qquad (10.7)$$

for t > 0.3 year (see Fig. 10.7). From the SDs listed in Table 10.4 and the coefficients of each term on the right-hand side of Eq. (10.6), the standard uncertainty of δS_A by Eq. (10.7) is estimated to be 0.0004 g kg^{-1}. The estimated uncertainty is well smaller than the measuring precision by the oscillation-type density meter (about 0.0014 g kg^{-1}, Uchida et al., 2011a).

Because DOC is not included in the Reference Composition, there is no need to make the DOC term an anomaly. However, in the relationship between density and salinity used to create the equation of the state of seawater, the effect of DOC should be included in the density. The DOC term is therefore expressed as an anomaly, like TA and DIC. Like DOC, silicate is not included in the Reference Composition, but because it is unknown how much silicate was in the SSW used to create the equation of state of seawater, silicate should not be an anomaly as Pawlowicz et al. (2011) did not treat it as an anomaly.

The δS_A values predicted from Eq. (10.7) were evaluated with δS_A derived from density measurements. The density measurements were conducted on 7–9 September 2020 (see Chap. 9 for S_P measurements). The density of SSW batch P161 was measured each day of measurement, and the density meter was calibrated with ultrapure water and SSW batch P161. The true density of P161 at the time of measurement was estimated to be 1024.7636 kg m^{-3} from the batch offset-corrected label S_P value (Uchida et al., 2020) and δS_A predicted from Eq. (10.7). The predicted δS_A for SSW batches P138 to P164 agreed well with the measured δS_A up to about 8 year of age (Fig. 10.10), but the older batches varied widely, probably because of the variation of silicate concentrations (Fig. 10.1).

In Eq. (10.7), it is assumed that there is no temporal variation of the S_R of SSW. However, changes in the concentrations of constituents of SSW may change the S_P

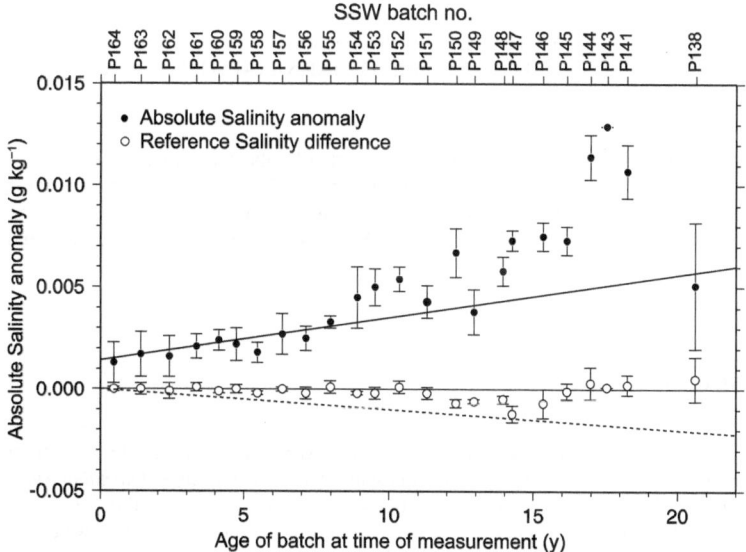

Fig. 10.10 Absolute Salinity anomalies (δS_A) (closed circles) from the batch offset-corrected label values of the Reference Salinity (S_R) plotted against age of batch at time of measurement for bottle-type SSW. S_R differences between measured values and the batch offset-corrected label values are also plotted (open circles). Vertical bars show their SDs. The solid line is the δS_A predicted from Eq. (10.7), and the dashed line indicates the decrease of S_R due to decrease of the DIC concentration (see text for details)

(and therefore the S_R) of the SSW. A DIC change of 100 μmol kg^{-1} will change S_P by 0.0007 (Pawlowicz, 2010). Because the DIC of SSW decreases at a rate of − 13.5 (μmol kg^{-1}) year^{-1}, the S_P of SSW would change at a rate of −0.0001 year^{-1}. The decrease of measured S_R might have been partly related to the decrease of DIC (Fig. 10.10). It is also likely that the presence of large concentrations of a nonionic constituent in seawater would tend to decrease the conductance slightly, possibly due to viscosity effects, and a silicate change of 100 μmol kg^{-1} will change S_P by −0.0007 (Brewer & Bradshaw, 1975). Because the silicate concentration of SSW increases at a rate of 5.47 (μmol kg^{-1}) year^{-1}, the S_P of SSW might decrease at a rate of −0.00004 year^{-1}. Although the change of DOC did not change S_P (Fig. 10.9), some DOC compounds are ionic and contribute to TA as organic alkalinity, which is typically deemed negligible and is not accounted for in conventional TA calculations (Kerr et al., 2021). Further investigation is needed of the changes of S_P associated with changes of the concentrations of other constituents of SSW.

Acknowledgements H. Uchida would like to thank the late Michio Aoyama for encouraging him to carry out this research. We thank Taketoshi Kodama of the University of Tokyo and the marine technicians of Marine Works Japan, Ltd., who analyzed the compositions of the SSWs. We also thank Richard Williams of Ocean Scientific International, Ltd., for providing information about SSW production. Comments by Masahito Shigemitsu of the Japan Agency for Marine-Earth Science and Technology (JAMSTEC), Rich Pawlowicz of the University of British Columbia, and Ryan Woosley of the Massachusetts Institute of Technology helped to improve the paper. The Red Sea samples were obtained from the first Joint King Abdullah University of Science and Technology (KAUST) and JAMSTEC cruise Kaja-001 by R/V *Thuwal* in February 2022, and we thank Ute Langner (KAUST) for sample collection and Vincent F. Saderne (KAUST) for preparation of regents. This paper is a contribution to the tasks of the Joint SCOR/IAPWS/IAPSO Committee on the Properties of Seawater (JCS).

Competing Interests Part of this work was supported by the Japan Society for the Promotion of Science KAKENHI Grants 18K03752, 21H01165 and 20H04349. The cruise Kaja-001 was supported by the Ministry of Economy, Trade and Industry (METI) Support Project for Strengthening Relations with Resource-rich Countries to Secure Oil and Natural Gas Interests and Stable Supply.

References

Bacon, S., Culkin, F., Higgs, N., & Ridout, P. (2007). IAPSO standard seawater: Definition of the uncertainty in the calibration procedure, and stability of recent batches. *Journal of Atmospheric and Oceanic Technology, 24*, 1785–1799. https://doi.org/10.1175/JTECH2081.1

Brewer, P. G., & Bradshaw, A. (1975). The effect of non-ideal composition of seawater on salinity and density. *Journal of Marine Research, 33*, 157–175.

Budéus, G. T. (2018). Potential bias in TEOS10 density of sea water samples. *Deep-Sea Research I, 134*, 41–47. https://doi.org/10.1016/j.dsr.2018.02.005

Culkin, F., & Smed, J. (1979). The history of standard seawater. *Oceanologica Acta, 2*, 355–364.

Duedall, I. W., Dayal, R., & Willey, J. D. (1976). The partial molal volume of silicic acid in 0.725 m NaCl. *Geochimica et Cosmochimica Acta, 40*, 1185–1189. https://doi.org/10.1016/0016-703 7(76)90153-8

Feistel, R., Weinreben, S., Wolf, H., Seitz, S., Spitzer, P., Adel, B., Nausch, G., Schneider, B., & Wright, D. G. (2010). Density and Absolute Salinity of the Baltic Sea 2006–2009. *Ocean Science, 6*, 3–24. https://doi.org/10.5194/os-6-3-2010

Gin, S., Delaye, J.-M., Angeli, F., & Schuller, S. (2021). Aqueous alteration of silicate glass: State of knowledge and perspectives. *Materials Degradation, 5*, 42. https://doi.org/10.1038/s41529-021-00190-5

Goyet, C., Poisson, A., Brunet, C., & Culkin, F. (1985). IAPSO Standard Seawater as a reference standard for alkalinity determinations. *Deep-Sea Research, 32*, 1437–1443. https://doi.org/10.1016/0198-0149(85)90058-5

Hahn, S. H., & van Duin, A. C. T. (2019). Surface reactivity and leaching of a sodium silicate glass under an aqueous environment: A reaxFF molecular dynamics study. *Journal of Physical Chemistry C, 123*, 15606–15617. https://doi.org/10.1021/acs.jpcc.9b02940

Hansell, D. A., Carlson, C. A., Amon, R. M. W., Álvarez-Salgado, X. A., Yamashita, Y., Romera-Castillo, C., & Bif, M. B. (2021). Compilation of dissolved organic matter (DOM) data obtained from the global ocean surveys from 1994 to 2020 (NCEI Accession 0227166). NOAA National Centers for Environmental Information. Dataset. https://doi.org/10.25921/s4f4-ye35. Accessed 10 November 2022.

Higano, R. (1977). On the calcium concentration of the Japanese Standard Sea Water. *Report of Hydrographic Researches, 12*, 121–124 (in Japanese with English abstract). https://www1.kaiho.mlit.go.jp/GIJUTSUKOKUSAI/KENKYU/report/rhr12/rhr12-07.pdf

IAPWS. (2018). *Revised release on the IAPWS formulation 1995 for the thermodynamic properties of ordinary water substance for general and scientific use*. http://iapws.org/relguide/IAPWS95-2018.pdf

IOC, SCOR, & IAPSO. (2010). The international thermodynamic equation of seawater—2010: Calculation and use of thermodynamic properties. Intergovernmental Oceanographic Commission, Manuals and Guides No. 56, UNESCO (English) (p. 196). http://teos-10.org/pubs/TEOS-10_Manual.pdf

Kawano, T. (2010). Method for salinity (conductivity ratio) measurement. The GO-SHIP repeat hydrography manual: A collection of expert reports and guidelines version 1 (IOCCP Rep. 14, ICPO Publ. Series 134, p. 13). http://www.go-ship.org/HydroMan.html

Kayukawa, Y., & Uchida, H. (2021). Absolute density measurements for standard sea-water by hydrostatic weighing of silicon sinker. *Measurement: Sensors, 18*. https://doi.org/10.1016/j.measen.2021.100200.

Kerr, D. E., Brown, P. J., Grey, A., & Kelleher, B. P. (2021). The influence of organic alkalinity on the carbonate system in coastal waters. *Marine Chemistry, 237*. https://doi.org/10.1016/j.marchem.2021.104050

Lewis, E., & Wallace, D. W. R. (1998). *Program developed for CO_2 system calculations* (Report 105, p. 33). Oak Ridge National Laboratory. http://cdiac.esd.ornl.gov/oceans/co2rprt.html

Li, X., Jiang, L., Liu, J., Wang, M., Li, J., & Yan, Y. (2021). Insight into the interaction between water and ion-exchanged aluminosilicate glass by nanoindentation. *Materials, 14*, 2959. https://doi.org/10.3390/ma14112959

Millero, F. J., Feistel, R., Wright, D. G., & McDougall, T. J. (2008). The composition of standard seawater and the definition of the reference-composition salinity scale. *Deep-Sea Research I, 55*, 50–72. https://doi.org/10.1016/j.dsr.2007.10.001

Millero, F. J., Huang, F., Woosley, R. J., Letscher, R. T., & Hansell, D. A. (2011). Effect of dissolved organic carbon and alkalinity on the density of Arctic Ocean waters. *Aquatic Geochemisty, 17*, 311–326. https://doi.org/10.1007/s10498-010-9111-2

Miyazaki, J., Kawagucci, S., Makabe, A., Takahashi, A., Kitada, K., Torimoto, J., Matsui, Y., Tasumi, E., Shibuya, T., Nakamura, K., Horai, S., Sato, S., Ishibashi, J., Kanzaki, H., Nakagawa, S., Hirai, M., Takaki, Y., Okino, K., Watanabe, H. K., … Chen, C. (2017). Deepest and hottest hydrothermal activity in the Okinawa Trough: The Yokosuka site at Yaeyama Knoll. *Royal Society Open Science, 4*, 171570. https://doi.org/10.1098/rsos.171570

Ogawa, H., Usui, T., & Koike, I. (2003). Distribution of dissolved organic carbon in the East China Sea. *Deep-Sea Research II, 50*, 353–366. https://doi.org/10.1016/S0967-0645(02)00459-9

Pawlowicz, R. (2010). A model for predicting changes in the electrical conductivity, practical salinity, and absolute salinity of seawater due to variations in relative chemical composition. *Ocean Science, 6*, 361–378. https://doi.org/10.5194/os-6-361-2010

Pawlowicz, R., Feistel, R., McDougall, T. J., Ridout, P., Seitz, S., & Wolf, H. (2016). Metrological challenges for measurements of key climatological observables part 2: Oceanic salinity. *Metrologia, 53*, R12–R25. https://doi.org/10.1088/0026-1394/53/1/R12

Pawlowicz, R., Wright, D. G., & Millero, F. J. (2011). The effects of biogeochemical processes on oceanic conductivity/salinity/density relationships and the characterization of real seawater. *Ocean Science, 7*, 363–387. https://doi.org/10.5194/os-7-363-2011

Picker, P., Tremblay, E., & Jolicoeur, C. (1974). A high-precision digital readout flow densimeter for liquids. *Journal of Solution Chemistry, 3*, 377–384. https://doi.org/10.1007/BF00646478

Poisson, A., Dauphinee, T., Ross, C. K., & Culkin, F. (1978). The reliability of standard seawater as an electrical conductivity standard. *Oceanologica Acta, 1*, 425–433.

Ridout, P. (2011). Developments in the standardisation of ocean salinity. *Science in Parliament, 68*, 7–9. https://www.scienceinparliament.org.uk/wp-content/uploads/2013/09/sip68-2-3.pdf

Romeo, R., Albo, P. A. G., & Lago, S. (2019). Density of standard seawater by vibrating tube densimeter: Analysis of the method and results. *Deep-Sea Research I, 154*, 103157. https://doi.org/10.1016/j.dsr.2019.103157

Schmidt, H., Seitz, S., Hassel, E., & Wolf, H. (2018). The density-salinity relation of standard seawater. *Ocean Science, 14*, 15–40. https://doi.org/10.5194/os-14-15-2018

Seitz, S., Spitzer, P., & Brown, R. J. C. (2010). CCQM-P111 study on traceable determination of practical salinity and mass fraction of major seawater compositions. *Accreditation and Quality Assurance, 15*, 9–17. https://doi.org/10.1007/s00769-009-0578-8

Steiner, Z., Sarkar, A., Liu, X., Berelson, W. M., Adkins, J. F., Achterberg, E. P., Sabu, P., Prakash, S., Vinaychandran, P. N., Byrne, R. H., & Turchyn, A. V. (2021). On calcium-to-alkalinity anomalies in the North Pacific, Red Sea, Indian Ocean and Southern Ocean. *Geochimica et Cosmochimica Acta, 303*, 1–14. https://doi.org/10.1016/j.gca.2021.03.027

Uchida, H., Kawano, T., Aoyama, M., & Murata, A. (2011a). Absolute salinity measurements of standard seawaters for conductivity and nutrients. *La Mer, 49*, 119–126. http://www.sfjo-lamer.org/la_mer/49-3_4/49-3-4-5.pdf

Uchida, H., Murata, A., & Doi, T. (2011b). WHP P21 Revisit Data Book. JAMSTEC. https://doi.org/10.17596/0000032

Uchida, H., Murata, A., Katsumata, K., Arulananthan, K., & Doi, T. (2021). WHP I08N Revisit/I07S in 2019/2020 Data Book. JAMSTEC. https://doi.org/10.17596/0002162

Uchida, H., Kawano, T., Nakano, T., Wakita, M., Tanaka, T., & Tanihara, S. (2020). An expanded batch-to-batch correction for IAPSO standard seawater. *Journal of Atmospheric and Oceanic Technology, 37*, 1507–1520. https://doi.org/10.1175/JTECH-D-19-0184.1

UNESCO. (1981). The practical salinity scale 1978 and the international equation of state of seawater 1980. In *UNESCO technical papers in marine science* (Vol. 36, p. 25).

Wakita, M., Honda, M. C., Matsumoto, K., Fujiki, T., Kawakami, H., Yasunaka, S., Sasai, Y., Sukigara, C., Uchimiya, M., Kitamura, M., Kobari, T., Mino, Y., Nagano, A., Watanabe, S., & Saino, T. (2016). Biological organic carbon export estimated from the annual carbon budget observed in the surface waters of the western subarctic and subtropical North Pacific Ocean from 2004 to 2013. *Journal of Oceanography, 72*, 665–685. https://doi.org/10.1007/s10872-016-0379-8

Wolf, H. (2008). Determination of water density: Limitations at the uncertainty level of 1×10^{-6}. *Accreditation and Quality Assurance, 13*, 587–591. https://doi.org/10.1007/s00769-008-0442-2

Wolf-Gladrow, D. A., Zeebe, R. E., Klaas, C., Körtzinger, A., & Dickson, A. G. (2007). Total alkalinity: The explicit conservation expression and its application to biogeochemical processes. *Marine Chemistry, 106*, 287–300. https://doi.org/10.1016/j.marchem.2007.01.006

Woosley, R., Huang, F., & Millero, F. J. (2014). Estimating absolute salinity (S_A) in the world's oceans using density and composition. *Deep-Sea Research I, 93*, 14–20, https://doi.org/10.1016/j.dsr.2014.07.009; Corrigendum in: *Deep-Sea Research I*, 142 (2018) 145. https://doi.org/10.1016/j.dsr.2018.09.007

Yamashita, M. (2011). Hydrolytic resistance of glass. *New Glass, 26*, 45–48 (in Japanese). https://www.newglass.jp/mag/TITL/maghtml/102-pdf/+102-p045.pdf

Chapter 11
On Japanese Standard Seawater for Salinity Measurements Used During and After World War II

Hiroshi Uchida

Abstract The Standard Seawater Service has provided Standard Seawater (SSW) since 1902. Although SSW is the only standard for salinity measurements around the world, for various reasons some countries have produced their own standards. Japanese Standard Seawater (JSSW) was produced and distributed in Japan for Japanese researchers from 1942 until the 1980s to ensure comparability of salinity measurements during and after World War II. The history of JSSW is reviewed based on the information available in the literature.

Keywords Secondary standard seawater · Japanese standard seawater · Salinity · Chlorinity

11.1 Introduction

Since its inception at the start of the twentieth century, the Standard Seawater Service (SSWS) has provided Standard Seawater (SSW) to ensure international comparability of salinity measurements (e.g., Chap. 8). However, some countries (the U.S.A., Japan, Russia, and China) have produced their own standards (secondary standard seawater) for salinity measurements for various reasons. The U.S.A. and Japan produced their own standards in the 1940s because availability of SSW to the oceanographic community was interrupted by World War II (WWII) (after batch P15 was produced in 1937, only one batch—P16 in 1943—was produced until P17 in 1948).

In the U.S.A., the U.S. Coast Guard and Woods Hole Oceanographic Institution briefly partnered to provide American Standard Seawater (ASSW) for American researchers at universities and government laboratories during WWII (Dawicki, 2022). Only one batch (batch 1 in 1941) seems to have been produced because batch 1 is the only batch that has been confirmed (tubes no. 49 and 52 in Dawicki (2022)

H. Uchida (✉)
Japan Agency for Marine-Earth Science and Technology, Yokosuka, Japan
e-mail: huchida@jamstec.go.jp

© The Author(s) 2025
M. Aoyama et al. (eds.), *Chemical Reference Materials for Oceanography*, Springer Oceanography, https://doi.org/10.1007/978-981-96-2520-8_11

and tube 537 in Japan Meteorological Agency [JMA], vide infra), and SSW was already available in the U.S.A. in 1947 (Dawicki, 2022; Richards, 1976).

In Russia, a seawater salinity standard has also been produced for distribution only in Russia (Ridout, 2011), but there is no detailed documentation about this standard seawater.

China has its own standard seawater (Chinese Primary Standard Seawater: CP SSW), but the international community has been largely unaware of its wide use by government and academic ocean scientists in China since the 1960s (Li et al., 2016). The source of the CP SSW is surface seawater collected in an area of the open ocean in the Western Pacific. Recent batches of CP SSW have been calibrated based on the definition of the Practical Salinity Scale 1978 (PSS-78) (UNESCO, 1981) and independently of SSW (Li et al., 2016).

In Japan, similar to the U.S.A., standard seawater (Japanese Standard Seawater: JSSW) has been produced and distributed for Japanese researchers at universities and government laboratories since 1942 (Hermann and Culkin (1978) briefly introduced it outside Japan). Unlike the standard seawater produced in the U.S.A., JSSW continued to be used in Japan until the 1980s. In this study, the history of JSSW is reviewed based on the information available in the literature.

11.2   Establishment of the Japanese Standard Seawater Committee

With the outbreak of WWII in 1939, disruption of transportation between Europe and Japan made SSW difficult to obtain in Japan. The efforts made by the oceanographic community in Japan to address this issue resulted in the establishment of the Japanese Standard Seawater Committee (JSSWC) under the Council for Scientific Research (now the Science Council of Japan) in 1940 (Tominaga et al., 1942). The establishment of the JSSWC led to the founding of the Oceanographic Society of Japan (OSJ). Kuroda (2020) has reviewed the details of the twists and turns of the history of the establishment of the OSJ.

Early technical members of the JSSWC were Takaharu Nomitsu, Masayoshi Ishibashi (Kyoto Univ.), Yuji Shibata (Univ. of Tokyo), Hitoshi Tominaga, Seiji Kokubo (Tohoku Univ.), Shinkichi Yoshimura (Tokyo Univ. of Literature and Science [now Univ. of Tsukuba]), Koji Hidaka, Yasuo Matsudaira (Marine Observatory [later Kobe Marine Observatory, JMA]), Shinichi Daito, Raizo Maruta (Naval Hydrographic Office [now Hydrographic and Oceanographic Department, Japan Coast Guard]), Michitaka Uda, Naraji Kakizaki (Fisheries Experiment Station, Ministry of Agriculture, Forestry and Fisheries [now Japan Fisheries Research and Education Agency]), Yasuo Miyake (Central Meteorological Observatory [CMO, now JMA]), and Mitsuyo Okada (Imperial Fisheries Institute [now Tokyo Univ. of Marine Science and Technology: TUMSAT]) (Tominaga et al., 1942).

Miyake (1941) concluded the discussions of the JSSWC with the following statement.

> In times of peace, it was indeed desirable for the world to have a single standard seawater, but this was not always the case in times of great international upheaval such as WWII. First, the status of the SSWS was unclear, and second, there was a question of whether it was really right from a national policy standpoint to use a large amount of SSW, which was terribly expensive because of foreign exchange and other factors (the price of one ampoule of SSW was equivalent to the monthly income of an average salaried worker before WWII (Hanzawa, 1987)). The answer was simple: Japan should use all of the capabilities of the oceanographic community to produce an excellent standard seawater with a higher level of quality assurance and quality control than the SSW produced in Denmark.

> This conclusion was considered important from a cultural-historical perspective because the Japanese oceanographic community was determined to break its dependence on foreign countries. This issue could therefore be used as an opportunity to establish close cooperation within that community (founding of the OSJ).

The basic policy established by the JSSWC was to divide the standard seawater into two types, one being the primary standard seawater and the other being the standard seawater for general use. Similarly to SSW, the chlorinity of the primary standard seawater was equated to 0.3285234 times the weight of pure silver equivalent to all the halides in the standard so that the result was independent of atomic weights (Jacobsen & Knudsen, 1940), and the general standard seawater was prepared with reference to the chlorinity of the primary standard seawater. Because it was necessary to immediately produce the standard seawater for general use, the first year was used to prepare for production of the primary standard seawater, and standard seawater for general use was temporarily calibrated with SSW and produced and distributed immediately (Tominaga et al., 1942) (However, there is no indication in the available literature that the primary standard seawater was subsequently produced).

The JSSWC seems to have continued its activities until about 1984 (Masuda, 1984). The Three-Governmental Oceanographic Operations Liaison Committee (current organizations are the JMA, the Japan Coast Guard [JCG], and the Fisheries Agency of Japan [FAJ]) was established in 1946, probably to coordinate the collection of source seawater for the JSSW (Seto, 2007) and to calibrate JSSW by joint comparison tests among responsible agencies (Shimano, 1982). This committee has continued to the present time with the addition of several organizations (Ministry of Defense in 1961, Ministry of Education, Culture, Sports, Science and Technology in 1962, Ministry of the Environment in 1989, Japan Agency for Marine-Earth Science and Technology in 1996, and the Japan Fisheries Research and Education Agency in 2001).

11.3   Early Production of JSSW

Seawater for the first batches of JSSW was collected in 1941 by the RV *Ryofu-maru* of the CMO (Matsudaira, 1943; Miyake, 1941) and calibrated in 1942 (Higano, 1977; Miyake & Yumura, 1943). Miyake (1941) has written a narrative of the cruise with many photographs of work on board the ship.

Matsudaira (1943) has provided a record of the collection of the source seawater for the JSSW and a detailed description of the methods and instruments that were used. A 20 L hard glass bottle for sampling was lowered to collect seawater from depths of 1–2 m in the western subtropical Pacific (25°N, 134°40'E). This site was chosen on the basis of the salinity and plankton content of the seawater. A total of 500 L of filtered seawater was stored in 25 hard-glass, 20 L, stoppered bottles. Temperature, salinity, and pH profiles to depths of 160 m at the sampling location as well as sea surface temperature and salinity data along with surface meteorological data obtained along the cruise track were also reported.

Miyake and Yumura (1943) reported details of preparation and calibration of the first batches of JSSW. The collected seawater was transported to the CMO, where it remained for about 2 months. The seawater from four hard-glass bottles was then transferred to a large, 80 L glass tank and stirred and mixed with a glass rod for 5 min. The seawater was then siphoned back into the original four 20 L glass bottles to make batches of the same composition. When the seawater was returned from the tank to the bottles, it was filtered through a glass filter plate (fineness no. 4). The mixed and filtered seawater was then sealed in hard-glass ampoules (volume, 200 mL). The first batch consisted of a total of about 2500 ampoules that were produced within a week.

A few ampoules were extracted from each batch, and the chlorinity of the seawater in the ampoules was determined. The chlorinity was first determined by the usual method for measuring chlorinity (silver nitrate titration [Mohr method]). This determination was followed by a precision method. The precision method was a potentiometric titration combined with a weight volumetric titration. A silver electrode was used as the indicator electrode for the precise determination of chlorinity. The electrode was immersed in seawater, and the change in electrode potential difference that occurred after adding a drop of silver nitrate (used in the silver nitrate titration) was measured with a potentiometer. At this time, the seawater and the silver nitrate solution to be added dropwise were all weighed. The error in the chlorinity of the JSSW calibrated relative to SSW (P15 for the first batches) was estimated to be ±0.002‰. Because the error of the silver nitrate titration was ±0.01–0.02‰, this precision method was an order of magnitude more precise (Miyake & Yumura, 1943).

All the JSSW ampoules were calibrated by the JSSWC and were manufactured and sold (under the supervision of the JSSWC) by Rigosha & Co. (Between 1933 and 1947, the company was renamed the Rikagaku Instrument Manufacturing Co.).

Matsudaira (1943) has discussed the effects of microorganisms on JSSW. The Marine Observatory had made a prototype of standard seawater more than a decade before the Matsudaira study. The prototype had been sterilized by boiling, but some bacterial colonies had developed in the standard seawater sealed within a glass ampoule. The chlorinity of the samples with bacterial colonies was lower than the chlorinity at the time of preparation, but the chlorinity of the samples without bacterial colonies was unchanged. Examination of the effects of microorganisms during long-term storage (culture experiments using several filtered seawater samples) was therefore conducted. The JSSW was filtered but not boiled because the JSSWC members felt that it was better not to boil the seawater. Matsudaira (1943) concluded that if the seawater was not sterilized by adding preservatives, the growth of microorganisms could not be stopped by any type of filtration. Because Kitamura (1952) detected microbial colonies on the glass walls in JSSW, a decision was made to add a preservative (mercuric chloride, vide infra) to future products.

11.4 Comparison Between JSSW and SSW

At the end of WWII, samples of JSSW were sent to the University of Washington's Department of Oceanography, the Scripps Institution of Oceanography, and the Woods Hole Oceanographic Institution for independent comparisons with SSW produced in Copenhagen and ASSW (JMA, 1975; Hanzawa, 1987; Richards, 1976). Norris Rakestraw conducted the comparisons at the Scripps Institution of Oceanography. The JSSW did not differ from the certified value by more than $\pm 0.02\%_o$ in chlorinity (Richards, 1976). Because the error of the usual method of measuring chlorinity at that time was $\pm 0.01-0.02\%_o$ (Miyake & Yumura, 1943), the chlorinity of the JSSW was confirmed to be highly accurate (JMA, 1975).

Because a measured ampoule of ASSW is stored at the JMA with the historical SSW and JSSW (Figs. 11.1 and 11.2), it is likely that Yasuo Miyake also participated in the comparisons.

The method for determination of chlorinity of JSSW using a potentiometric titration combined with a weight volumetric titration was apparently the first use of this highly precise method for preparing standard seawater. The same method was later adopted for SSW in 1969 (Hermann & Culkin, 1978), although Hermann and Culkin (1978) mentioned only the existence of JSSW. JSSW was carefully calibrated with SSW using this advanced methodology (Richards, 1976). This is one of the reasons why the chlorinities of the JSSW and SSW agreed so well (vide supra).

Fig. 11.1 A photograph of JSSWs stored at the JMA. From left to right, calibration year (chlorinity) is 1950 (19.390‰), 1950 (19.394‰), 1948 (19.410‰), 1956 (19.387‰), 1951 (19.349‰), 1953 (19.372‰), 1943 [2603 in Japan's Imperial era] (19.39‰), 1956 (19.386‰), 1950 (19.395‰), and 1951 (19.355‰). All ampoules state that they were calibrated by the JSSWC, except for the 1948 ampoule, which states that it was calibrated by the CMO. Unlike SSW, which is sealed in ampoules made of soda glass, JSSW is sealed in borosilicate glass ampoules, and hence there are few suspended particles resembling glass flakes and crystals on the inner walls of the borosilicate glass

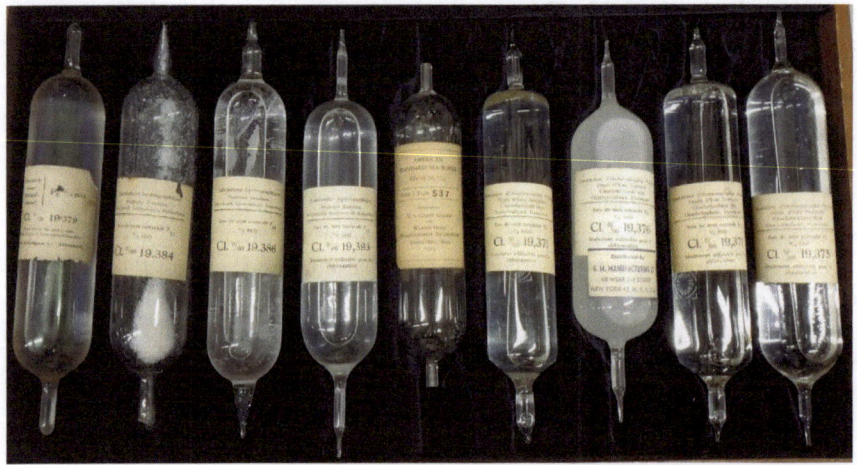

Fig. 11.2 A photograph of the historical SSWs and ASSW stored with the JSSWs shown in Fig. 11.1 at the JMA. From left to right, batch numbers (chlorinity, calibration date) are P8 (19.379‰, 15 April 1913), P13 (19.384‰, 28 June 1929) (broken ampoule without seawater), P14 (19.386‰, 21 September 1932), P15 (19.393‰, 20 June 1937), ASSW batch 1 tube 537 (19.380‰, 1941) (used ampoule without seawater), P19 (19.371‰, 6 April 1950), P18 (19.376‰, 4 December 1949) (distributed by G. M. Manufacturing Co., N.Y., U.S.A.), P19 (19.371‰, 6 April 1950), and P21 (19.375‰, 29 March 1953). The ampoule of P18 contained white suspended material, as noted by Park (1964)

11.5 Production and Use of JSSW During the Post-WWII Period

Only a few records about production of JSSW exist after the early production described in Sect. 11.3. Shimano (1982) has written a retrospective about the determination of the chlorinity of JSSW by the three governmental oceanographic organizations (JMA, JCG, and FAJ) commissioned by the JSSWC. Seto (2007) has also written a retrospective about the production and determination of chlorinity of JSSW. Collections of source seawater were conducted by the three governmental oceanographic organizations at around 30°N, south of the Kuroshio Current, through the fire hydrants on ships. The collected seawater (more than thirty 20 L glass bottles) was transported to the Rigosha & Co. laboratory. It remained there for about a year to allow remineralization of the organic matter in the water.

Francis A. Richards of the Office of Naval Research, U.S.A., visited Rigosha & Co., Ltd., around 1976 and reported details of JSSW at that time to the Office of Naval Research as follows:

> Seawater for the Japanese standards is collected near the Ogasawara Shoto (Bonin Islands), because the salinity in this area is close to the desired value of 35‰. Various ships, primarily the *Yofu Maru* and *Takuyo Maru* of the Hydrographic Division Maritime Safety Agency [now JCG], or the *Umitaka Maru* of the Tokyo University of Fisheries [now TUMSAT], collect the samples. Seventy or eighty 20-liter polyethylene bottles of water are collected at a time; these are set aside for five or six years to allow remineralization of their organic content. After this time the water is filtered through a fine pore sintered glass filter and transferred to either 100 L enameled steel or 300 L polyethylene containers and mixed….

> So that it will compare with the international standards, the final chlorinity of the secondary standards should be between 19.35 and 19.40‰, a good "average" value for the Atlantic Ocean. As the collected water is less saline than this, the chlorinity is increased by removing 10 L from every 100 L and evaporating the 10 L to a calculated volume. The concentrated water is then mixed back with the remaining 90 L. At this point the water is bottled in 20 L glass-stoppered glass containers to prevent further evaporation and poisoned with mercuric chloride (1 mL of a saturated solution for each 100 L of seawater. Of course, this causes a very small chlorinity change).

> Things are then ready for introducing the water into ampoules. The contents of $2^1/_2$ of the 20 L bottles are mixed to make a lot. While the lot is constantly stirred, the water is forced by air pressure into the open ampoules (either 100 or 200 mL…) through a thin glass tube connected to the reservoir by a rubber tube. This is when the inspections begin.

> The preparation of Japanese standard seawater is directed by a committee appointed by the Japan Science Council. It consists of some 10 members representing the Council, the Japan Meteorological Agency, the Meteorological Research Institute, the Hydrographic Division of the Maritime Safety Agency, the Ocean Research Institute of Tokyo University, and the Tokai Fisheries Research Institute, with a representative of the Rigosha Company attending meetings as an observer. Dr. Yasuo Miyake, former director of the Meteorological Research Institute and of its Geochemical Laboratory and until the end of 1977 a member of the Japan Science Council, has been Chairman of the Committee since its first formation in 1941.

> All the procedures that the company follows have been approved by the supervisory committee. Cleaning the ampoules is very important. The adopted process is simple and appears to be effective. Ampoules are filled $^1/_2$ to $^3/_4$ full of distilled water and shaken on a mechanical shaker for 20–30 min. After draining, the insides are steamed on a manifold

for another 30 min in a covered container kept a little above room temperature…. They are then oven dried at 110 °C. The filling and sealing of the ampoules are inspected by three or four representatives of the organizations supplying the membership of the supervisory committee; these inspectors are present at all times during the filling and sealing, although they rotate during the several days of the preparation of a new supply….

Following the filling of an ampoule, an inspector checks to see that no droplets have been left in the neck of the ampoule; one such droplet is cause for rejection. The rejected ampoule is emptied (the water is saved for re-use) and recleaned for another try. The neck of an accepted filled ampoule is than flamed off and again inspected. Any sign of solid material in the neck of the ampoule or in the water is cause for rejection, although if evidence of solid material in the neck disappears on tilting water into it and the neck drains clean, the ampoule is considered acceptable.

The ampoules themselves are made of special borosilicate hard glass. The glass is checked by leaching pulverized sample ampoules with distilled water and checking the salinity by potentiometric titration (after some concentration). There is little if any difference from the leachings of the ampoules used by the International Standard Seawater Service.

The chlorinity (salinity) of every 15th ampoule is determined either by a potentiometric titration in the chemical laboratory of the Japan Meteorological Agency or by inductive salinometer in the laboratories of the Hydrographic Division or the Tokai Fisheries Research Institute; titration is considered more accurate and disagreement within a lot of more than 0.003‰ in chlorinity is cause for discarding the lot. (A lot could consist of up to 260 200 mL ampoules or 530 100 mL ampoules, depending on the number of rejected.)

Ordinarily Rigosha & Co., Ltd., prepares around 6000 100 mL ampoules (for standardization by titration) and around 2000 200 mL ampoules (for standardization by conductivity) every two years. For standardization against the international standard Rigosha buys some 1000 ampoules of Normal Water each year from the Institute of Oceanographic Sciences in England, where the international Standard Seawater Service is now located, so the (secondary) Japanese Standard Seawater is well tied to the international standard…. (Richards, 1976)

With the introduction of the salinometer in the 1960s, the relationship between electrical conductivity and salinity was internationally recognized; the accuracy of salinity measurements with a salinometer was ±0.003‰, comparable to the variation in the relationship between chlorinity and conductivity (Higano, 1977; see also Fig. 9.5 of Chap. 9). When glass bottles were used to store seawater for several years before calibration, it was imagined that the pH was high near the boundary between the seawater and glass, and considerable calcium carbonate was precipitated, but this precipitation should not have been a problem at all if JSSW was used only as a chlorinity standard. However, the conductivity method has now been adopted to measure salinity, and if the calcium in seawater is precipitated as calcium carbonate, the conductivity should decrease even though the chlorinity remains constant (Higano, 1977; see also Chap. 10). In the 1970s, polyethylene bottles were used to store JSSW, but precipitation of calcium carbonate could occur even during storage in polyethylene bottles. In 1975, the JSSWC therefore decided to study the possible decrease in conductivity if the calcium in JSSW precipitated as calcium carbonate (Higano, 1977). Higano (1977) analyzed two bottles of seawater used to prepare JSSW (sixty-three 20 L polyethylene bottles) that had been collected and stored in 1972 and analyzed in 1976. The amount of calcium loss over the four-year period was estimated to be 10–20% of the change of calcium that would naturally occur in the ocean. The change during storage was less than the sensitivity of the salinometer measurement (0.0012‰ in salinity).

JSSW is believed to have been widely used in Japan during the post-WWII period until Japan's economy developed, but detailed records could not be found. Hanzawa (1986) reported that two types of JSSW (100 and 200 mL) were sold. Hirakawa and Takeishi (1985) used JSSW produced in 1980 in their laboratory study of sedimentation potential. However, during the cruise of the RV *Hakuho-maru* conducted in 1968 by the Ocean Research Institute, University of Tokyo, about 53 ampoules of SSW were used to calibrate an inductive salinometer (Horibe, 1970). JMA conducted a batch-to-batch comparison of many batches of SSW between P32 (1961) and P128 (1995) in 1996 (Aoyama et al., 2002). Recently, it was found that dozens of P51 (1969) ampoules were stored at the National Institute of Radiological Sciences (Chap. 9). Despite the paucity of records, it can be inferred that Japanese researchers at universities and government laboratories have been using SSW rather than JSSW since the 1960s in synchrony with the spread of the salinometer.

11.6 Epilogue

In 1943 during WWII, a Japanese submarine *I-8* (2231 ton) returned from Germany with a cargo manifest that included "Standard Seawater (5 ampoules)". Because it was impossible to transport goods and people between Europe and Japan by land, air, or sea during that time, submarine missions to Germany were conducted five times between 1942 and 1944, and the *I-8*, which conducted the second mission, was the only submarine to successfully complete its mission, a six-month round trip between Japan and Germany. Japanese naval experts in Germany had gathered weapons from the military and munitions companies, and the amount of equipment they had gathered was so vast that a single submarine could not transport it all to Japan. Even after carefully selecting the best of the equipment, the loading of the weapons still required ingenuity, and every conceivable space was used. Some of the bullets and torpedoes were not loaded on the submarine, and other cargo was packed in the torpedo tubes. After the death of Tadao Yokoi, a naval officer stationed in Germany who returned to Japan aboard the *I-8*, a cargo manifest was found among his belongings. The manifest listed 56 weapons and drawings (Niinobe & Sato, 1997). Toshitaka Gamo of the University of Tokyo found a fourth entry, "Standard Seawater (5 ampoules)" and speculated that the five ampoules of SSW produced in Copenhagen may have been used by Japanese oceanographers to accurately calibrate JSSW so that their oceanographic observations during WWII would not be compromised (Gamo, 2021).

In the fall of 2022, Michio Aoyama, who had planned the publication of this book, passed away suddenly. He belonged to the Meteorological Research Institute (MRI). Yasuo Miyake, who was an important contributor to the creation of JSSW, worked hard to establish the MRI. Aoyama succeeded in Miyake's spirit and devoted himself to research on the batch offset evaluation of SSW (the author has since taken over that study, see Chap. 9) and the creation of Reference Materials for Nutrients in Seawater, which originated in Japan. It is necessary to continue to convey to the world the importance of standards in oceanographic measurements. Recognition of that importance has been handed down from generation to generation in Japan. My thoughts and prayers are with Michio Aoyama.

Acknowledgements I thank Atsushi Kojima of the JMA for providing photographs of the JSSWs and historical SSWs. I also thank Shinji Masuda of the JMA for providing information about JSSW.

Competing Interests The author has no conflicts of interest to declare that are relevant to the content of this chapter.

References

Aoyama, M., Joyce, T. M., Kawano, T., & Takatsuki, Y. (2002). Standard seawater comparison up to P129. *Deep-Sea Research, I, 49*, 1103–1114. https://doi.org/10.1016/S0967-0637(02)00018-3

Dawicki, S. (2022). *Standard seawater? Yes, there is such a thing!* Northeast Fisheries Science Center, National Oceanic and Atmospheric Administration. https://www.fisheries.noaa.gov/feature-story/standard-seawater-yes-there-such-thing

Gamo, T. (2021). [The identity of the submarine that secretly submarined in the Indian Ocean during the war. What important materials did they bring back?] Senjikano indoyowo gokuhi senkoshita sensuikanno shoutai. Kareraga mochikaetta jyuuyou na busshitutowa? Gendai Business (in Japanese). https://gendai.media/articles/-/86939

Hanzawa, M. (1986). [The latest technology for ocean observation] Kaiyo kansoku no saishin gijyutu. *Journal of the Marine Engineering Society in Japan, 21*, 32–36 (in Japanese). https://doi.org/10.5988/jime1966.21.86

Hanzawa, M. (1987). A retrospective description of the famous Japanese survey and oceanographic vessels before and during the WWII. *Annual Bulletin of Maritime Museum, 15*, 27–36 (in Japanese). https://doi.org/10.24546/81005749

Hermann, F. E., & Culkin, F. (1978). The preparation and chlorinity calibration of standard seawater. *Deep-Sea Research, 25*, 1265–1270. https://doi.org/10.1016/0146-6291(78)90020-6

Higano, R. (1977). On the calcium concentration of the Japanese Standard Sea Water. In *Report of hydrographic researches, No. 12*, (pp. 121–124) (in Japanese with English abstract). https://www1.kaiho.mlit.go.jp/GIJUTSUKOKUSAI/KENKYU/report/rhr12/rhr12-07.pdf

Hirakawa, H., & Takeishi, T. (1985). Measurement of sedimentation potential of sea water. *La Mer, 23*, 118–122 (in Japanese with English abstract). http://www.sfjo-lamer.org/lamer.html

Horibe, Y. (1970). *Preliminary report of the Hakuho Maru Cruise KH-68-4 (Southern Cross Cruise)* (p. 170). Ocean Research Institute, University of Tokyo.

Jacobsen, J. P., & Knudsen, M. (1940). Urnormal 1937 or primary standard sea-water 1937. *Association d'Océanographie Physique, Publication. Scientifique, 7*, 5–38. https://iapso-ocean.org/images/stories/pdf/IAPSO_publications/Publications_Scientifiques/Pub_Sci_No_7.pdf

Japan Meteorological Agency. (1975). [A centennial history of meteorology] Kishou Hyakunenshi. *II, Section, 15*, 466 (in Japanese).

Kitamura, H. (1952). The microbes in the standard sea water (1). *Journal of Meteorological Research, 4*, 113–115 (in Japanese).

Kuroda, K. (2020). The process to the establishment of "the Oceanographic Society of Japan" from its mother "the Colloquium on Oceanography" and the first decade afterwards. *Oceanography in Japan, 29*, 37–53 (in Japanese with English abstract). https://doi.org/10.5928/kaiyou.29.2_37

Li, Y., Luo, Y., Kang, Y., Yu, T., Wang, A., & Zhang, C. (2016). Chinese primary standard seawater: Stability checks and comparisons with IAPSO standard seawater. *Deep-Sea Research, I*, 101–106. https://doi.org/10.1016/j.dsr.2016.04.005

Masuda, Y. (1984). Report on the 96th general meeting of the science council of Japan. *Tenki, 12*, 46 (in Japanese). https://www.metsoc.jp/tenki/pdf/1984/1984_12_0756.pdf

Matudaira, Y. (1943). The preparation of the standard sea water in Japan. Part 3. The micro-organisms remained in the filtered sea-water. *Journal of Oceanographic Society of Japan, 1*, 1–9 (in Japanese). https://doi.org/10.5928/kaiyou1942.2.2_1

Miyake, Y. (1941). Tokyo−Chichijima−Ioujima [Record of sampling of the standard seawater] Hyoujun kaisui saisuiki. *Kaiyo no Kagaku, 1*, 41–47 (in Japanese).

Miyake, Y., & Yumura, Y. (1943). The preparation of the standard sea water in Japan. Part 2. The determination of the chlorinity. *Journal of Oceanographic Society of Japan, 2*, 29–33 (in Japanese). https://doi.org/10.5928/kaiyou1942.2.29

Niinobe, A., & Sato, H. (1997). [Missing submarine I-52] Kieta sensuikan I52 (p. 251). NHK Publishing, Inc. ISBN: 4-14-080307-X (in Japanese).

Park, K. (1964). Reliability of standard sea water as a conductivity standard. *Deep-Sea Research, 11*, 85–87. https://doi.org/10.1016/0011-7471(64)91084-8

Richards, F. A. (1976). Japanese standard seawater. *Office of Naval Research Scientific Bulletin, 1*, 602–604.

Ridout, P. (2011). Developments in the standardization of ocean salinity. *Science in Parliament, 68*, 7–9. http://www.scienceinparliament.org.uk/wp-content/uploads/2013/09/sip68-2-3.pdf

Seto, Y. (2007). [Methods for measuring salinity of seawater] Kaisui-chu no enbun sokuteihou. *Bulletin of the Japan Marine Surveys Association, 87*, 10–17 (in Japanese). https://www.jamsa.or.jp/images/technical/pdf/087.pdf

Shimano, T. (1982). Salinity determination and standard sea water in oldtimes. *The Suiro (Hydrography), 11*(3), 38–40 (in Japanese). https://www.jha.or.jp/jp/shop/products/suiro/pdf/suiro043.pdf

Tominaga, H., Isibasi, M., Hidaka, K., Matudaira, Y., & Miyake, Y. (1942). On the preparation of the standard sea-water in Japan. Part 1. The course of sampling of water. *Journal of Oceanographic Society of Japan, 1*, 95–102 (in Japanese). https://doi.org/10.5928/kaiyou1942.1.95

UNESCO. (1981). The practical salinity scale 1978 and the international equation of state of seawater 1980. In *UNESCO technical papers in marine science* (Vol. 36, p. 25).

Chapter 12
Development of Multiparametric Standard Seawater (MSSW) for CO_2 Parameters, Dissolved Oxygen, and Density of Seawater

Hiroshi Uchida, Akihiko Murata, Masahide Wakita, Hitoshi Mitsuda, Yasuhiro Nagasawa, Tatsuya Tanaka, Yohei Kayukawa, Kazuhiko Takeda, Kazuaki Ito, Takeshi Yoshimura, and Daisuke Sasano

Abstract Multiparametric Standard Seawater (MSSW) is being developed based on the technology used to manufacture the reference material for nutrients in seawater (RMNS), without adding mercuric chloride to sterilize it, but by adopting aluminum bottles and plastic inner caps with high impermeability to gas and water vapor. The

H. Uchida (✉) · A. Murata
Global Ocean Observation Research Center, Japan Agency for Marine-Earth Science and Technology, Yokosuka, Japan
e-mail: huchida@jamstec.go.jp

A. Murata
e-mail: murataa@jamstec.go.jp

M. Wakita
Mutsu Institute for Oceanography, Japan Agency for Marine-Earth Science and Technology, Mutsu, Japan
e-mail: mwakita@jamstec.go.jp

H. Mitsuda
Laboratory for Instrumentation and Analysis, KANSO TECHNOS Co., Ltd., Katano, Japan
e-mail: rminfo@kanso.co.jp

Y. Nagasawa
JFE Advantech Co., Ltd., Nishinomiya, Japan
e-mail: nagasawa@jfe-advantech.co.jp

T. Tanaka
Marine Works Japan Ltd., Yokosuka, Japan
e-mail: tanakats@mwj.co.jp

Y. Kayukawa
National Metrology Institute of Japan, National Institute of Advanced Industrial Science and Technology, Tsukuba, Japan
e-mail: kayukawa-y@aist.go.jp

K. Takeda
Graduate School of Integrated Sciences for Life, Hiroshima University, Higashi-Hiroshima, Japan

M. Aoyama et al. (eds.), *Chemical Reference Materials for Oceanography*, Springer Oceanography, https://doi.org/10.1007/978-981-96-2520-8_12

history, current status, and future plans for the development of MSSW for measurements of Practical Salinity, density, dissolved inorganic carbon, total alkalinity, pH, dissolved oxygen (DO), and dissolved organic matter are discussed. Substances that interfere with the Winkler method for DO determination (nitrite, iodate, and hydrogen peroxide) were evaluated for MSSW. The values of the parameters of interest were relatively homogeneous, but the concentrations of dissolved organic carbon depended on the serial number in the lot tested. Long-term stabilities were good in most cases, but DO concentrations began to gradually decrease immediately after production and appeared to stabilize a few years later. The Practical Salinity also tended to decrease (-0.00015 year^{-1}), and the cause of the decreasing trend urgently needs to be clarified. One possibility is an increasing trend (about $+0.00015$ year^{-1}) of the Practical Salinity of IAPSO standard seawater for salinity measurements.

Keywords Reference seawater · Non-toxic · CO$_2$ parameters · Dissolved oxygen · Salinity · Density

12.1 Introduction

The accuracy of oceanographic measurements depends on calibration against reference materials using standard equipment to ensure comparability over time and among laboratories (National Research Council of the National Academies, 2002). However, recent scientific challenges in evaluating changes in the ocean caused by global climate change from observational data suggest that the continued development of reference seawaters, both for higher precision in intercomparisons and for use in comparing a wider variety of measurement parameters is essential. The following seawater reference materials have long been used: the International Association for the Physical Sciences of the Oceans (IAPSO) Standard Seawater (SSW) (Ocean Scientific International, Ltd., Havant, Hampshire, United Kingdom) for Practical Salinity (S_P) measurements (Chap. 8), a certified reference material (Dickson-RM) provided by Prof. A. G. Dickson (Scripps Institution of Oceanography, University of California, San Diego) for measurements of dissolved inorganic carbon (DIC) and total alkalinity (TA) (Dickson, 2010), and a consensus reference material provided

e-mail: takedaq@hiroshima-u.ac.jp

K. Ito
Seawater Assessment Technologies Research Institute, Hiroshima, Japan
e-mail: itok1481@gmail.com

T. Yoshimura
Faculty of Fisheries Sciences, Hokkaido University, Sapporo, Japan
e-mail: yoshimura-t@fish.hokudai.ac.jp

D. Sasano
Japan Meteorological Agency, Tokyo, Japan
e-mail: daisuke_sasano@met.kishou.go.jp

by Prof. D. A. Hansell (University of Miami) for measurements of dissolved organic carbon (DOC) (Chap. 15). In addition to these sources, reference materials for nutrients in seawater (RMNS) (KANSO TECHNOS Co., Ltd., Osaka, Japan), which originated in Japan, have recently become popular worldwide (Aoyama et al., 2010a).

In contrast, it has been difficult to discuss the comparability of measurements of dissolved oxygen (DO) and pH, which have been foci of attention in recent years because of concerns about ocean hypoxia and acidification, because no standard seawater exists for these parameters (Stendardo et al., 2009; Velo et al., 2010). There are also problems with extant standard seawaters. For example, because mercuric chloride is added to Dickson-RM to sterilize it for DIC measurements, international distribution of that seawater may be at risk if mercury regulations are tightened in the future. It has also been pointed out that the certified values of IAPSO SSW for conductance-based S_P measurements are not traceable to the International System of Units (SI) (Pawlowicz et al., 2016; Seitz et al., 2011) and that there are small but measurable offsets between batches of certified values (Chap. 9).

Partly to address the issues with conductance-based salinity measurements, there has been a proposal to ensure the traceability of salinities derived instead from density measurements to the SI (Pawlowicz et al., 2016). However, although the oscillation-type density meter used to routinely measure seawater density is usually calibrated with air and pure water (e.g., Wolf, 2008), measured densities of SSW tend to be lower than the densities calculated by the Thermodynamic Equation of Seawater 2010 (TEOS-10) (Intergovernmental Oceanographic Commission [IOC] et al., 2010), and the deviations vary widely from paper to paper. The density discrepancy has been reported to be about -0.004 kg m^{-3} at 20 °C by Feistel et al. (2010), -0.006 kg m^{-3} at 20 °C by Uchida et al. (2011a), -0.001 kg m^{-3} at 25 °C by Woosley et al. (2014), -0.016 kg m^{-3} at 20 °C by Schmidt et al. (2018), and -0.003 kg m^{-3} at 20 °C and -0.008 kg m^{-3} at 25 °C by Romeo et al. (2019). Most of these inconsistencies may have been due to differences in the nonlinearities of the density meters as their use for seawater densities requires an extrapolation of the calibration. We have also confirmed a case where the density discrepancy derived from the density meter used by Uchida et al. (2011a) changed from -0.006 to -0.011 kg m^{-3} after four years of use. It may therefore be that oscillation-type density meters should be calibrated with pure water and an appropriate standard seawater to reduce or eliminate the extrapolation, but there is a problem with the SSW density's changing over time due to compositional changes (increase of silicic acid and decrease of DIC, Chap. 10).

The RMNS production technique (sterilization by giant autoclave) can be used to stop biological activity in standard seawater without the addition of mercuric chloride. If the exchange of gas and water vapor with the ambient air can be prevented, a standard seawater that maintains the matrix of natural seawater and whose nutrient concentrations, carbonate system parameters, salinity, and other components are consistent with each other can be realized. We are therefore developing a new reference material, Multiparametric Standard Seawater (MSSW), based on the technology used to manufacture RMNS.

In this study, we describe the history, current status, and future plans for the development of MSSW.

12.2   History of Development of MSSW

The development of non-toxic reference seawater for carbonate system parameters (DIC and TA) which began in 2003 with lots Pre1–11, the predecessor of MSSW, was conducted based on the technology for manufacturing RMNS (sterilization by large-volume autoclave) (Ota et al., 2010), through contract research with the Kansai Electric Power Co., Inc. (Osaka, Japan) to KANSO TECHNOS (2003–2004, principal investigator [PI]: Hidekazu Ota) and a study "A study on international reference materials for carbon and nutrients cycles" (2005–2008, PI: Michio Aoyama of Meteorological Research Institute, Japan Meteorological Agency [JMA]) supported by the Ministry of Education, Culture, Sports, Science and Technology (MEXT) KAKENHI Grant 17310015.

We initially considered glass bottles or polypropylene (PP) bottles as containers, the same bottles used for RMNS (Fig. 12.1a–d). However, the results were not satisfactory and we subsequently achieved stability of the DIC for about 300 days by: connecting three stainless steel (SUS316) 230-L reaction chambers (Fig. 12.2) to maintain a constant water level (i.e., head space) in the lead chamber of the initial seawater sample, by improving the bottling method, by using 500-mL aluminum beverage bottles (New Bottle Can, Daiwa Can Company, Chiyoda, Tokyo, Japan) to contain the reference seawater, and finally by enclosing the bottles in degassed and sealed aluminum bags (HRS-1725 S, Meiwa Pax Co., Ltd., Kashiwara, Osaka, Japan) (Fig. 12.1e). In these containers, the standard deviation (SD) of the DIC concentration increased by only 3 $\mu mol\ kg^{-1}$ and the DIC concentration decreased by about 4 $\mu mol\ kg^{-1}$ after 300 days (Murata, 2010).

We then considered an extension of this approach to produce non-toxic reference seawater for DO measurements with lots Pre12–13 through contract research from Kansai Electric Power to KANSO TECHNOS (2008–2009, PI: Hitoshi Mitsuda). The method of sterilizing aluminum bottles was changed from sterilization with heat to sterilization with gamma rays. The DO concentration was stable for two months after production, but after five months, the DO concentration decreased by 3% (Mitsuda et al., 2010).

The development of non-toxic reference seawater for DO measurements, lots Pre14–18, was continued through cooperative research: "Development of standard dissolved oxygen sensors and reference materials for dissolved oxygen measurement and for dissolved oxygen sensor calibration" (2010–2013) and "Improvement of standard dissolved oxygen sensors and evaluation of long-term stability of reference materials for dissolved oxygen measurement" (2014) by KANSO TECHNOS (PI: Hitoshi Mitsuda), JFE Advantech Co., Ltd. (Nishinomiya, Hyogo, Japan) (PI: Yasuhiro Nagasawa), and the Japan Agency for Marine-Earth Science and Technology (JAMSTEC) (PI: Hiroshi Uchida).

Because it was suspected that the stability of DO was reduced because of changes in integrity of the screw cap liner caused by opening aluminum caps for empty and pre-capped bottles, bottling the seawater, and re-capping by hand (Mitsuda et al., 2010), we introduced a capping machine (Fig. 12.2d) beginning with the production

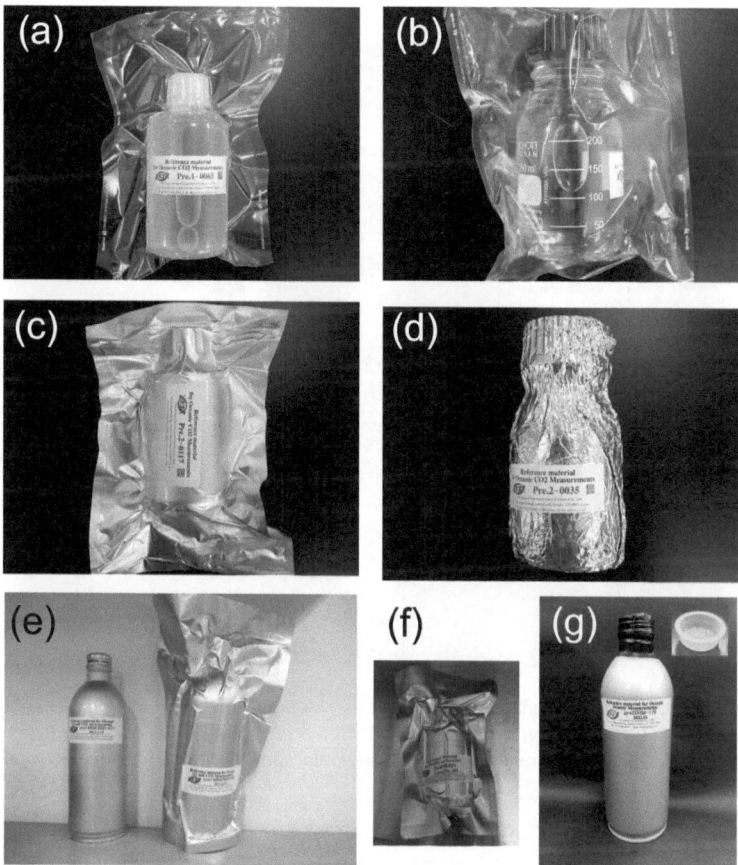

Fig. 12.1 Photographs of prototypes of non-toxic reference seawater. **a** Pre1 (a 100 mL polypropylene [PP] bottle sealed with a two-layer [nylon and polyethylene] film bag), **b** Pre1 (a 200 mL borosilicate glass bottle [Schott Duran] sealed with the two-layer film bag), **c** Pre2 (a 100 mL PP bottle sealed with an aluminum bag), **d** Pre2 (a 200 mL borosilicate glass bottle [Schott Duran] sealed with aluminum foil), **e** Pre14 (a 500 mL aluminum bottle [left] sealed with an aluminum bag [right]), **f** DnRM-1 (a 100 mL perfluoroalkoxy [PFA] fluoroplastic bottle sealed with double-wrapped aluminum bags), and **g** MSSW Pre20 (a 500 mL aluminum bottle with a polychlorotrifluoroethene [PCTFE] inner cap [upper right corner] sealed with butyl rubber tape)

of lot Pre14 in 2010, but this change did not improve the stability of the DO. From lot Pre16 (produced in 2012), UV-sterilized plastic inner caps with high gas and water vapor impermeability (Fig. 12.1g) were therefore adopted for the aluminum bottles. After autoclaving the initial seawater stored in large containers in a giant autoclave (Fig. 12.2b), the containers were transferred into a clean room and cooled. By connecting three (two 320-L and one 230-L) chambers (Fig. 12.2), seawater was stirred with circulation (open to the atmosphere) for 24 h before bottling. The filtered and sterilized seawater was dispensed into the aluminum bottles from the leading

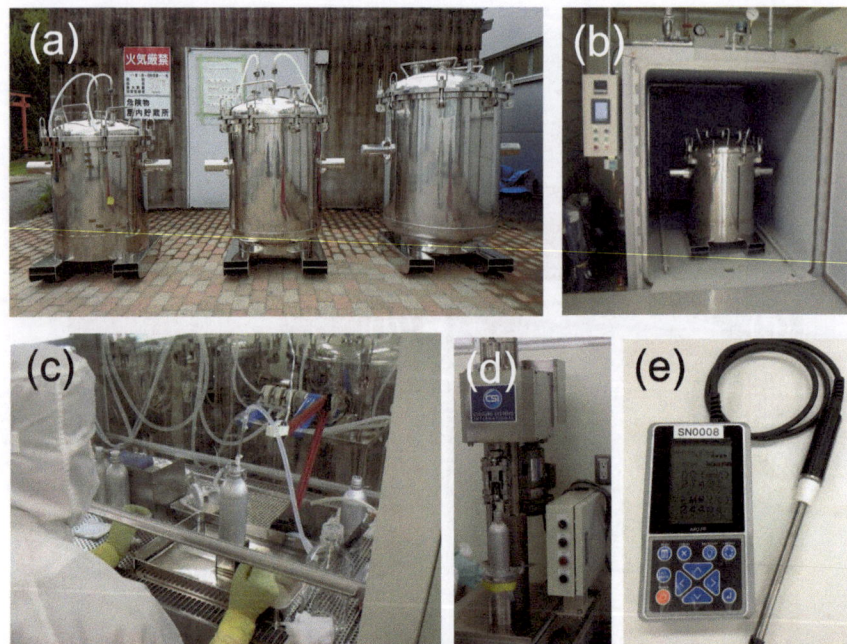

Fig. 12.2 Photographs of **a** stainless steel (SUS316) reaction chambers (left: 230-L, middle: 320-L, right: 500-L), **b** the reaction chamber (230-L) in the giant autoclave, **c** a clean bench in the clean room, **d** a capping machine for the aluminum bottle, and **e** the portable optical oxygen sensor (model ARO-PR)

chamber on a clean bench (Fig. 12.2c). After overflowing 500 mL of seawater, the aluminum bottle filled with seawater was sealed with a plastic inner cap without air space and capped with an aluminum cap. A total of 310 bottles of Pre16 and 57 bottles of DnRM-2 (300 mL aluminum bottle for density measurements described below) were produced. With the new caps, the stability of DIC and DO was greatly improved (Sects. 12.4 and 12.7) without the additional use of degassing-packaging aluminum bags.

A portable, optical DO sensor (model ARO-PR, JFE Advantech) (Fig. 12.2e) was also developed through the cooperative studies mentioned above and used to monitor the stability of the DO in the reference seawater. The DO sensor was calibrated with SI-traceable standard gases. Unlike electrode DO sensors, the optical DO sensor we developed had the advantage of not consuming DO in the sample at the DO-sensing element and could be inserted directly into the aluminum bottle to measure the DO of the reference seawater. Unlike DO measurements using the Winkler method, the DO measurements made by the sensor were unaffected by the substances that interfere with the Winkler method such as nitrite, iodate, and hydrogen peroxide (see Sect. 12.6).

The development of reference seawater for density measurements was conducted in parallel through cooperative research: "Development of seawater density reference materials" (2011–2012) by Marine Works Japan, Ltd. (MWJ) (PI: Tatsuya Tanaka) and JAMSTEC (PI: Hiroshi Uchida). Based on the technology used to manufacture RMNS, we examined various containers (plastic and aluminum bottles) and various packaging bags (plastic and aluminum bags) suitable for density measurements of reference seawater. Based on the results of the examination, we produced lot DnRM-1 by adopting 100-mL perfluoroalkoxy (PFA) fluoroplastic bottles for containers with degassing packaging double aluminum bags (MAL-1320 S, Meiwa Pax) (Fig. 12.1f). However, the homogeneity of silicic acid concentrations was poor, and salinity increased because of evaporation of water. We then produced lot DnRM-2 at the same time as lot Pre16. The only difference between Pre16 and DnRM-2 was the volume of the container (500 mL for Pre16 and 300 mL for DnRM-2). There was little change in the S_P of Pre16 and DnRM-2 during storage for seven years (Uchida et al., 2020). Uchida et al. (2020) did not distinguish between Pre16 and DnRM-2 but referred to both as Pre16.

Finally, from lot Pre16, instead of making separate reference seawaters for different parameters in independent projects, the goal was set to produce a single reference seawater, MSSW (stored in 500 mL aluminum bottles), which could be used for measurement of multiple analytical properties such as DIC, TA, pH, DO, nutrient concentrations, S_P, and density (or Absolute Salinity: S_A).

To establish traceability of the density of MSSW to SI with small uncertainty, we developed a hydrostatic weighing apparatus using a silicon single-crystal sinker through contract research: "Research on absolute measurement of seawater density" from JAMSTEC to the National Metrology Institute of Japan (NMIJ), National Institute of Advanced Industrial Science and Technology (AIST) (2012, PI: Yohei Kayukawa). The density of that water is directly traceable to the density of the Japanese National Standard for seawater density measurements. Currently, the NMIJ is offering density-calibration services for aqueous solutions, including seawater, using the developed hydrostatic weighing apparatus.

For lot Pre18, which was produced in 2013 in the same way as lot Pre16, the DO had become less homogeneous than the DO of lot Pre16, probably because of the connection of three reaction chambers, which may have been associated with changes of the internal pressure of the leading chamber during bottling. Three hundred thirty bottles of lot Pre19 were therefore produced using only one large-capacity (500 L) chamber (Fig. 12.2a) to improve homogeneity by reducing the amount of overflowing water from 500 mL to 250 mL as part of the study "Clarification of global warming based on changes in the deep ocean" (2018–2020, PI: Hiroshi Uchida of JAMSTEC) supported by the Japan Society for the Promotion of Science (JSPS) KAKENHI Grant 18K03752. In addition, the size of the plastic inner cap was slightly modified, and the aluminum screw cap was wrapped with butyl rubber to improve stability (Fig. 12.1g). JAMSTEC, KANSO TECHNOS, JMA, NMIJ, and Hiroshima University collaborated to evaluate the homogeneity and stability of the DO, DIC, TA, pH, nutrient concentrations, S_P, density (or SA), DOC, and concentrations of iodate and hydrogen peroxide.

Most recently, to evaluate the certified S_P values of IAPSO SSWs based on the definition of PSS-78, we produced 333 bottles of lot Pre20 in 2022. Some of those bottles were produced as a part of the study "Understanding climate change from freshening of deep ocean" (2021–2025, PI: Hiroshi Uchida of JAMSTEC) supported by the JSPS KAKENHI Grant 21H01165.

12.3 Materials and Methods

12.3.1 Production of MSSWs

Table 12.1 summarizes the reference seawaters produced since 2011, with lots Pre16 and later being considered prototypes for MSSW, although some of the stability results described below also involved measurements in earlier lots. The source seawater for these references was obtained from the Ogasawara Seaweed Research Association (https://sea-water.jp) for lot Pre14 and Suruga Bay Deep Water Intake and Supply Facility (https://www.pref.shizuoka.jp/sangyoshigoto/suisan/shinsosui/) for lots Pre15 and thereafter. This water has a typical TEOS-10 Salinity Anomaly of 0.008 g kg^{-1} (see below).

The process used to manufacture the most recent lot (Pre20) is described below as an example. The source seawater was collected in 20 L low-density polyethylene (LDPE) bags (Union Container, As One, Co., Ltd., Osaka, Japan) and transported to the RMNS production facility (KANSO TECHNOS) on the same day (18 November 2021). The seawater collected in the LDPE bags was filtered (0.45 μm membrane filter) into new LDPE bags (18–19 November 2021) and stored at room temperature (18–25 °C). The filtered seawater was placed in a 500-L stainless steel reaction chamber (Fig. 12.2a), and a magnetic stir bar (polytetrafluoroethylene [PTFE]) was also placed in the chamber before closing the lid.

The reaction chamber was autoclaved two times in a giant autoclave (Fig. 12.2b) at 120 °C for two hours (22 and 24 November 2021). The reaction chamber was then

Table 12.1 List of lots of reference seawater produced, with lots Pre16 and later being considered as MSSW. The asterisk indicates lot DnRM-2 (300 mL version of lot Pre16)

Lot no.	Production date	No. of bottles	Source seawater
Pre14	19 January 2011	350	0 m (Chichijima, Ogasawara)
Pre15	4 August 2011	60	397 m (Suruga Bay, Shizuoka)
Pre16	7 September 2012	310 + 70*	397 m (Suruga Bay, Shizuoka)
Pre17	28 February 2013	60	397 m (Suruga Bay, Shizuoka)
Pre18	4 October 2013	390	397 m (Suruga Bay, Shizuoka)
Pre19	20 February 2019	334	397 m (Suruga Bay, Shizuoka)
Pre20	12 January 2022	333	397 m (Suruga Bay, Shizuoka)

transferred to a clean room (Fig. 12.2c) and cooled to room temperature. The seawater in the reaction chamber was mixed with the magnetic stirrer and a mixing inline peristaltic pump with an inline filter (0.22 μm pore size) connected with autoclaved tubes (see Mitsuda et al., 2010) for 24 h (11–12 January 2022) until just before the bottling of MSSW (12 January 2022). The reaction chamber was open to the atmosphere during the mixing and bottling, and a silicon tube was used to draw the inlet air from a class 100 clean bench (Fig. 12.2c).

The aluminum bottles, aluminum caps, and inner caps (Fig. 12.1g) to be used were washed with ultrapure water. After the aluminum bottles had been sterilized with gamma rays and the aluminum caps and inner caps sterilized with ultraviolet (UV) light, they were transferred to the clean room.

Before bottling, a new inline filter was installed, and the seawater was pumped from the reaction chamber into the aluminum bottles on the clean bench in the clean room (Fig. 12.2c). The seawater used to fill an aluminum bottle was allowed overflow the bottle by half the volume of the bottle (250 mL). The inner cap was then emplaced in the aluminum bottle without an air space. The capping machine was then used to emplace the aluminum cap over the inner cap (Fig. 12.2d), and the cap was wrapped with butyl rubber (Fig. 12.1g).

12.3.2   Analytical Methods and Instruments

The S_P was measured with a salinometer (Autosal model 8400B; Guildline Instruments, Ltd., Smiths Falls, Ontario, Canada). We used P-series IAPSO SSW to calibrate the salinometer. The S_P was measured at 24 °C following the method of Kawano (2010) or Uchida et al. (2020).

We measured the density with an oscillation-type density meter (model DMA 5000M with an Xsample 122 sample changer; Anton-Paar GmbH, Graz, Austria). We calibrated the density meter with pure water and P-series SSW, and we measured the density at 20 °C following the method of Uchida et al. (2011a). We used ultrapure water (Milli-Q water from Millipore Sigma) or pure water (Pure Water [water hardness 0], Ako Kasei Co. Ltd., Ako, Hygo, Japan) made from seawater collected from a depth of 344 m off Muroto, Kochi, Japan, by filtering twice with a reverse osmosis membrane. In addition, we deionized the Pure Water using an ion exchange resin (Pure Maker, Sanei Corp., Arao, Kumamoto, Japan).

Concentrations of most of the nutrients in our prototype reference seawaters (nitrate, nitrite, silicic acid, phosphate, and ammonium), TA, DIC, pH, and DO were measured on the Global Ocean Ship-based Hydrographic Investigations Program (GO-SHIP) cruises by R/V *Mirai* (e.g., Uchida et al., 2011b; Uchida et al., 2021) or on cruises equivalent to GO-SHIP by JAMSTEC or JMA. Some of the data were measured in a laboratory at JAMSTEC or KANSO TECHNOS by using systems identical to or similar to those on R/V *Mirai*.

Nutrient concentrations were measured with a continuous flow analyzer (TRAACS 800 system [in 2010]; BRAN + LUEBBE, or model QuAAtro 2-HR

system [since 2011]; BL TEC K.K., Osaka, Japan). These concentrations were calibrated by using reference material for nutrients in seawater (RMNS) and certified
reference material (CRM) for nutrients (KANSO TECHNOS).

The TA was measured based on spectrophotometry with a custom-made system
(Nippon ANS, Inc., Tokyo, Japan), and the DIC was measured with a coulometric
system that measured total CO_2 (model 3000; Nippon ANS, Inc.). The TA and DIC
values were calibrated by using certified reference material (Dickson-RM) provided
by Prof. A. G. Dickson (Scripps Institution of Oceanography, University of California San Diego) or CO_2 in seawater reference material (KANSO-RM), which was
prepared by KANSO TECHNOS using a method similar to the method used to
produce the Dickson-RM, and the values of which were determined by measuring
the TA and DIC relative to the Dickson-RM.

We measured the pH with a pH-measuring system (Nippon ANS, Inc.) based on
spectrophotometry. The hydrogen ion concentration exponent at 25 °C was determined on the total hydrogen ion concentration scale with the spectrophotometric
technique using the indicator dye m-cresol purple (Clayton & Byrne, 1993).

The DO was measured based on a modified Winkler titration method (Carpenter,
1965a, 1965b). We used an automatic photometric titrator DOT-01X or DOT-15X
(Kimoto Electronics Co., Ltd., Osaka, Japan). The DO concentration was determined
with the addition of azide to remove the effect of nitrite interference only at KANSO
TECHNOS.

We used a portable optical DO sensor (Fig. 12.2e) to measure DO in MSSW. We
calibrated the DO sensor using O_2/N_2 Gravimetric Mixture Standard Gases (Specialty
grade gas, Taiyo Nippon Sanso CO., Tokyo, Japan) (see Uchida et al., 2021 for more
details).

The concentrations of DOC and total dissolved nitrogen (TDN) were measured
with a total organic carbon analyzer (model TOC-V or TOC-L; Shimadzu Co., Kyoto,
Japan). The DOC and TDN values were calibrated following the method of Ogawa
et al. (2003) (see also Wakita et al., 2016) using consensus reference material provided
by Prof. D. A. Hansell (University of Miami). The dissolved organic phosphorus
(DOP) concentration was estimated as the difference between total dissolved phosphorus (TDP) and phosphate (PO_4^{3-}) concentrations. Samples for TDP analysis were
autoclaved in an acid potassium persulfate solution at 123 °C for 120 min (Hansen &
Koroleff, 1999; Ridal & Moore, 1990). We measured the concentration of PO_4^{3-}
with the molybdenum blue method (Hansen & Koroleff, 1999).

The concentration of iodate (IO_3^-) was determined by two ion chromatographic
methods with UV detection. One concentration was equated to the difference between
the total inorganic iodine ($IO_3^- + I^-$) and iodide (I^-) in seawater. Total inorganic
iodine was determined after conversion from iodate to iodide. Iodide was directly
determined (Ito et al., 1991, 2003). The other concentration was determined with
a dodecylammonium-coated ODS column (Takeda et al., 2017) with UV detection (UV-2075 plus; JASCO Co., Hachioji, Tokyo, Japan). We used chemicals and
solvents of reagent grade or high-performance liquid chromatography-grade.

The concentration of hydrogen peroxide (H_2O_2) was measured by a sensitive and
simple method proposed by Takeda et al. (2018). Terephthalate (TP), which reacts

with the hydroxyl radical generated from H_2O_2 and Fe(II), was used to quantitatively form a strongly fluorescent 2-hydroxyterephthalate (HTP), and the HTP formed was analyzed with an isocratic high performance liquid chromatography (HPLC) system (LC-980 series; JASCO). We used chemicals and solvents of reagent grade or high-performance liquid chromatography-grade, and hydrogen peroxide (30 %) acquired from Santoku Chemical Industries Co., Ltd., Chuo, Tokyo, Japan, was used as a reference.

12.4 Improvement of MSSW

12.4.1 Homogeneity

The homogeneities of the DO, DIC, and TA of lots Pre14, Pre16, Pre18, and Pre19 were examined initially (Table 12.2). Coefficients of variation for the concentrations of DO were as high as 0.34% for all lots beginning with lot Pre14, except for lot Pre18, with a value of 0.85%. For lots prior to Pre19, three reaction chambers were connected and used for the production of MSSW. The large coefficient of variation for lot Pre18 might have been caused by pressure changes in the leading chamber during production. This problem might have been solved by the use of a single, large-capacity (500-L) chamber. Although the coefficient of variation of DO for lot Pre14 was small (0.28%), the DO concentration increased by 0.8% after production, judging from comparisons with the DO concentration of samples collected into oxygen flasks during the course of production of MSSW (Table 12.2). This problem was solved by using plastic inner caps with high impermeability to gas and water vapor beginning with lot Pre16.

Coefficients of variation of the initial concentrations of DIC and TA were as much as 0.09% and 0.06%, respectively, for all lots beginning with lot Pre14.

For lot Pre19, however, there were intra-lot concentration gradients in silicic acid (Fig. 12.3) and DOC (Fig. 12.4) that depended on serial number. The silicic acid

Table 12.2 Homogeneity of initial concentrations of DO, DIC, and TA for lots Pre14, Pre16, Pre18, and Pre19. Averages ± SDs are shown. Coefficients of variation (%) and numbers of bottles measured are shown in parentheses. DO concentrations with asterisks were not concentrations in MSSW but in samples collected into oxygen flasks periodically during production of MSSW

Lot no.	DO [μmol kg^{-1}]	DIC [μmol kg^{-1}]	TA [μmol kg^{-1}]
Pre14	214.66 ± 0.66 (0.28, 14) 212.95 ± 0.32* (0.15, 14)	1578.19 ± 1.03 (0.07, 14)	1705.03 ± 0.77 (0.05, 14)
Pre16	215.87 ± 0.74 (0.34, 10) 215.71 ± 0.69* (0.32, 25)	2226.06 ± 1.58 (0.07, 10)	2303.05 ± 1.29 (0.06, 10)
Pre18	219.76 ± 1.88 (0.85, 23)	2213.14 ± 1.42 (0.06, 10)	2310.85 ± 1.01 (0.04, 10)
Pre19	212.95 ± 0.32* (0.32, 15)	2140.54 ± 1.84 (0.09, 5)	2303.16 ± 0.62 (0.03, 5)

concentration gradient might have been eliminated by accelerated testing because some of the bottles in lot Pre19 were stored at 40 °C for one month. The DOC concentration gradient appeared to remain the same over time.

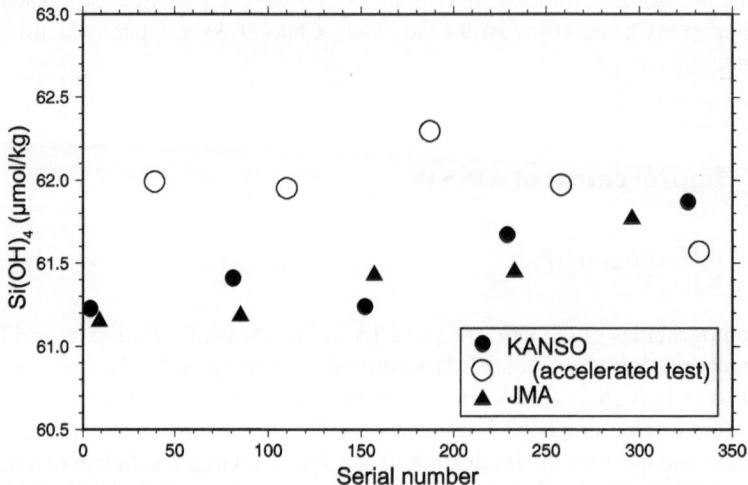

Fig. 12.3 Silicic acid ($Si(OH)_4$) concentrations in MSSW lot Pre19 plotted against serial number. 334 bottles were produced for lot Pre19

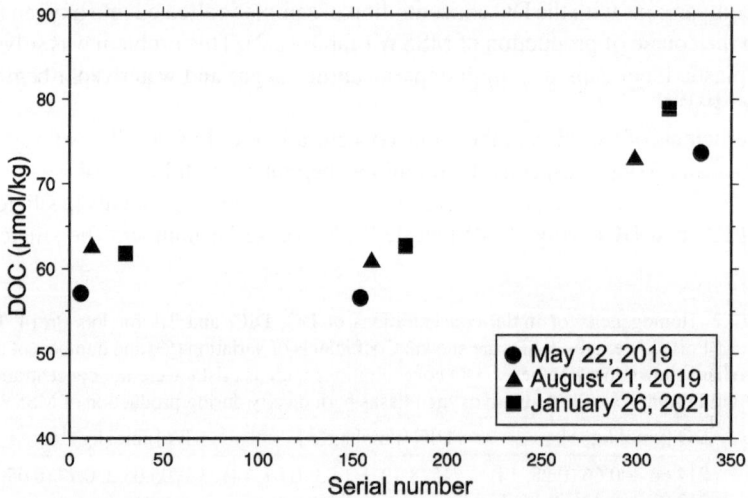

Fig. 12.4 Dissolved organic carbon (DOC) concentrations in MSSW lot Pre19 plotted against serial number. 334 bottles were produced for lot Pre19

12.4.2 Accelerated Testing

For lots Pre14 and Pre16, we conducted storage tests (1 month) for DO, DIC, and TA at low (4 °C) and high (50 °C) temperatures as well as shaking tests (24 h) and low-pressure tests (24 h) with transportation in mind. We observed the following significant changes before and after these tests. For lot Pre14, DIC concentrations increased from 1578 ± 1.0 to 1582.1 ± 1.9 µmol kg^{-1} during storage at high temperature. The DO concentration of lot Pre14 decreased from 214.7 ± 0.6 to 199.1 ± 2.3 µmol kg^{-1} and increased to 218.6 ± 2.3 µmol kg^{-1} during storage tests at high and low temperature, respectively. For lot Pre16, the DO concentration decreased from 213.9 ± 0.7 to 209.9 ± 0.4 µmol kg^{-1} during storage at high temperature.

Considering that the air-saturated solubilities of DO at an S_P of 34 are 143, 208, and 317 µmol kg^{-1} at temperatures of 50, 25, and 4 °C, respectively, some of the changes in lot Pre14 could be attributed to gas exchange with ambient air because of failure to use plastic inner caps with high impermeability to gas and water vapor. In contrast, for lot Pre16, the DO concentration after the storage test at high temperature was close to the equilibrium value of DO after long-term storage, as described in Sect. 12.7.3.

The change of S_P after an accelerated storage test at high temperature (60 °C) for 10 days was 0.0001 ± 0.0002 for lot Pre16 (Uchida et al., 2020).

For lot Pre19, accelerated storage tests (one month) for many parameters were conducted only at high temperature (Table 12.3) based on the results of storage tests for lots Pre14 and Pre16. The storage temperature was reduced to 40 °C because the plastic inner caps of some bottles blew off because of increased internal pressure during previous storage tests. Differences between before and after the accelerated storage tests were significant for concentrations of silicic acid (see also Fig. 12.3) and DO. The DO concentration after the test was close to the result for lot Pre16 and much larger than the solubility of DO at the S_P and storage temperature (165 µmol kg^{-1} at S_P of 34 and temperature of 40 °C). These changes might be due to compositional changes of the bottle interior at 40 °C, because the DO concentration after the storage test at high temperature was close to the equilibrium value of DO after long-term storage, as in the case of lot Pre16.

12.5 Changes of Properties Due to Autoclaving

As with RMNS, the source water for MSSW was autoclaved to stop biological activity. Lot Pre14, which was intended to be a standard seawater for DO, was surface seawater from Chichijima Island (Ogasawara Islands) south of Japan. Seawater has low concentrations of nutrients and is nearly saturated with DO. However, when that surface seawater was autoclaved, the DIC and TA decreased significantly, and the S_P increased (Table 12.4). The fact that white crystals remained in the reaction chamber after the MSSW was dispensed suggested that calcium carbonate precipitated from

Table 12.3 Changes of properties during accelerated storage tests (1 month) at high temperature (40 °C) for lot Pre19. Averages ± SDs are shown

Parameter	Control value	Value after acceleration test
S_P	34.2744 ± 0.0002	34.2744 ± 0.0001
NO_3^-	27.21 ± 0.06	27.31 ± 0.09 µmol kg^{-1}
NO_2^-	0.52 ± 0.02 µmol kg^{-1}	0.55 ± 0.03 µmol kg^{-1}
$Si(OH)_4$	61.48 ± 0.28 µmol kg^{-1}	61.96 ± 0.26 µmol kg^{-1}
PO_4^{3-}	1.977 ± 0.003 µmol kg^{-1}	1.980 ± 0.002 µmol kg^{-1}
NH_4^+	0.82 ± 0.05 µmol kg^{-1}	0.87 ± 0.07 µmol kg^{-1}
DIC	2140.54 ± 1.84 µmol kg^{-1}	2139.50 ± 1.32 µmol kg^{-1}
TA	2303.16 ± 0.62 µmol kg^{-1}	2302.43 ± 0.92 µmol kg^{-1}
DO	212.55 ± 1.45 µmol kg^{-1}	209.70 ± 0.77 µmol kg^{-1}

the source seawater, and the compositional ratios of the seawater changed. For lot Pre15, we therefore used intermediate water from a depth of 397 m in Suruga Bay, Shizuoka, Japan as the source seawater, and we produced 60 bottles on a trial basis. We found that there was no significant change of the DIC or TA before and after autoclaving when we used that intermediate water. We therefore decided to use the Suruga Bay intermediate water as the source seawater for MSSW beginning with lot Pre15.

Table 12.4 shows how the following properties changed as a result of autoclaving the source seawater.

The DO concentration increased and became closer to the concentration in equilibrium with the atmosphere because the seawater was stirred while open to the atmosphere for 24 h before bottling.

The nitrite and ammonium concentrations increased for some unknown reason. The increase of the nitrite concentration was likely due to the effect of gamma-ray sterilization of the aluminum bottles because the nitrite concentration in the seawater that remained in the reaction chamber after the production of MSSW was not different from the concentration before autoclaving.

We believe that the concentration of H_2O_2 increased because the concentration of H_2O_2 in the source seawater taken from a depth of 397 m was expected to be small. In natural water, H_2O_2 is produced by natural sunlight in surface waters (tens of nmol kg^{-1} to >0.1 µmol kg^{-1}) and rapidly decreases with depth within the euphotic zone because of decomposition due to biological activity (Takeda et al., 2018; Yuan & Shiller, 2004). In fact, the concentration of H_2O_2 in bottles not sterilized with gamma rays was half that of bottles sterilized with gamma rays. The implication is that H_2O_2 may have been decomposed by microorganisms attached to the inner wall of the aluminum bottles and not by gamma ray sterilization.

Table 12.4 Concentrations of some constituents of prototype reference seawaters before and after autoclaving. Averages ± SDs are shown. The H_2O_2 concentrations with asterisks are for samples in aluminum bottles not sterilized with gamma rays

Property	Before autoclaving	After autoclaving	Source seawater
Lot Pre14			0 m (Chichijima, Ogasawara)
DIC [μmol kg^{-1}]	2006.9	1578.2 ± 1.0	
TA [μmol kg^{-1}]	2267.5	1705.0 ± 0.8	
S_P	34.734	34.853	
Lot Pre15			397 m (Suruga Bay, Shizuoka)
DIC [μmol kg^{-1}]	2213.2	2216.1 ± 1.5	
TA [μmol kg^{-1}]	2308.2	2308.5 ± 0.7	
H_2O_2 [μmol kg^{-1}]	<0.01	0.23	
H_2O_2 [μmol kg^{-1}]	<0.01*	0.11*	
Lot Pre16			397 m (Suruga Bay, Shizuoka)
DIC [μmol kg^{-1}]	2266.6 ± 3.0	2224.7 ± 2.5	
TA [μmol kg^{-1}]	2309.2 ± 0.7	2303.1 ± 1.3	
S_P	34.349	34.351	
DO [μmol kg^{-1}]	180.9 ± 4.7	215.9 ± 0.7	
Lot Pre19			397 m (Suruga Bay, Shizuoka)
NO_3^- [μmol kg^{-1}]	27.30	27.2 ± 0.06	
NO_2^- [μmol kg^{-1}]	0.05	0.52 ± 0.02	
$Si(OH)_4$ [μmol kg^{-1}]	61.41	61.48 ± 0.28	
PO_4^{3-} [μmol kg^{-1}]	1.96	1.98 ± 0.003	
NH_4^+ [μmol kg^{-1}]	0.56	0.82 ± 0.05	

12.6 Substances that Interfere with the Winkler Method

Dissolved oxygen in seawater is usually determined by the Winkler method, which is based on the iodometric reaction scheme (Carpenter, 1965a, 1965b). The chemical reactions involved in the Winkler scheme are as follows:

$$Mn^{2+} + 2OH^- \rightarrow Mn(OH)_2 \tag{12.1}$$

$$4Mn(OH)_2 + O_2 + 2H_2O \rightarrow 4Mn(OH)_3 \tag{12.2}$$

$$4Mn(OH)_3 + 12H^+ + 4I^- \rightarrow 2I_2 + 4Mn^{2+} + 12H_2O. \tag{12.3}$$

Each mole of oxygen is thus equivalent to 2 mol of iodine (I_2), and the concentration of dissolved oxygen can be determined by iodometric titration. Because of the presence of substances in seawater that interfere with the Winkler method, it is necessary to know the magnitude of the error due to these substances when comparing sensor readings with values determined by the Winkler method.

Nitrite (NO_2^-) is well known to interfere with iodometry (Wong, 2012). Because nitrite reacts with iodide under acidic conditions to form molecular iodine, each mole of nitrite can result in an apparent presence of 0.25 mol of dissolved oxygen as follows:

$$NO_2^- + 4H^+ + 2I^- \rightarrow I_2 + 2NO + 2H_2O. \tag{12.4}$$

In addition, if atmospheric oxygen enters the sample during analysis, a sequence of semi-cyclic reactions may occur as follows:

$$2NO + O_2 \rightarrow 2NO_2 \tag{12.5}$$

$$2NO_2 + H_2O \rightarrow NO_2^- + NO_3^- + 2H^+. \tag{12.6}$$

Because 1 mol of nitrite may give rise to a theoretical maximum of 0.5 mol of apparent oxygen, each mole of nitrite may result in the apparent presence of between 0.25 and 0.5 mol of dissolved oxygen (Wong, 2012).

The nitrite in MSSW lot Pre16 was stable for about 6 year ($0.59 \pm 0.15 \, \mu mol \, kg^{-1}$) (Sect. 12.7), although it might have been produced by the interaction of gamma rays with the sterilized aluminum bottles (Sect. 12.5). The error of oxygen determined by the Winkler method due to nitrite interference could therefore range from 0.15 to $0.3 \, \mu mol \, kg^{-1}$ for the MSSW produced using Suruga Bay intermediate water. The interference caused by nitrite can be readily eliminated by the azide modification of the Winkler method (Wong, 2012).

Iodate (IO_3^-) is also known to interfere with the Winkler method (Wong & Li, 2009). Each mole of iodate can result in an apparent presence of 1.5 mol of dissolved oxygen as follows:

$$IO_3^- + 6H^+ + 5I^- \rightarrow 3I_2 + 3H_2O. \tag{12.7}$$

Because the iodate concentration in the open ocean ranges from $0.34 \pm 0.1 \, \mu mol \, kg^{-1}$ in the surface mixed layer to $0.42 \pm 0.03 \, \mu mol \, kg^{-1}$ in the deep ocean, it can lead to an overestimation of 0.52 ± 0.15 to $0.63 \pm 0.05 \, \mu mol \, kg^{-1}$ for oxygen estimated by the Winkler method (Wong & Li, 2009).

We measured the iodate in MSSW within a few months after production (Table 12.5). The iodate concentrations in the MSSWs were comparable to the open ocean surface water and deep water values (Huang et al., 2005; Wong & Li, 2009).

Hydrogen peroxide is also known to interfere with the Winkler method (Wong et al., 2010). Each mole of H_2O_2 can result in an apparent presence of 0.5 mol of dissolved oxygen as follows:

Table 12.5 Iodate concentrations of lots Pre14, Pre16, and Pre19. Averages ± SDs are shown. The number of bottles measured is shown in parentheses. The error of oxygen determined by the Winkler method due to iodate interference is also shown

Lot no.	IO_3^- [µmol kg^{-1}]	Error in oxygen [µmol kg^{-1}]	Source seawater
Pre14	0.366 ± 0.002 (4)	0.549 ± 0.003	0 m (Chichijima, Ogasawara)
Pre16	0.415 ± 0.002 (3)	0.622 ± 0.003	397 m (Suruga Bay, Shizuoka)
Pre19	0.445 ± 0.008 (5)	0.668 ± 0.012	397 m (Suruga Bay, Shizuoka)

$$H_2O_2 + 2H^+ + 2I^- \rightarrow I_2 + 2H_2O. \tag{12.8}$$

Hydrogen peroxide might have been produced in MSSW by autoclaving. The H_2O_2 concentration was about 1 µmol kg^{-1} immediately after autoclaving, but it then decreased with time (Fig. 12.5). Its decay rate estimated from the regression line in Fig. 12.5 was 0.006 day^{-1}, and its half-life was 116 days. Hydrogen peroxide could therefore lead to an overestimation of oxygen by about 0.5 µmol kg^{-1} at the time of production of MSSW by the Winkler method. This overestimation decreased with time and reached as little as 0.1 µmol kg^{-1} of oxygen eight months after production. At that time, it would be difficult to detect this overestimation with the Winkler method.

The decay rate of the H_2O_2 in MSSW was one order magnitude smaller than the decay rate that has been observed in the deep North Pacific (0.07 day^{-1}, Yuan & Shiller, 2004). Because the decomposition of H_2O_2 is thought to be a process

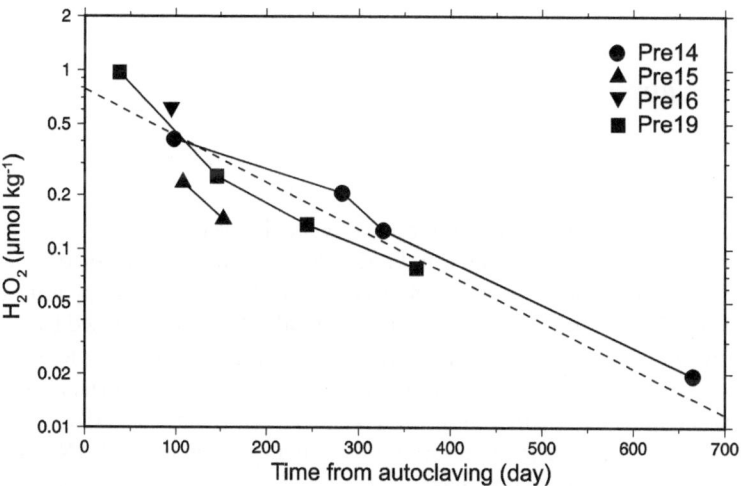

Fig. 12.5 Natural logarithm of the measured concentrations of hydrogen peroxide (H_2O_2) in prototype reference seawaters. The decay rate (0.006 day^{-1}) was estimated by fitting a straight line (dashed line) to the data for lots Pre14 and Pre19

primarily mediated by microorganisms (Petasne & Zika, 1997), the biological activity in MSSW, which had been autoclaved and used to fill aluminum bottles sterilized with gamma rays, was likely quite small.

12.7 Current Status of MSSW

This section discusses the current status of MSSW and focuses on the results from Pre16 and Pre19, in which the stabilities of many properties were examined over a relatively long period of time.

12.7.1 S_p

Uchida et al. (2020) have reported the stability of the S_P of MSSW lot Pre16 over seven years. The SD of the S_P was estimated to be 0.00026 by application of the batch offset correction for IAPSO SSW. Because the salinometer used to measure the S_P had a measurement resolution of 0.0002, the S_P of the MSSW was stable over a long period. However, we found that there was a slight decreasing trend (-0.00015 year^{-1}) in the S_P of Pre19 (Fig. 12.6). Improvement of the homogeneity and stability of the MSSW contributed to the detection of the decreasing trend of the S_P. The decreasing trend of the S_P of the MSSW is likely an apparent result of an increasing trend of the S_P of IAPSO SSW, which we discuss in Sect. 12.8.1.

12.7.2 Nutrients

The stability of nutrients was evaluated only for lot Pre16. The results are summarized in Table 12.6 and shown in Figs. 12.7 and 12.8. Nitrate, nitrite, phosphate, and ammonium concentrations were relatively stable over time, but the silicic acid concentration varied by ±2%. Part of the variability of the silicic acid concentration might have been caused by inhomogeneity within the lot, as was apparent for lot Pre19 (Fig. 12.3). Some of this variability might also have been caused by the relatively low comparability of the silicic acid measurements. For example, the SD of silicic acid measurements in an inter-laboratory comparison of nutrient concentrations was larger than those for nitrate and phosphate (Aoyama et al., 2010b). Because this variability in silicic acid concentrations was not apparent in the RMNS produced by a similar method, there may have been a cause specific to the MSSW stored in aluminum bottles with plastic inner caps.

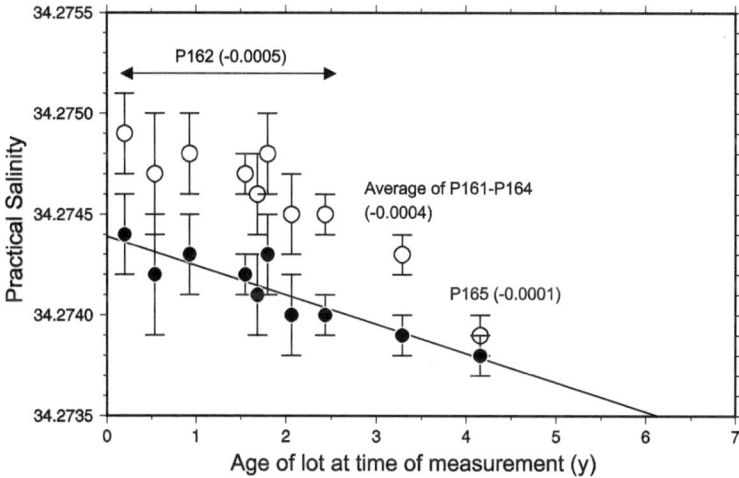

Fig. 12.6 Practical Salinity (S_P) of MSSW lot Pre19 plotted against age of lot at time of measurement. Open circles indicate original S_P values and closed circles indicate S_P values after the application of the batch offset correction for IAPSO SSW (Chap. 9). The batch number of the SSW used as a reference is also shown, and the offset value is shown in parentheses following the batch number. Vertical bars show the SDs of the S_P values. The straight line is the ordinary least squares regression line for the batch offset corrected S_P values

Table 12.6 NO_3^-, NO_2^-, $Si(OH)_4$, PO_4^{3-}, and NH_4^+ concentrations in MSSW lot Pre16. Averages ± SDs are shown. The number of evaluations is shown in parentheses

NO_3^- [μmol kg⁻¹]	NO_2^- [μmol kg⁻¹]	$Si(OH)_4$ [μmol kg⁻¹]	PO_4^{3-} [μmol kg⁻¹]	NH_4^+ [μmol kg⁻¹]
Lot Pre16				
11.56 ± 0.09 (7)	0.59 ± 0.15 (7)	44.58 ± 0.80 (7)	1.054 ± 0.028 (7)	2.27 ± 0.14 (5)

12.7.3 DO

The stability of DO was evaluated for lots Pre16 and Pre19. The DO concentrations gradually decreased immediately after production and then appeared to stabilize a few years after production (Fig. 12.9). The stabilized value was close to the solubility of DO at the storage temperature (e.g., 208 μmol kg⁻¹ at a temperature of 25 °C and salinity of 34).

We also used the portable optical DO sensor (Fig. 12.2e) to evaluate MSSW lot Pre16 (triangles in Fig. 12.9). We applied a bias of 0.7 μmol kg⁻¹ to the DO sensor data and plotted the corrected sensor data in Fig. 12.9 to compare the sensor data with the data determined by the Winkler method to assess the error in the Winkler method due to interfering substances. The comparability of the data from the DO sensor and the Winkler method suggested that comparison with a DO sensor

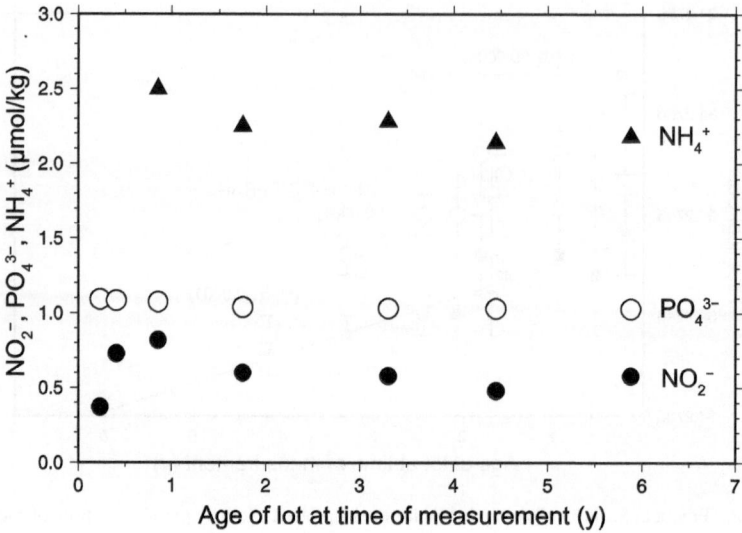

Fig. 12.7 Concentrations of nitrite (NO_2^-), phosphate (PO_4^{3-}), and ammonium (NH_4^+) in MSSW lot Pre16 plotted against age of lot at time of measurement

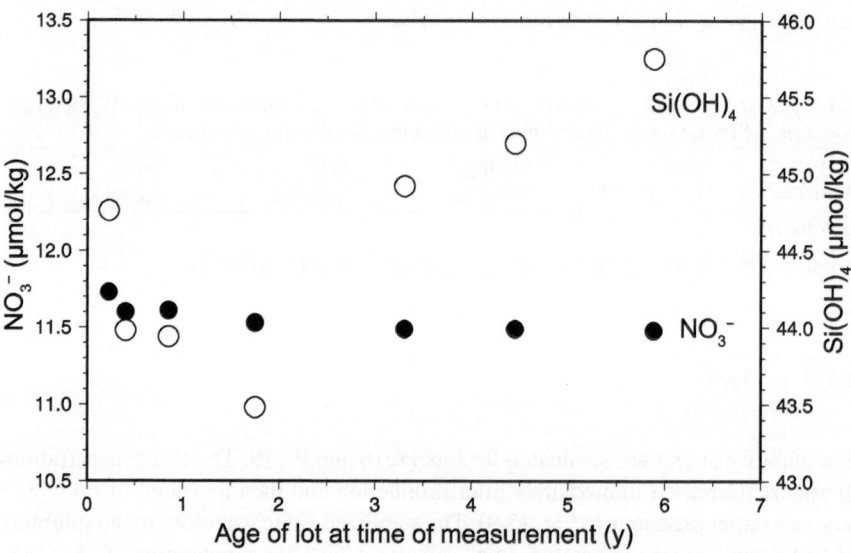

Fig. 12.8 Concentrations of nitrate (NO_3^-) and silicic acid ($Si(OH)_4$) in MSSW lot Pre16 plotted against age of lot at time of measurement

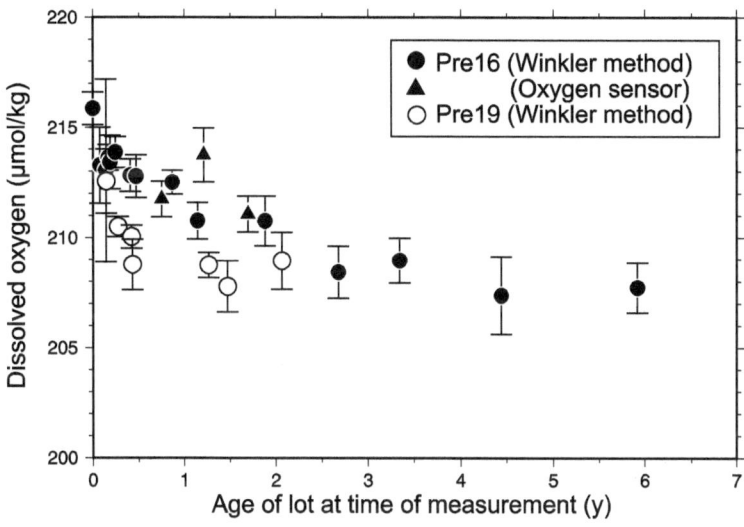

Fig. 12.9 Concentrations of dissolved oxygen (DO) in MSSW lots Pre16 and Pre19 plotted against age of lot at time of measurement. Vertical bars show SDs

calibrated with standard gases may be a useful way to ensure the comparability of DO measurements.

12.7.4 DIC and TA

We evaluated the stability of DIC and TA for lots Pre16 and Pre19. The results are summarized in Table 12.7 and shown in Figs. 12.10 and 12.11 for DIC and TA, respectively. The DIC and TA were relatively stable over time, and the TA for lot Pre19 was the least stable. This relatively large variability of the TA was caused by systematic differences between different laboratories. The average $\pm$ SD of data measured by KANSO TECHNOS (labeled as K in Fig. 12.11) was 2301.9 $\pm$ 1.2 μmol kg^{-1}. It was 2305.1 $\pm$ 0.6 μmol kg^{-1} for the other data measured by JAMSTEC and JMA.

Table 12.7 DIC and TA concentrations of MSSW lots Pre16 and Pre19. Averages $\pm$ SDs are shown. The number of evaluations is shown in parentheses

Lot no.	DIC [μmol kg^{-1}]	TA [μmol kg^{-1}]
Pre16	2222.0 $\pm$ 1.3 (11)	2302.9 $\pm$ 0.7 (10)
Pre19	2139.9 $\pm$ 1.6 (5)	2303.7 $\pm$ 1.9 (7)

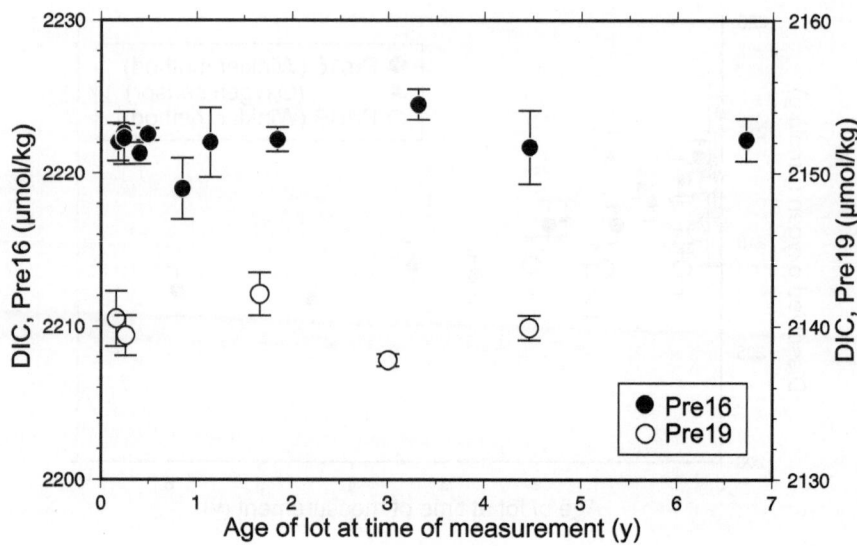

Fig. 12.10 Dissolved inorganic carbon (DIC) of MSSW lots Pre16 and Pre19 plotted against age of lot at time of measurement. Vertical bars show SDs

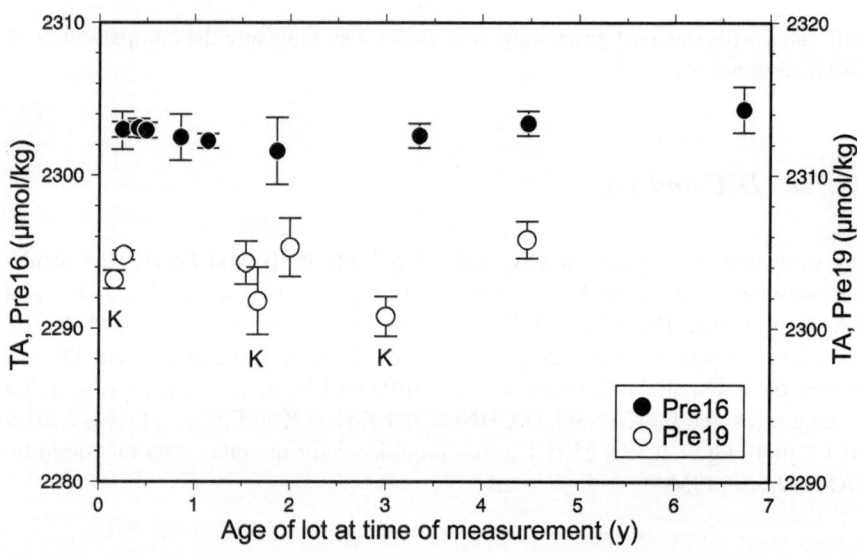

Fig. 12.11 Total alkalinity (TA) of MSSW lots Pre16 and Pre19 plotted against age of lot at time of measurement. Vertical bars show SDs. TA data labeled K were measured by KANSO TECHNOS

12.7.5 pH

The variability of pH was large, although the SD at each time of measurement was small (Fig. 12.12). The pH might tend to rapidly decrease just after production. Other sources of variation could be systematic differences between measurement laboratories rather than changes in MSSW over time. In fact, the pH data for lot Pre19 were measured only by JMA, and the data were relatively stable over time, except for the rapid decrease just after production.

However, even if the pH was measured in the same laboratory using the same equipment and methods, there were large, systematic differences (about 0.01 in pH) between cruises MR12-05 legs 2/3 and MR14-04/MR15-05 (Fig. 12.12). For cruises MR12-05 legs 2/3, the pH data were calibrated with respect to the in-house Tris buffer measurements based on the difference between the measured pH and the pH calculated from the DIC and TA data using the CO2SYS program developed by Lewis and Wallace (1998) for seawater samples (Murata et al., 2015) (Table 12.8). If the pH value for lot Pre16 was assigned based on the measurements obtained during cruises MR14-04 and MR15-05, the difference between the measured and calculated pH values for the seawater samples calibrated with MSSW was closer to zero than for those calibrated with the in-house Tris buffer (Table 12.8).

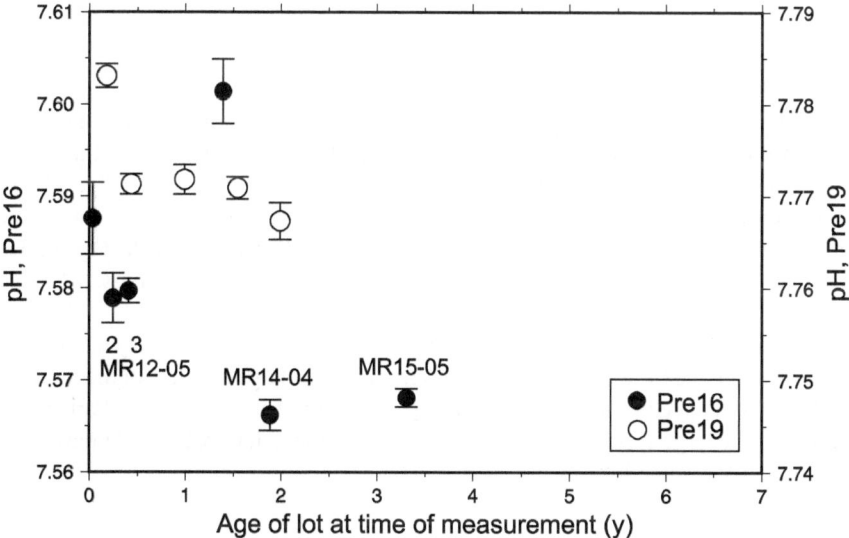

Fig. 12.12 pH of MSSW lots Pre16 and Pre19 plotted against the age of lot at time of measurement. Vertical bars show SDs. Data measured on cruises MR12-05 legs 2 and 3, MR14-04, and MR15-05 are labeled

Table 12.8 Difference between measured and assigned pH values for MSSW lot Pre16 and the in-house Tris buffer obtained during cruises MR12-05 legs 2 and 3. The assigned pH value for lot Pre16 (7.5672) was estimated from the average of pH measurements made during cruises MR14-04 (7.5662 ± 0.0017) and MR15-05 (7.5681 ± 0.0010). The differences between the measured pH and the pH calculated from DIC and TA data for the seawater samples are also shown for the original data, for samples calibrated with the in-house Tris buffer, and for those calibrated with MSSW

Measurement	Leg 2	Leg 3
MSSW	0.0117 ± 0.0027	0.0125 ± 0.0013
In-house Tris buffer	0.0089 ± 0.0011	0.0140 ± 0.0012
Seawater samples (original)	0.0148 ± 0.0054	0.0121 ± 0.0056
Seawater samples calibrated with Tris buffer	0.0059	−0.0019
Seawater samples calibrated with MSSW	0.0031	−0.0004

Table 12.9 DOC, TDN, and DOP concentrations of MSSW lots Pre16 and Pre19. Averages ± SDs are shown. The number of evaluations is shown in parentheses

Lot no.	DOC [μmol kg^{-1}]	TDN [μmol kg^{-1}]	DOP [μmol kg^{-1}]
Pre16	95.2 ± 5.9 (8)	17.4 ± 0.5 (4)	0.10 ± 0.06 (3)
Pre19	65.0 ± 3.1 (3)	30.5 ± 0.3 (3)	

12.7.6 DOM

The stability of DOP was evaluated for lot Pre16, and the stabilities of DOC and TDN were evaluated for lots Pre16 and Pre19. The results are summarized in Table 12.9 and shown in Figs. 12.13 and 12.14. DOP was stable for one year (Fig. 12.13). Dissolved organic nitrogen, which is defined as TDN minus dissolved inorganic nitrogen (DIN), was also expected to be stable over time because TDN (Fig. 12.13) and DIN (sum of NO_3^-, NO_2^- and NH_4^+) were stable over time (Figs. 12.7 and 12.8). However, the DOC of lot Pre16 was variable compared to the SD at the time of each measurement (Fig. 12.14). The DOC of lot Pre19 was relatively stable over time, although the SD at each time of measurement was large because of the increase of the DOC concentration with increasing serial number (Fig. 12.4). Part of the reason for the large variation of the DOC concentration in lot Pre16 might have been inhomogeneity of the lot.

12.7.7 Absolute Salinity (S_A)

We directly estimated the stability of the S_A of lot Pre19 from its measured density by back-calculating from the TEOS-10 equations using the temperature and pressure at which the density was measured (Fig. 12.15). Given that the resolution of the density

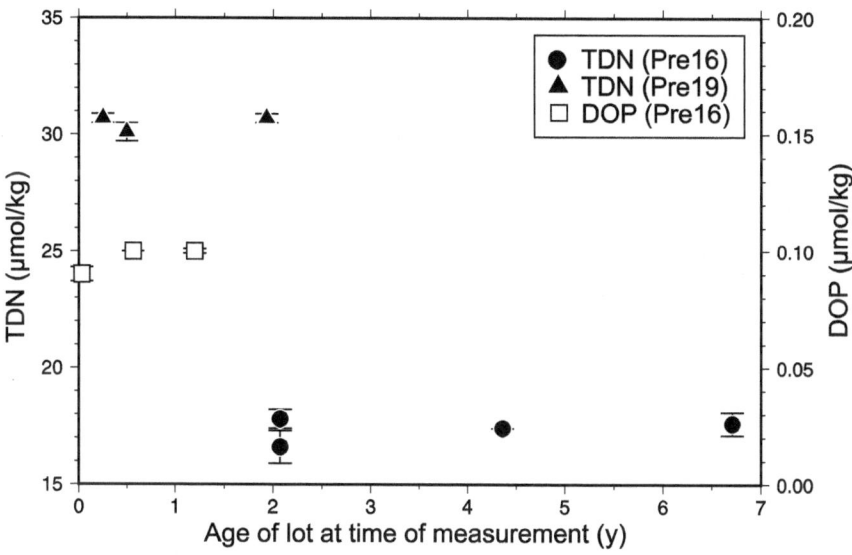

Fig. 12.13 Total dissolved nitrogen (TDN) and dissolved organic phosphorus (DOP) plotted against age of lot at time of measurement for MSSW lots Pre16 and Pre19. Vertical bars show SDs

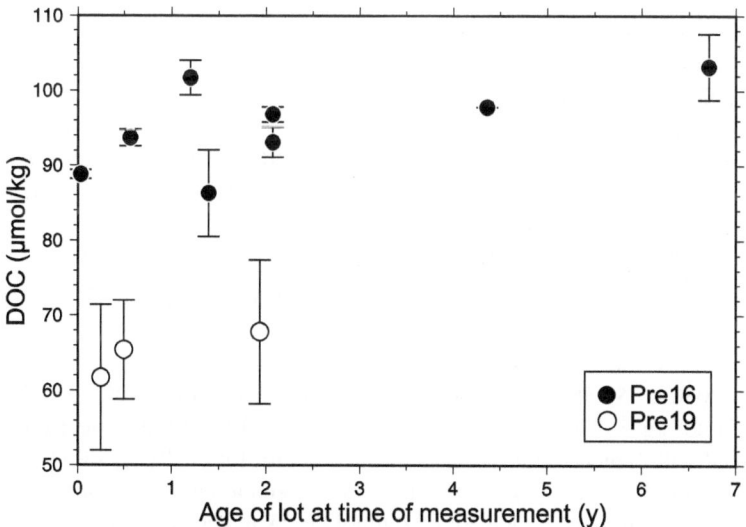

Fig. 12.14 Dissolved organic carbon (DOC) of MSSW lots Pre16 and Pre19 plotted against age of lot at time of measurement. Vertical bars show SDs

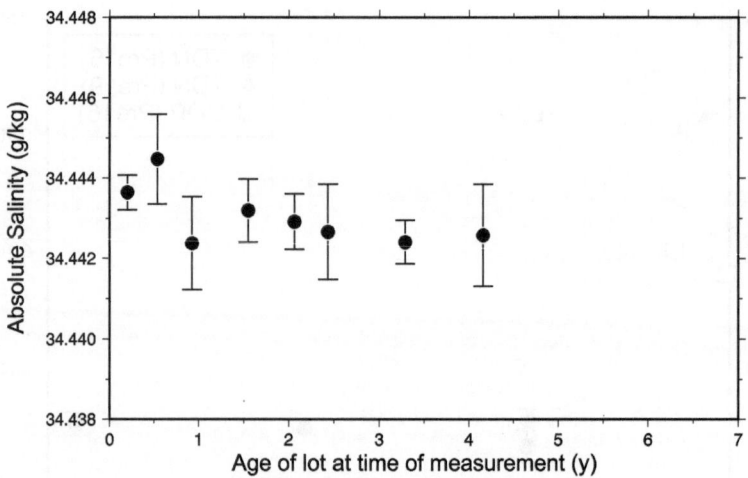

Fig. 12.15 Absolute salinity (S_A) of MSSW lot Pre19 plotted against age of lot at time of measurement. Vertical bars show SDs

meter used to measure the S_A was 0.0013 g kg^{-1}, we concluded that the S_A (34.4430 ± 0.0007 g kg^{-1}) of the MSSW was stable over a long period.

The S_A introduced in TEOS-10 can also be linked to the conductance used to determine the S_P. The relevant equation is $S_A = S_R + \delta S_A$, where S_R is the Reference-Composition Salinity and is calculated from the S_P ($S_R = 35.16504$ g kg^{-1}/35 × S_P), and δS_A is the Absolute Salinity anomaly. The δS_A can be independently estimated from the concentrations of silicic acid, nitrate, TA, DIC, and DOC by using the model of Pawlowicz et al. (2011) with modification for DOC (Chap. 10) as follows:

$$\delta S_A \left[\text{g kg}^{-1} \right] = (5.07\text{Si(OH)}_4 + 3.89 \text{ NO}_3^- + 5.56\Delta\text{NTA}$$
$$+ 0.47\Delta\text{NDIC} + 1.76\Delta\text{NDOC}) \times 10^{-5}, \qquad (12.9)$$

where ΔNTA is the TA anomaly (TA − 2300 × S_P/35), ΔNDIC is the DIC anomaly (DIC − 2080 × S_P/35), and ΔNDOC is the DOC anomaly (DOC − 54.3 × S_P/35), in μmol kg^{-1}. The value of S_A (34.4433 g kg^{-1}) from the S_R and Eq. (12.9) calculated with typical values of these parameters for lot Pre19 (S_P of 34.274, DIC of 2139.9 μmol kg^{-1}, TA of 2303.7 μmol kg^{-1}, silicic acid of 61.48 μmol kg^{-1}, nitrate of 27.2 μmol kg^{-1}, DOC of 65.0 μmol kg^{-1}) agreed well with the measured S_A (34.4430 g kg^{-1}).

12.8 Future Development Plans for MSSW

12.8.1 Density (Absolute Salinity) and Practical Salinity

Among the properties of MSSW, the long-term stability of S_P has been most precisely examined (Uchida et al., 2020). However, the cause of the decreasing trend found in recent MSSWs (Fig. 12.6 for Pre19 and not shown in this manuscript for Pre20) urgently needs to be clarified. There are three possible causes of the decreasing trend: (1) composition change of MSSW, (2) an evaporation of water from IAPSO SSW, and (3) composition change of IAPSO SSW. The second possible cause might be unlikely because the change of S_P after an accelerated storage test at high temperature (60 °C) for 10 days was −0.0002 ± 0.0002 for IAPSO SSW (batch P153) (Uchida et al., 2020). The third possible cause might be more likely than the first possible cause: for IAPSO SSW, silicic acid increased with time (5.47 μmol kg⁻¹ year⁻¹), DIC decreased with time (−13.5 μmol kg⁻¹ year⁻¹), and increase of Na⁺ and decrease of Ca²⁺ are suggested associated with the changes of silicic acid and DIC (Chap. 10), although long-term stabilities were good in most cases for MSSW (Sect. 12.7).

If we assume that there has been an increasing trend of S_P (0.00015 year⁻¹) in IAPSO SSW and reevaluate the batch offset correction table for IAPSO SSW (Chap. 9) and correct the estimated initial batch offset and the increasing trend of S_P at time of measurement of IAPSO SSW, we can eliminate the decreasing trend of S_P in Pre19.

Density-related parameters (DIC, TA, silicic acid, and nitrate) are stable in time, as described in Sect. 12.7, in terms of δS_A (see Eq. (12.9)). The density of MSSW is therefore expected to be stable over time or decrease slightly with time at a rate of −0.00008 kg m⁻³ year⁻¹ because of the decrease with time of S_P. Yohei Kayukawa of the NMIJ is currently improving the hydrostatic weighing apparatus for seawater density measurements (Kayukawa & Uchida, 2021). His goal is to improve the reproducibility of density measurements by making it possible to adjust the pressure in the measurement system. Current plans are to certify the density of MSSW (Pre19 and Pre20) with traceability to SI with a small uncertainty (the expanded uncertainty of 0.0014 kg m⁻³ [equivalent to 0.0018 g kg⁻¹ in S_A]).

Within-batch differences of density of MSSW can be evaluated with an ultra-high-resolution seawater density sensor based on a refractive index measured using the spectroscopic interference method (Uchida et al., 2019), which was developed as a part of the research project "Development of a high-accuracy absolute sali-nometer" (2014–2015, PI: Hiroshi Uchida of JAMSTEC) supported by the Japan Society for the Promotion of Science KAKENHI Grant 26610154, and "The challenge of measuring ultra-deep sea salinity using an innovative method" (2017–2018, PI: Hiroshi Uchida of JAMSTEC) supported by the Japan Agency for Marine-Earth Science and Technology (JAMSTEC) Specially Promoted Program.

The S_A of MSSW can be directly estimated from its certified density by back-calculating from the TEOS-10 equations using the temperature and pressure at which the density was certified. Because the S_A is linked to a conductance determination as noted in Sect. 12.7.7, certification of the S_P of MSSW can be made using the S_A estimated from density measurements and the δS_A estimated from Eq. (12.9) as an alternative way to trace S_P to the SI. In this case, it may be necessary to note whether the silicic acid changed over time (e.g., Fig. 12.8), although any effect on the calculated δS_A is expected to be small (less than 0.0001 g kg^{-1} year^{-1}).

Our plan is to evaluate label values of multiple IAPSO SSWs based on the certified S_P of a common MSSW (Pre20) over the next several decades.

12.8.2 Carbonate System Parameters

For DIC and TA, we will examine the use of the same method used to certify the Dickson-RM and aim to use MSSW as a secondary RM. Unlike the Dickson-RM, to which mercuric chloride is added, MSSW is a non-toxic standard and can be easily distributed internationally. In addition, whereas the Dickson-RM is made from surface seawater, MSSW is based on intermediate water. The DIC and TA values therefore differ from the Dickson-RM values (typical values of DIC and TA are 2030 and 2200 μmol kg^{-1} for the Dickson-RM and 2220 and 2300 μmol kg^{-1} for MSSW, respectively). The MSSW can therefore be used effectively in combination with the Dickson-RM, for example, to confirm the linearity of DIC and TA analyses (Bockmon & Dickson, 2015).

For the present, evaluation of the long-term stability of MSSW lot Pre19 will be continued by using the value preliminarily assigned through comparative measurements with a certain batch of Dickson-RM. The usefulness of MSSW will be evaluated by comparing the stability of deep-water DIC and TA values for seawater samples collected at time-series station K2 (47°N, 160°E; water depth 5215 m) or in the Chishima Trench south of Hokkaido, Japan, when calibrated with Dickson-RM and when calibrated with MSSW.

No standard currently exists for pH (Bockman & Dickson, 2015). It is not easy to evaluate the stability of the pH of MSSW, although it could be done by using a Tris buffer (Nemzer & Dickson, 2005) or by comparing it with the pH calculated from the DIC and TA of the Dickson-RM. However, as indicated in Sect. 12.7.5, the pH of MSSW may be more stable than that of Tris buffers. Alternatively, the determination of pH for MSSW by Harned cell measurements is a primary method in metrology (Maksimov et al., 2008), and it is straightforward to maintain the SI traceability. It would be useful to measure the pH with a Harned cell for MSSW lot Pre19 and to compare the results with those measured by spectrophotometric determinations using the indicator dye *m*-cresol purple (ISO, 2015), by a glass pH electrode, or by an ion-selective field effect transistor (ISFET) non-glass pH electrode.

12.8.3 DO

Extreme DO outliers (about 30 μmol kg^{-1} difference) were frequently observed in the Pre19 evaluation by the JMA, but such a frequency of outliers was not observed in the evaluations by KANSO and JAMSTEC. It is possible that there were individual differences in the preparative method used to separate a sample from an MSSW bottle into a sample flask for measurement by the Winkler method, and it may be necessary to consider a preparative method with high reproducibility. In particular, the lack of airspace in MSSW bottles makes it difficult to stir before preparative sampling. The outliers may therefore have been caused by non-uniform DO concentrations in MSSW bottles during storage, and a highly reproducible pre-preparation stirring method should therefore be considered.

Part of the change over time in the measured DO values (Fig. 12.9) was due to the change over time of the H$_2$O$_2$ concentration (Fig. 12.5), one of the chemicals that interferes in the Winkler method. It may therefore be better to evaluate the DO about eight months after the MSSW is produced. However, the change over time of the DO concentration cannot be explained entirely by the DO error estimated from the change of the H$_2$O$_2$ concentration over time (initial value of ~1 μmol kg^{-1}: Winkler error of ~0.5 μmol kg^{-1}), and some other factors should be considered. Because the stabilized values are close to the solubility of DO at the storage temperature (Sect. 12.7.3), exchange of O$_2$ with the ambient air may have occurred. However, this hypothesis seems unlikely because the DO concentrations did not increase after a long-term storage test (1.4 years) at low (4 °C) temperature for lot Pre19 (207.43, 197.31, 207.10, and 207.72 μmol kg^{-1}) (see Fig. 12.9).

12.8.4 Organic Matter

The tendency of the concentration of DOC to increase as the serial number increased (Fig. 12.4) was probably due to a vertical gradient of DOC in the stainless-steel reaction chamber when the MSSW was produced. To improve the homogeneity of DOC in the MSSW, the seawater in the chamber should be stirred thoroughly.

As with DIC and TA, evaluation of the long-term stability of DOC for MSSW lot Pre19 will be continued by using the value assigned preliminarily through comparative measurements with a certain batch of consensus reference material provided by Prof. D. A. Hansell. The usefulness of MSSW will be evaluated by comparing the stability of deep-water DOC values for seawater samples collected at time-series station K2 or in the Chishima Trench when calibrated with the consensus reference material and when calibrated with MSSW.

In addition to DOC, fluorescent dissolved organic matter (FDOM) will be evaluated for stability in MSSW (lot Pre20) as part of the study "Acceleration of ocean acidification due to shutdown of winter CO$_2$ efflux in the subarctic western North Pacific Ocean" (2020–2024, PI: Masahide Wakita of JAMSTEC) supported by JSPS

KAKENHI Grant 20H04349. In FDOM analysis, benchtop fluorometers used in laboratories are usually calibrated using pure water (Shigemitsu et al., 2020), but the existence of systematic errors in seawater measurements has been suggested by comparison between different analytical laboratories (Masahito Shigemitsu of JAMSTEC, 2023, personal communication). An inter-laboratory comparison of FDOM is underway using lot Pre20. MSSW is expected to be used as a reference seawater in FDOM analysis and for laboratory calibration of fluorescent organic matter (FOM) sensors for field use (Shigemitsu et al., 2020).

Acknowledgements H. Uchida thanks the late Michio Aoyama for encouraging him to carry out this research. Multiparametric Standard Seawater was developed through cooperative research by KANSO TECHNOS (author Mitsuda), JFE Advantech (author Nagasawa) and JAMSTEC (authors Uchida, Murata, and Wakita), and its salinity was evaluated through cooperative research by MWJ (author Tanaka) and JAMSTEC (author Uchida). We thank the marine technicians of MWJ, who analyzed the composition of the MSSW. Comments by Rich Pawlowicz of the University of British Columbia helped to improve the paper. This paper is a contribution to the tasks of the Joint SCOR/IAPWS/IAPSO Committee on the Properties of Seawater (JCS).

Competing Interests Part of this work was supported by MEXT/JSPS KAKENHI Grants 17310015, 26610154, 18K03752, 21H01165, and 20H04349.

References

Aoyama, M., Dickson, A. G., Hydes, D. J., Murata, A., Oh, J. R., Roose, P., & Woodward, E. M. S. (Eds.). (2010a). *Comparability of nutrients in the world's ocean*. Mother Tank.

Aoyama, M., et al. (2010b). 2008 Inter-laboratory comparison study of a reference material for nutrients in seawater. In *Technical reports of the Meteorological Research Institute*, Tsukuba, Japan (Vol. 60, p. 134). https://www.mri-jma.go.jp/Publish/Technical/DATA/VOL_60/tec_rep_mri_60_1.pdf

Bockmon, E., & Dickson, A. G. (2015). An inter-laboratory comparison assessing the quality of seawater carbon dioxide measurements. *Marine Chemistry, 171*, 36–43. https://doi.org/10.1016/j.marchem.2015.02.002

Carpenter, J. H. (1965a). The accuracy of the Winkler method for dissolved oxygen analysis. *Limnology and Oceanography, 10*, 135–140. https://doi.org/10.4319/lo.1965.10.1.0135

Carpenter, J. H. (1965b). The Chesapeake Bay Institute technique for the Winkler dissolved oxygen method. *Limnology and Oceanography, 10*, 141–143. https://doi.org/10.4319/lo.1965.10.1.0141

Clayton, T. D., & Byrne, R. H. (1993). Spectrophotometric seawater pH measurements: Total hydrogen ion concentration scale calibration of m-cresol purple and at-sea results. *Deep-Sea Research, 40*, 2115–2129. https://doi.org/10.1016/0967-0637(93)90048-8

Dickson, A. G. (2010). Standards for ocean measurements. *Oceanography, 23*(3), 34–47. https://doi.org/10.5670/oceanog.2010.22

Feistel, R., Weinreben, S., Wolf, H., Seitz, S., Spitzer, P., Adel, B., Nausch, G., Schneider, B., & Wright, D. G. (2010). Density and absolute salinity of the Baltic Sea 2006–2009. *Ocean Science, 6*, 3–24. https://doi.org/10.5194/os-6-3-2010

Hansen, H. P., & Koroleff, F. (1999). Determination of nutrients. In K. Grasshoff, K. Kremling, & M. Ehrhardt (Eds.), *Methods of seawater analysis* (3rd ed., pp. 159–228). Wiley-VCH Verlag GmbH. https://doi.org/10.1002/9783527613984.ch10

Huang, Z., Ito, K., Morita, I., Yokota, K., Fukushi, K., Timerbaev, A. R., Watanabe, S., & Hirokawa, T. (2005). Sensitive monitoring of iodine species in sea water using capillary electrophoresis:

Vertical profiles of dissolved iodine in the Pacific Ocean. *Journal of Environmental Monitoring, 7,* 804–808. https://doi.org/10.1039/B501398D

IOC, SCOR, & IAPSO. (2010). *The international thermodynamic equation of seawater—2010: Calculation and use of thermodynamic properties* (p. 196). Intergovernmental Oceanographic Commission, Manuals and Guides No. 56, UNESCO (English). http://teos-10.org/pubs/TEOS-10_Manual.pdf

ISO. (2015). *Water quality—Determination of pHt in seawater—Method using the indicator dye m-cresol purple* (ISO 18191:2015).

Ito, K., Ichihara, T., Zhuo, H., Kumamoto, K., Timerbaev, A. R., & Hirokawa, T. (2003). Determination of trace iodide in seawater by capillary electrophoresis following transient isotachophoretic preconcentration: Comparison with ion chromatography. *Analytica Chimica Acta, 497,* 67–74. https://doi.org/10.1016/j.aca.2003.08.052

Ito, K., Shoto, E., & Sunahara, H. (1991). Ion chromatography of inorganic iodine species using C_{18} reversed-phase columns coated with cetyltrimethylammonium. *Journal of Chromatography, 549,* 265–272. https://doi.org/10.1016/S0021-9673(00)91438-9

Kawano, T. (2010). Method for salinity (conductivity ratio) measurement. In *The GO-SHIP repeat hydrography manual: A collection of expert reports and guidelines version 1* (IOCCP Rep. 14, ICPO Publ. Series 134, p. 13). http://www.go-ship.org/HydroMan.html

Kayukawa, Y., & Uchida, H. (2021). Absolute density measurements for standard sea-water by hydrostatic weighing of silicon sinker. *Measurement: Sensors, 18.* https://doi.org/10.1016/j.measen.2021.100200

Lewis, E., & Wallace, D. W. R. (1998). *Program developed for CO_2 system calculations* (Report 105, p. 33). Oak Ridge National Laboratory. http://cdiac.esd.ornl.gov/oceans/co2rprt.html

Maksimov, I., Ohata, M., Nakamura, S., Hioki, A., Chiba, K., & Spitzer, P. (2008). PH determination on a carbonate buffer by Harned cells of different designs. *Accreditation and Quality Assurance, 13,* 381–387. https://doi.org/10.1007/s00769-008-0399-1

Mitsuda, H., Kimura, M., Murao, A., Ishii, R., Takeda, Y., Ota, H., & Aoyama, M. (2010). Development of a reference material for dissolved oxygen in seawater. In M. Aoyama, A. Dickson, D. Hydes, A. Murata, P. Roose, & E. Woodward (Eds.), *Comparability of nutrients in the world's ocean* (pp. 101–112). Mother Tank.

Murata, A. (2010). The development of non-toxic reference materials for oceanic CO_2 measurements: Current status and future plans. In M. Aoyama, A. Dickson, D. Hydes, A. Murata, P. Roose, & E. Woodward (Eds.), *Comparability of nutrients in the world's ocean* (pp. 34–41). Mother Tank.

Murata, A., Ishikawa, Y., Watai, T., Deguchi, E., Ono, A., & Tsubata, K. (2015). Carbon items (C_T, A_T and pH). In H. Uchida, K. Katsumata, & T. Doi (Eds.), *WHP P14S, S04I revisit in 2012 data book* (pp. 90–94). JAMSTEC. https://doi.org/10.17596/0000030

National Research Council of the National Academies. (2002). *Chemical reference materials, setting the standards for ocean sciences* (p. 130). The National Academies Press.

Nemzer, B. V., & Dickson, A. G. (2005). The stability and reproducibility of Tris buffers in synthetic seawater. *Marine Chemistry, 96,* 237–242. https://doi.org/10.1016/j.marchem.2005.01.004

Ogawa, H., Usui, T., & Koike, I. (2003). Distribution of dissolved organic carbon in the East China Sea. *Deep-Sea Research II, 50,* 353–366. https://doi.org/10.1016/S0967-0645(02)00459-9

Ota, H., Mitsuda, H., Kimura, M., & Kitao, T. (2010). Reference materials for nutrients in seawater: Their development and present homogeneity and stability. In M. Aoyama, A. Dickson, D. Hydes, A. Murata, P. Roose, & E. Woodward (Eds.), *Comparability of nutrients in the world's ocean* (pp. 11–30). Mother Tank.

Pawlowicz, R., Feistel, R., McDougall, T. J., Ridout, P., Seitz, S., & Wolf, H. (2016). Metrological challenges for measurements of key climatological observables part 2: Oceanic salinity. *Metrologia, 53,* R12–R25. https://doi.org/10.1088/0026-1394/53/1/R12

Pawlowicz, R., Wright, D. G., & Millero, F. J. (2011). The effects of biogeochemical processes on oceanic conductivity/salinity/density relationships and the characterization of real seawater. *Ocean Science, 7,* 363–387. https://doi.org/10.5194/os-7-363-2011

Petasne, R. G., & Zika, R. G. (1997). Hydrogen peroxide lifetimes in south Florida coastal and offshore waters. *Marine Chemistry, 56*, 215–225. https://doi.org/10.1016/S0304-4203(96)000 72-2

Ridal, J. J., & Moore, R. M. (1990). A re-examination of the measurement of dissolved organic phosphorus in seawater. *Marine Chemistry, 29*, 19–31. https://doi.org/10.1016/0304-4203(90)900 03-U

Romeo, R., Albo, P. A. G., & Lago, S. (2019). Density of standard seawater by vibrating tube densimeter: Analysis of the method and results. *Deep-Sea Research I, 154*, 103157. https://doi.org/10.1016/j.dsr.2019.103157

Schmidt, H., Seitz, S., Hassel, E., & Wolf, H. (2018). The density-salinity relation of standard seawater. *Ocean Science, 14*, 15–40. https://doi.org/10.5194/os-14-15-2018

Seitz, S., Feistel, R., Wright, D. G., Weinreben, S., Spitzer, P., & De Bievre, P. (2011). Metrological traceability of oceanographic salinity measurement results. *Ocean Science, 7*, 45–62. https://doi.org/10.5194/os-7-45-2011

Shigemitsu, M., Uchida, H., Yokokawa, T., Arulananthan, K., & Murata, A. (2020). Determining the distribution of fluorescent organic matter in the Indian Ocean using in situ fluorometry. *Frontiers in Microbiology, 11*, 589262. https://doi.org/10.3389/fmicb.2020.589262

Stendardo, I., Gruber, N., & Körtzinger, A. (2009). CARINA oxygen data in the Atlantic Ocean. *Earth System Science Data, 1*, 87–100. https://doi.org/10.5194/essd-1-87-2009

Takeda, K., Nojima, H., Kuwahara, K., Chidya, R. C., Adesina, A. O., & Sakugawa, H. (2018). Nanomolar determination of hydrogen peroxide in coastal seawater based on the Fenton reaction with terephthalate. *Analytical Sciences, 34*, 459–464. https://doi.org/10.2116/analsci.17p536

Takeda, K., Yamane, K., Horioka, Y., & Ito, K. (2017). The iodide and iodate distribution in the Seto Inland Sea, Japan. *Aquatic Geochemistry, 23*, 315–330. https://doi.org/10.1007/s10498-017-9324-8

Uchida, H., Kawano, T., Aoyama, M., & Murata, A. (2011a). Absolute salinity measurements of standard seawaters for conductivity and nutrients. *La Mer – Bulletin de la Société franco-japonaise d'océanographie, 49*, 119–126. http://www.sfjo-lamer.org/la_mer/49-3_4/49-3-4-5.pdf

Uchida, H., Kawano, T., Nakano, T., Wakita, M., Tanaka, T., & Tanihara, S. (2020). An expanded batch-to-batch correction for IAPSO standard seawater. *Journal of Atmospheric and Oceanic Technology, 37*, 1507–1520. https://doi.org/10.1175/JTECH-D-19-0184.1

Uchida, H., Kayukawa, Y., & Maeda, Y. (2019). Ultra high-resolution seawater density sensor based on a refractive index measurement using the spectroscopic interference method. *Scientific Reports, 9*, 15482. https://doi.org/10.1038/s41598-019-52020-z

Uchida, H., Murata, A., & Doi, T. (2011b). *WHP P21 revisit data book*. JAMSTEC. https://doi.org/10.17596/0000032

Uchida, H., Murata, A., Katsumata, K., Arulananthan, K., & Doi, T. (2021). *WHP I08N revisit/I07S in 2019/2020 data book*. JAMSTEC. https://doi.org/10.17596/0002162

Velo, A., Pérez, F. F., Lin, X., Key, R. M., Tauhua, T., de la Paz, M., Olsen, A., van Heuven, S., Jutterström, S., & Ríos, A. F. (2010). CARINA data synthesis project: pH data scale unification and cruise adjustments. *Earth System Science Data, 2*, 133–155. https://doi.org/10.5194/essd-2-133-2010

Wakita, M., Honda, M. C., Matsumoto, K., Fujiki, T., Kawakami, H., Yasunaka, S., Sasai, Y., Sukigara, C., Uchimiya, M., Kitamura, M., Kobari, T., Mino, Y., Nagano, A., Watanabe, S., & Saino, T. (2016). Biological organic carbon export estimated from the annual carbon budget observed in the surface waters of the western subarctic and subtropical North Pacific Ocean from 2004 to 2013. *Journal of Oceanography, 72*, 665–685. https://doi.org/10.1007/s10872-016-0379-8

Wolf, H. (2008). Determination of water density: Limitations at the uncertainty level of 1×10^{-6}. *Accreditation and Quality Assurance, 13*, 587–591. https://doi.org/10.1007/s00769-008-0442-2

Wong, G. T. F. (2012). Removal of nitrite interference in the Winkler determination of dissolved oxygen in seawater. *Marine Chemistry, 130–131*, 28–32. https://doi.org/10.1016/j.marchem.2011.11.003

Wong, G. T. F., & Li, K.-Y. (2009). Winkler's method overestimates dissolved oxygen in seawater: Iodate interference and its oceanographic implications. *Marine Chemistry, 115*, 86–91. https://doi.org/10.1016/j.marchem.2009.06.008

Wong, G. T. F., Wu, Y.-C., & Li, K.-Y. (2010). Winkler's method overestimates dissolved oxygen in natural waters: Hydrogen peroxide interference and its implications. *Marine Chemistry, 120*, 83–90. https://doi.org/10.1016/j.marchem.2010.07.006

Woosley, R., Huang, F., & Millero, F. J. (2014). Estimating absolute salinity (S_A) in the world's oceans using density and composition. *Deep-Sea Research Part I: Oceanographic Research Papers, 93*, 14–20. https://doi.org/10.1016/j.dsr.2014.07.009; Woosley, R., Huang, F., & Millero, F. J. (2018). Estimating absolute salinity (S_A) in the world's oceans using density and composition. *Deep-Sea Research Part I: Oceanographic Research Papers, 142*, 145. https://doi.org/10.1016/j.dsr.2018.09.007

Yuan, J., & Shiller, A. M. (2004). Hydrogen peroxide in deep waters of the North Pacific Ocean. *Geophysical Research Letters, 31*, L01310. https://doi.org/10.1029/2003GL018439

Chapter 13
Future Strategy for a Resilient Production and Certification of Seawater Reference Materials for the Carbonate System

Maribel I. García-Ibáñez and Regina A. Easley-Vidal

Abstract Ocean inorganic carbon research is crucial to quantify the global ocean uptake of atmospheric carbon dioxide (CO_2), understand its spatiotemporal variability and mechanisms that control this process, and monitor ocean acidification. This requires high-quality measurements of the seawater carbonate system that rely on the availability of reference materials (RMs). The COVID-19 pandemic highlighted the fragility of the production system of the seawater RMs for the carbonate system, currently depending on one single laboratory. With that in mind, a new model for seawater RMs for the carbonate system, centered on regional hubs, is being developed to create a more resilient system. Challenges associated with establishing new production centers, such as funding their startup and ongoing costs, ensuring the quality and stability of the materials, and staff training are discussed. Opportunities to minimize the cost of these RMs and to supply certified or indicative values for currently uncertified carbonate system variables are explored. Additionally, a vision to integrate the new model into the global metrology landscape whereby the materials are comparable and metrologically traceable to the International System of Units is highlighted. As more laboratories are seeking to undertake seawater carbonate system measurements, access to these RMs is ever more critical.

Keywords Ocean carbon uptake · Reference materials · Seawater · Seawater CO_2 system

M. I. García-Ibáñez (✉)
Oceanographic Center of the Balearic Islands (COB-IEO), CSIC, Palma, Spain
e-mail: maribel.garcia@ieo.csic.es

R. A. Easley-Vidal
Chemical Sciences Division, National Institute of Standards and Technology (NIST),
Gaithersburg, MD, USA
e-mail: regina.easley@nist.gov

© The Author(s) 2025

M. Aoyama et al. (eds.), *Chemical Reference Materials for Oceanography*, Springer
Oceanography, https://doi.org/10.1007/978-981-96-2520-8_13

13.1 Introduction on Seawater Inorganic Carbon Research and Reference Materials (RMs) for the Seawater Carbonate System

The global ocean has absorbed around 20% to 30% of atmospheric carbon dioxide (CO_2) emissions accumulated since the industrial revolution and serves as an overall sink for excess anthropogenic atmospheric CO_2 (Gruber et al., 2019; Friedlingstein et al., 2022). The increased reservoir of oceanic CO_2 modifies ocean chemistry leading to changes collectively known as ocean acidification that are mainly traced by decreasing seawater pH and saturation states of biologically essential minerals such as the calcium carbonate polymorphs, calcite, and aragonite (Riebesell et al., 2000; Kroeker et al., 2013; Pörtner et al., 2014; Mostofa et al., 2016; Doney et al., 2020).

The increase in the oceanic CO_2 reservoir and its consequences (i.e., ocean acidification) are monitored by measuring four key variables of the seawater carbonate system: total dissolved inorganic carbon (DIC, the sum of the concentrations of all inorganic carbon species), total alkalinity (TA, a measure of the buffering capacity of seawater), pH (pH = $-\log_{10} a[H^+]$), and partial pressure of CO_2 (pCO_2, a measure of dissolved CO_2 in seawater) (Fig. 13.1). Measuring any two of these variables, along with auxiliary determinations of salinity, temperature, pressure, and dissolved inorganic nutrient concentrations, allows for the complete characterization of the seawater carbonate system, which includes the amount of carbonate and bicarbonate ions, and calcium carbonate saturation states. Measurements of DIC and TA are routinely performed by coulometric titration and potentiometric titration, respectively (Dickson et al., 2007), and are well-established within the monitoring efforts, with accuracies of around 2 μmol kg^{-1}. Discrete samples of seawater pH are typically measured with spectrophotometric pH indicator dyes calibrated over the range of salinities and temperatures expected in the open ocean, with pH accuracies of 0.01 (Clayton & Byrne, 1993). Spectroscopic methods (e.g., cavity ring-down spectroscopy or infrared spectroscopy) or gas chromatography are used to quantify pCO_2 with accuracies of around 2 μatm for surface waters (Wanninkhof & Thoning, 1993; Neill et al., 1997).

Efforts to monitor changes in the oceanic inventory of CO_2 began in the 1970s using large-scale repeat hydrography cruises starting with the GEOSECS (Geochemical Ocean Sections Study) program (Takahashi et al., 1982), which laid the groundwork for later programs such as WOCE (World Ocean Circulation Experiment) and JGOFS (Joint Global Ocean Flux Study) in the late 1980s and 1990s, respectively. Nowadays, GO-SHIP (Global Ocean Ship-based Hydrographic Investigations Program) maintains the global survey of select hydrographic sections with modern standard practices and methods (Sloyan et al., 2019). On the other hand, in recognition that an understanding of the variability over a range of time scales from seasonal to interannual was required to better understand the connection between climate and ocean biogeochemistry, sustained ocean time series were promoted around the world in the late 1980s and early 1990s as part of the international WOCE/

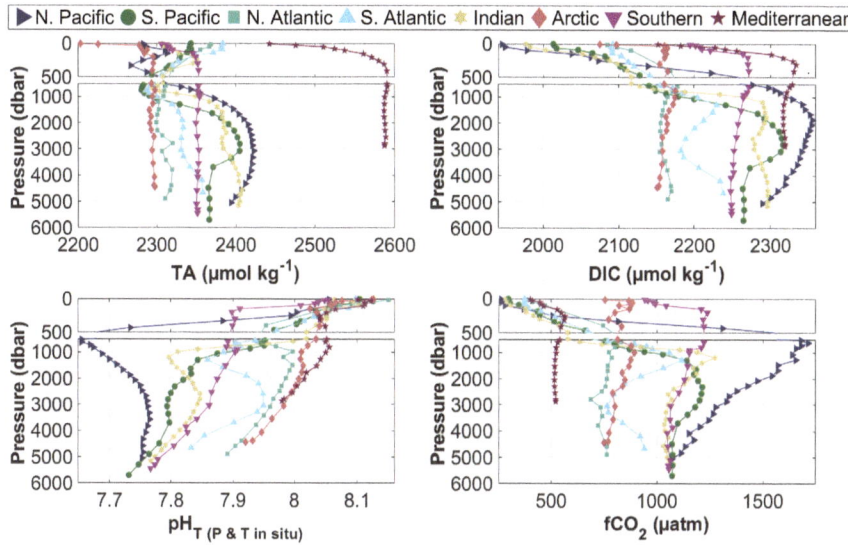

Fig. 13.1 Representative profiles of the seawater carbonate system variables in different ocean basins. pH is presented on the total hydrogen ion scale at in situ conditions of temperature and pressure ($pH_{T\,(P\,\&\,T\,in\,situ)}$). The fugacity of CO_2 (fCO_2) is the partial pressure of CO_2 corrected for the non-ideality of gases. Data from GLODAPv2.2022 (Lauvset et al., 2022): North Pacific Ocean (N. Pacific, navy right-pointing triangles) profile from station 324 of cruise 49NZ20071008 (expocode; expocode refers to the Expedition Code of the cruise and is a series of numbers and letters guaranteed to be unique and constructed by combining the country code and platform code from the ICES (International Council for the Exploration of the Sea) library (https://vocab.ices.dk/) with the date of departure in the format YYYYMMDD), South Pacific Ocean (S. Pacific, dark green circles) profile from station 86 of cruise 096U20160426, North Atlantic Ocean (N. Atlantic, teal squares) profile from station 83 of cruise 06MT20010507, South Atlantic Ocean (S. Atlantic, cyan upward-pointing triangles) profile from station 66 of cruise 33RO20110926, Indian Ocean (Indian, yellow six-pointed stars) profile from station 503 of cruise 49NZ20031209, Arctic Ocean (Arctic, pink diamonds) profile from station 76 of cruise 06AQ20150817, Southern Ocean (Southern, purple downward-pointing triangles) profile from station 24 of cruise 06AQ20141202, and Mediterranean Sea (Mediterranean, rosy five-pointed stars) profile from station 324 of cruise 06MT20110405

JGOFS effort where designated stations such as the Bermuda Atlantic Time Series (BATS) and Hawaii Ocean Time series (HOT) were sampled monthly (Sabine et al., 2010). Currently, seawater carbonate chemistry monitoring focuses on increasing the spatiotemporal coverage of observations to also improve our understanding of short-scale variability. For this, autonomous systems have been developed to routinely measure seawater carbonate system variables (Bushinsky et al., 2019). Global ocean monitoring has expanded through the development of regional programs and the founding of the Global Ocean Acidification Observing Network (GOA-ON) (Newton et al., 2015). Ultimately, measurements of the seawater carbonate system variables are populated into quality-controlled data products, such as SOCAT (Surface Ocean CO_2 Atlas) (Bakker et al., 2016) and GLODAP (Global Ocean Data Analysis Project) (Olsen et al., 2020), that inform bodies such as the Intergovernmental Panel on Climate Change (IPCC), an international cooperative established by the World Meteorological Organization (WMO), and the United Nations Environment Program (UNEP), which helps to shape global environmental policies.

Thanks to the collection of systematic measurements of the seawater carbonate system, it has been observed that, with an increased reservoir of oceanic CO_2, the pH of the surface ocean is decreasing between 0.0013 and 0.0026 per year depending on the study area and the frequency of the observations (Bates et al., 2014). While these massive efforts have greatly expanded our understanding of the seawater carbonate system and have initiated ocean acidification as one of the global indicators of ocean health (World Meteorological Organization, 2021), the oceanic reservoir of CO_2 still represents one of the largest uncertainties in the global carbon budgets (Friedlingstein et al., 2022). For this reason, accurate and precise seawater carbonate system measurements are essential to monitoring the ongoing progression of climate change and assessing the health of marine ecosystems.

Despite numerous technological advances over the last several decades, ship-based hydrography remains the only method for obtaining high-quality measurements over the full water column, especially for the deep ocean below 2 km (52% of global ocean volume). The precision and accuracy of ship-based measurements of the seawater carbonate system variables have improved by an order of magnitude since the 1990s (García-Ibáñez et al., 2022; Lauvset et al., 2022) due to the creation and adoption of standard operating procedures (SOPs) for measurements of DIC, TA, pH, and pCO_2 (Dickson et al., 2007), and the availability of reference materials (RMs) for TA and DIC (Dickson, 2010). Over these past three decades, a single laboratory managed by Prof. Andrew Dickson at Scripps Institution of Oceanography (SIO) in the United States has provided over 150,000 bottles of well-characterized seawater RMs for TA and DIC to the international oceanographic community. These seawater RMs are produced from modified local seawater (Pacific Ocean) with lower than average salinity and characterized for TA and DIC (Dickson, 2010), with values for salinity and dissolved inorganic nutrients. The demand for the seawater carbonate system RMs grew to a point where by 2015 production capacity exceeded 10,000 bottles annually, with approximately 70% of the RMs being shipped outside the U.S. (A. Dickson, pers. comm).

The COVID-19 pandemic highlighted the fragility of the current production and distribution system of the seawater RMs for the carbonate system (Catherman, 2021). The dramatic reduction in the supply of RMs during the pandemic forced many groups to produce in-house or working RMs to overcome the shortage of RMs. This is a challenging task since there are no standard protocols for their production, thus resulting in non-uniform production and compliance measures and uncertainty. In Europe, ICOS (Integrated Carbon Observation System) performed initial attempts for regional distribution of in-house or working RMs within the ICOS network. The ICOS-lead production showed that it is possible to produce stable in-house or working RMs. However, it still needs work to assign values to these working RMs and not compromise the quality of the seawater carbonate system measurements. Most, if not all, of these working RMs are supplied with reported repeatability of measurement values for their only uncertainty source.

With the expected increase in the use of RMs for the seawater carbonate system by the current users (Acquafredda et al., 2022) and the additional needs coming from the increased focus on ocean and coastal observing and the introduction of mitigation measures to curb ocean acidification, the demand is there for seawater RMs for the carbonate system. Additionally, the increasing need to serve national targets and actions in the global assessment (e.g., Sustainable Development Goals and national action plans to limit carbon and greenhouse gas emissions within the Paris Agreement) requires trustworthy data collected using global Best Practices that incorporate RMs in the workflow. Therefore, a robust and sustained production and a global supply of seawater RMs for the carbonate system are crucial. In this chapter, we propose a model of production and certification of those seawater RMs based on discussions among the seawater carbonate system research community.

13.2 Community Needs

The term RM is used to broadly categorize materials that are both sufficiently homogeneous and stable with reference to specified property values. The current seawater RMs for the carbonate system are produced at Prof. Andrew Dickson's laboratory at SIO in the U.S. using Pacific seawater, with an average salinity of 33.449 $\pm$ 0.135, average DIC of 2038 μmol kg^{-1} $\pm$ 39 μmol kg^{-1}, and average TA of 2226 μmol kg^{-1} $\pm$ 27 μmol kg^{-1} (based on properties from batch #119 onwards, which ensures the same seawater source and calibration methodologies (OCADS website 2023), with $\pm$ indicating one standard deviation; Fig. 13.2). However, those seawater RMs are used to ensure measurement accuracy on seawater samples not only from the Pacific Ocean but from the global ocean (Acquafredda et al., 2022), which presents a wider range of salinity, DIC, and TA values (Fig. 13.2). In some instances, for example, in the high-salinity and high-TA Mediterranean waters, seawater carbonate system properties vary from the RM by more than 370 μmol kg^{-1} for TA and 280 μmol kg^{-1} for DIC (Fig. 13.2), quantities 100 times larger than the uncertainty of those measurements. The mismatch of properties of the seawater RMs for the carbonate system and the seawater samples analyzed can pose some issues, especially for laboratories that use the seawater RMs to calibrate their equipment due to the lack of other calibration techniques (see Sect. 13.4). Therefore, having multiple RM compositions will increase the measurement accuracy of those groups by providing more than one point to confirm their calibrations.

In 2021, the current uses of RMs by the seawater carbonate system research community were assessed through surveys issued by the Interagency Working Group on Ocean Acidification (IWGOA), a U.S. federal subcommittee under the Office of Science and Technology Policy (Acquafredda et al., 2022), and the GOA-ON Mediterranean Hub, a network that connects Mediterranean scientists who are working and are interested in ocean acidification in the Mediterranean Sea (Hassoun et al., 2022). The results from both surveys emphasized the need for better and sustained access to the widely used seawater RMs for the carbonate system with

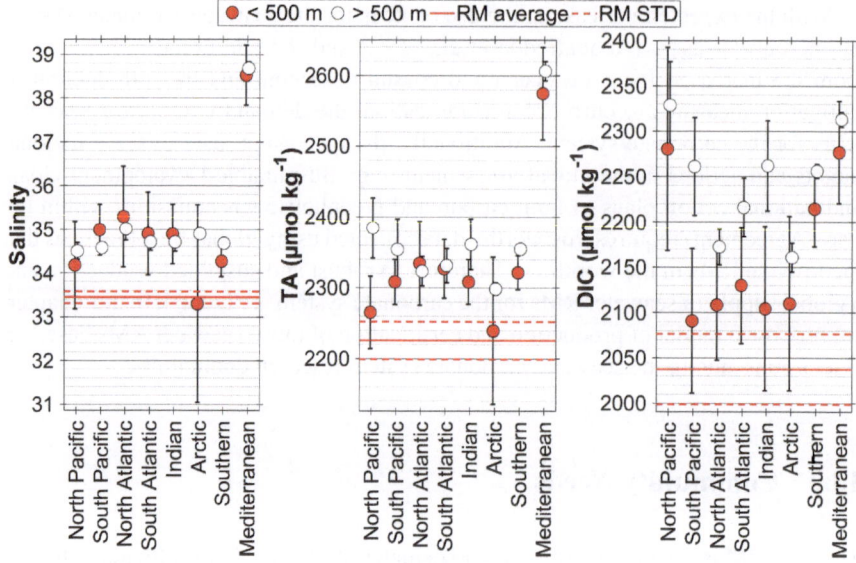

Fig. 13.2 Average seawater properties and standard deviations (STD) of different ocean basins and the seawater reference materials (RMs; properties from batch #119 onwards (OCADS website, 2023)). Oceanographic data from GLODAPv2.2022 (Lauvset et al., 2022), with the following sample distribution: North Pacific Ocean < 500 m water depth: 65,940 samples, > 500 m water depth: 74,381 samples; South Pacific Ocean < 500 m water depth: 28,860 samples, > 500 m water depth: 40,369 samples; North Atlantic Ocean < 500 m water depth: 44,941 samples, > 500 m water depth: 58,629 samples; South Atlantic Ocean < 500 m water depth: 17,593 samples, > 500 m water depth: 28,843 samples; Indian Ocean < 500 m water depth: 22,781 samples, > 500 m water depth: 32,541 samples; Arctic Ocean < 500 m water depth: 39,477 samples, > 500 m water depth: 17,776 samples; Southern Ocean < 500 m water depth: 21,388 samples, > 500 m 23,127 samples; Mediterranean Sea < 500 m water depth: 617 samples, > 500 m water depth: 528 samples. The Arctic and Southern Oceans are defined as North of 60 °N and South of 60 °S, respectively

values for DIC and TA, with the majority of research groups surveyed expecting to increase their consumption of these RMs over time (Acquafredda et al., 2022). Furthermore, higher demand is expected from the increased focus on ocean and coastal observing (Dobson et al., 2022) and the introduction of marine carbon dioxide removal (mCDR) applications (National Academies of Sciences, Engineering, & Medicine, 2022).

Community surveys (Acquafredda et al., 2022; Hassoun et al., 2022) also highlight the demand for seawater RMs for pH through the high number of research groups using the current seawater RMs to check the accuracy of their pH measurements, even though those RMs only include values for TA and DIC. Some laboratories routinely calculate RM pH from DIC and TA and use these values to quality control their discrete pH measurements. Despite the high number of seawater pH measurements, currently, there is no single recommended measurement procedure, nor is there an internationally accepted RM for seawater pH measurement that enables different laboratories to reliably achieve comparable measurements. This can jeopardize the detection of the long-term anthropogenically-driven changes in ocean pH over multi-decadal timescales. This is especially important as pH was designated within the United Nations Sustainable Development Goals (SDGs) Target 14.3 to "Minimize and address the impacts of ocean acidification, including through enhanced scientific cooperation at all levels" as a climate indicator 14.3.1 (United Nations, 2015).

To increase the number of laboratories performing seawater carbonate system measurements with a known uncertainty, not only will the community need increased production of existing seawater RMs, but to satisfy the goals of the SDGs, developing countries need access to more affordable RMs. This objective can be achieved in multiple ways, for example, by producing RMs with a higher uncertainty, lowering the costs associated with the production of RMs, or subsidizing a portion of the RM costs. Additionally, there is a general need for SOPs on the production and value assignment of in-house or working RMs that are often used to check equipment drift, thus reducing the number of RMs used. SOPs for in-house or working RMs would ensure higher measurement quality from the global community. Prof. Andrew Dickson and the EuroGO-SHIP infrastructure are starting to draft those SOPs, which will likely include a protocol for their preparation and guidance on the determination of their long-term stability, value assignment, and uncertainty quantification (A. Dickson and T. Steinhoff, pers. comm.).

13.3   Integrating Seawater RM Production and Certification into the Global Metrology System

The improved quality of seawater carbonate system measurements is directly attributed to the availability of carefully characterized seawater batches of RMs using measurement capabilities grounded in metrological principles to minimize uncertainties and guarantee consistent traceability between batches. Recent discussions, such as the organization of the BIPM-WMO Metrology for Climate Action workshop (October 2022; Metrology for Climate Action, 2022), have highlighted the need to merge expertise from the oceanographic monitoring community and the global metrological community to ensure that these critical measurements are adequately optimized to quantify changes in the environment, are reproducible over long timescales, and are ideally metrologically traceable to the International System of Units (SI). One of the goals moving forward is to connect existing RM programs, such as the seawater RM carbonate system program, to the global metrology system.

The International Bureau of Weights and Measures (BIPM: abbreviated from the French: Bureau International des Poids et Mesures) is a global body established by the Metre Convention, an international treaty signed in 1875 by 17 nations, which established the basis for an internationally agreed-upon system of measurements (Fig. 13.3). As of June 2024, the BIPM consists of 64 Member States and 36 Associate States and Economies (see Fig. 13.4, where dark-shaded countries are BIPM Member States and medium-shaded countries are BIPM Associate States

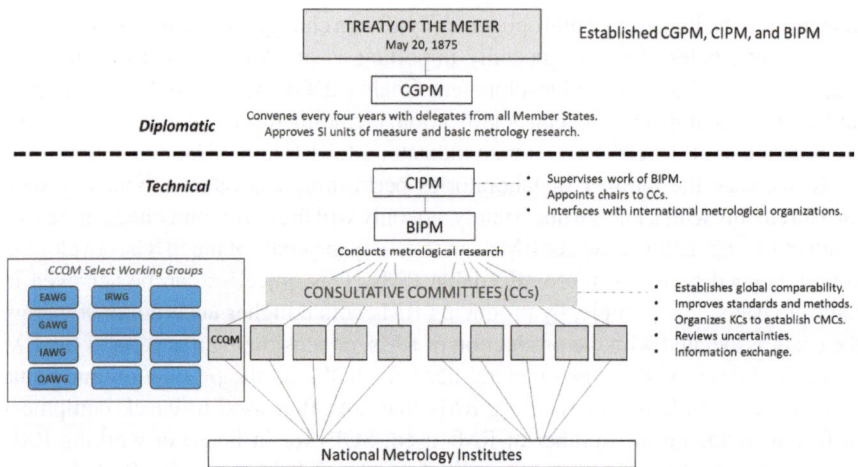

Fig. 13.3 Organization of the CGPM (General Conference on Weights and Measures; abbreviated CGPM from the French: Conférence Générale des Poids et Mesures), CIPM, and BIPM

and Economies). Within this structure, the International Committee for Weights and Measures (CIPM: abbreviated from the French: Comité International des Poids et Mesures), managed by 18 representatives from Member States, promotes global uniformity of units of measure by supervising the research of the BIPM. This metrological research is actualized through 10 Consultative Committees (CCs). Most relevant to the seawater carbonate system community is the work conducted by the CC for Amount of Substance: Metrology in Chemistry and Biology (CCQM), which houses 12 working groups consisting of representatives from National Metrology Institutes (NMIs) specialized in various areas of metrology, such as the Electrochemical Analysis Working Group (EAWG), the Gas Analysis Working Group (GAWG), the Inorganic Analysis Working Group (IAWG), the Organic Analysis Working Group (OAWG), or the Isotope Ratio Working Group (IRWG).

CCQM working groups convene to discuss measurement challenges, assess measurement capabilities, and prove their performance on high-quality certification methods through their participation in interlaboratory comparisons called key comparisons (KCs). Ideally, every five years, each working group cycles through KCs specifically designed to assess Calibration and Measurement Capability (CMC) claims for a given analyte in a specified matrix within a designated range of amount content. The KCs are designed to reflect typical RMs issued by the participating NMIs. Obtaining CMCs for a specified measurand is subject to an international peer-assessed approval process, which requires the review of the institute's quality management system and usually requires participation in KCs. In the design of KCs, metrologists recognize that demonstration of capabilities to support CMC claims can be performed independent of certification method or technique. In some comparisons, for example, within the EAWG, analysts test their expertise in measuring the activity of hydrogen ion in buffers using techniques ranging from primary pH measurements

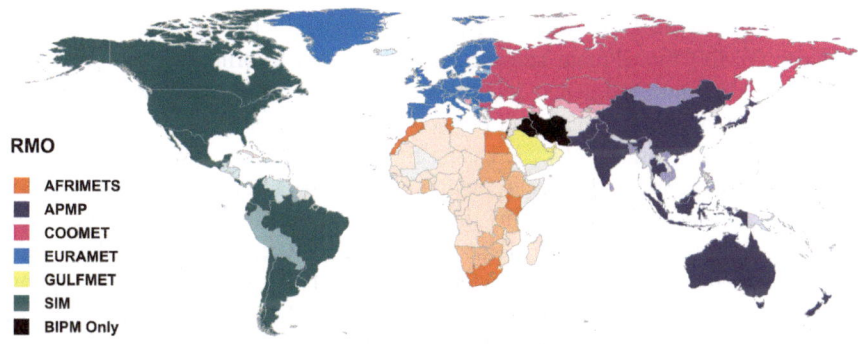

Fig. 13.4 The world's regional metrology organizations (RMOs) as of June 2024 where dark-shaded countries represent BIPM Member States, medium-shaded countries represent BIPM Associate States and Economies, and light-shaded countries represent non-BIPM members participating in the RMO

(Harned cells) to secondary differential cells and glass electrodes (Spitzer et al., 2013). From a metrological viewpoint, while the true value of a material can never be known, the use of high-quality measurements whereby all sources of uncertainty have been accounted for will likely minimize errors between the measurement result and the true value. The KCs are critical for maintaining quality in RM production as they shed light on potential biases and uncertainties in the measurement process.

Six Regional Metrological Organizations (RMOs) have been established (Fig. 13.4) to implement regional strategies in addition to serving the BIPM by organizing supplemental comparisons, monitoring quality management systems, and providing guidance to the BIPM CCs for KCs. These RMOs consist of BIPM Member States and Associate States and Economies, other countries and stakeholders within the region, and entities including NMIs beyond the region who hold invested economic interests in helping to strengthen measurement capabilities for the RMO. For example, within AFRIMETS, its "Ordinary" voting members consist of three groups: BIPM Member States (South Africa, Kenya, Egypt, and Morocco), BIPM Associate States and Economies (Tanzania, Mauritius, Namibia, Botswana, Ethiopia, and Ghana), and legal metrology institutes (LMIs) not associated with BIPM. AFRIMETS Associate Members (nonvoting) include NMIs such as Physikalisch-Technische Bundesanstalt (PTB, Germany) and LMIs outside of Africa, which may aid in strengthening the regional measurement capabilities and/or provide traceability to their national standards. The structure enables the RMOs to organize their comparisons and, at times, provide guidance to help associate members improve their capabilities such that they qualify for full BIPM membership. The organization and structure of the global metrology system are established to ensure confidence that the standards needed to drive the global economy have the highest quality and lowest uncertainties.

One example of regional coordination is through the European Association of National Metrology Institutes (EURAMET), a collective of representatives from 38 states as of June 2024. In 2014, EURAMET established the European Metrology Network (EMN) for Climate and Ocean Observation to bring together expertise from 44 NMIs and Designated Institutes (DIs) along with academic groups and businesses to collaborate on a set of projects to improve, for example, the quality of Essential Climate Variables by increasing linkages of the variables to SI units and improving the accuracy of measurements (Woolliams et al., 2019). The group to-date has been successful in outlining the key stakeholders in the areas of terrestrial, atmospheric, and oceanographic monitoring and has worked with these stakeholders to understand the critical need to provide certified RMs (CRMs) and RMs to support the monitoring of Essential Ocean Variables (EOVs). In addition to the supply of RMs, this group is promoting the idea that laboratories providing data to national and global repositories should seek accreditation for their measurement capabilities in line with the guidance outlined in ISO/IEC 17025 (International Organization for Standardization and International Electrotechnical Commission, 2017) or ISO 17034 (International Organization for Standardization, 2016). This network is a leading example of the integration of metrological perspectives with global environmental monitoring.

Within the BIPM, CCQM has a range of activities closely related to the development of seawater standards and, particularly, for the determination of both TA and DIC. The determination of TA requires the accurate standardization of a hydrochloric acid titrant in a background of sodium chloride. The CCQM EAWG organized a KC on the standardization of hydrochloric acid titrant (CCQM-K73), with the last reported comparison conducted in 2020 (Pratt et al., 2013; Bastkowski et al., 2020). For DIC measurements, a seawater sample of known mass is acidified to convert all the inorganic carbon species (H_2CO_3, HCO_3^-, and CO_3^{2-}) to CO_2 gas. The gas is extracted from seawater using a vacuum line, which isolates and purifies CO_2 through a series of sublimation steps. The quantity of CO_2 evolved from the original seawater sample is determined using digital manometry, which relies on accurate volume, pressure, and temperature determinations. The final steps in this measurement process to quantify CO_2 are quite similar to the determinations of CO_2 in air or gas mixtures performed by members of the CCQM GAWG. In 2019, the GAWG conducted a Pilot Study CCQM-P188 in parallel with KC CCQM-K120 on the determination of CO_2 in air. This study provided comparisons of CO_2 values obtained by FTIR (Fourier-transform infrared spectroscopy) and GC-FID (gas chromatography with flame ionization detection) (Flores et al., 2019). In both instances, the KCs give insight into uncertainties associated with these measurements and help assess the participants' abilities to provide certified values on similar materials. We expect that to envelop ocean carbon RMs in the global metrology system, studies such as these will be evermore important in solidifying the ability of NMIs or other laboratories to certify seawater carbonate system RMs. Ideally, future KCs directly examining CO_2 in seawater should be conducted; however, the challenge is gaining support within the NMIs to prioritize these measurement capabilities.

13.4 Classification of RMs and Their Uses

While the linkages between the agreed-upon SI units and national standards are established through the work of NMIs and DIs, the connection between the SI units and users can be understood through a hierarchy of RMs (Fig. 13.5). Metrological traceability provides connections between the SI units and materials routinely adopted by end-users (International Organization for Standardization, 2007). For the seawater carbonate system research community, this can be understood as a chain progressing from primary CRMs to RMs and then to Quality Control Materials (QCMs). National governments (NMIs) are confirmed as authorized bodies through their participation in the BIPM where their CMC claims are established; thus, they can ensure standardization to the top level of the traceability chain. Additionally, they routinely adhere to guidance outlined in ISO 17034 (International Organization for Standardization, 2016) and ISO/IEC 17025 (International Organization for Standardization & International Electrotechnical Commission, 2017) for RMs producers and testing and calibration laboratories, respectively.

Within chemical metrology, at the highest level, direct linkages to the SI units are created using primary reference measurement procedures (RMPs) to value-assign primary CRMs (high purity materials). Here we use the term 'primary CRM' in the same sense in which the term 'primary measurement standard' has been defined within the VIM–International Vocabulary of Metrology–(BIPM et al., 2012). The criterion for producing primary CRMs is strict. Primary RMPs, used to value-assign

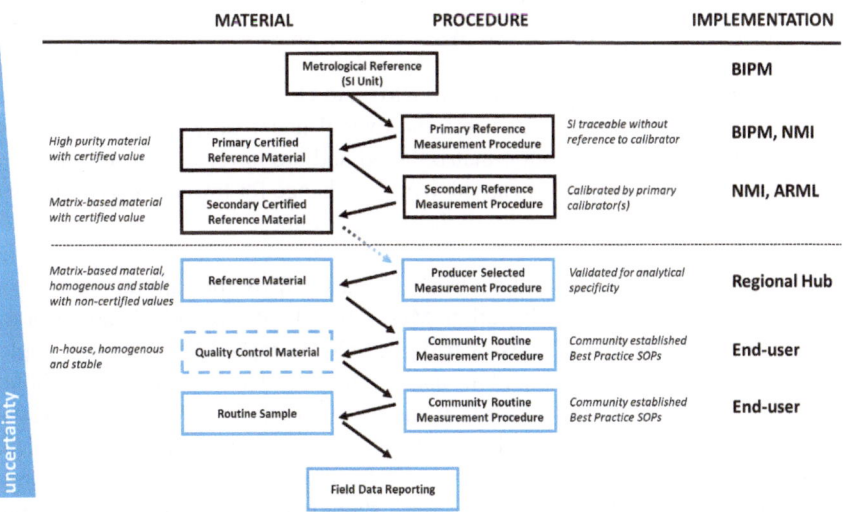

Fig. 13.5 General traceability chain of the seawater carbonate system measurements illustrating the relationship between calibrator type, measurement procedures, and entity responsible for implementation where NMI is the national metrology laboratory and ARML is an accredited reference (calibration) laboratory. Solid lines represent known links in the traceability chain, while dotted lines represent contingent linkages and usage by the community

primary CRMs, are performed without reference to a calibrator having assigned values of the same kind being measured. For example, to establish metrological traceability to the amount of substance, there are only a few measurement procedures, such as gravimetry and coulometry, and freezing point depression, which qualify as primary RMPs (Milton & Quinn, 2001). Secondary CRMs, which are often useful for customers who require a matrix-based material with a similar composition to their samples, are often calibrated against a primary calibrator using secondary RMPs. In cases where the primary RMPs are applicable, the certification of secondary CRMs can be performed solely with the high-caliber method. However, when the high-caliber method is not an option, secondary RMPs, which combine at least two different approaches, and which can include measurements performed by interlaboratory comparison of expert laboratories, can be used to establish certified values. The CRM label (or SRM for NIST and ERM for EURAMET) requires that the material is not only homogeneous and stable, but also that the true value lies within the stated uncertainty to a specified level of confidence as given in a certificate supplied to the user. At NIST (National Institute of Standards and Technology; U.S.), certified values have "the highest confidence in that all known or suspected sources of bias and imprecision have been considered and any contributions they may make to measurement uncertainty have been quantified and are expressed in the reported uncertainty."

RMs lacking certified values consist of a broader category whose only specifications are that the material is sufficiently homogeneous and stable with reference to specified property values. The values for the materials are assigned using well-established methods, which have been calibrated sufficiently based on the user community's fitness-for-purpose (Magnusson & Örnemark, 2014), and full uncertainty budgets are not required. As the traceability chain progresses from the SI units (or other agreed-upon standards) to RMs and QCMs, the uncertainty for each successive level gets incorporated into the overall uncertainty budget at the next level in the chain. RMs are often matrix-based and are lower in cost as compared to CRMs, while the CRM prices are higher in exchange for lower uncertainties and greater confidence in the reliability of the certified values.

The main purpose of using RMs is to obtain the correct analysis result. Valid analysis results can be obtained by using primary CRMs for calibration, method validation, and instrument qualification. Matrix-based CRMs and RMs are ideal for recovery studies. The lower costs of RMs and QCMs make them ideal for implementing quality control throughout the analysis process. Ideally, calibrations should be conducted using high-purity CRMs. In the absence of these materials, the void can be filled by calibrating with a matrix-based material. The accuracy and validity of the calibration should be checked with a material that is different from the material used for calibration. The CRM for quality control should preferably not be equal to one of the calibration points and should not be too close to the calibration upper-lower limit values.

The current seawater RMs for the carbonate system were designed to improve quality control and to provide harmonization of analysis across time and space. These materials allowed laboratories worldwide to determine the accuracy and validity of the analysis results as compared to the single RM producer and have enabled the community to assess change over time (where change is seen in the environments studied and even changes in laboratory personnel). The methods used by Prof. Andrew Dickson's laboratory at SIO were developed to bottle and characterize the materials to ensure the stability and homogeneity of every batch produced. One reality is that even though the materials are often referred to by the community as "CRMs", based on practices defined within the field of metrology and based on the ISO definitions, these materials do not fit the current criteria for CRM. They are indeed RMs proven to be stable and homogenous but with non-certified values. To qualify as CRMs, they would require a somewhat more rigorous value assignment process, including the development of uncertainty budgets that incorporate all significant components of uncertainty. The SIO RMs are currently reported with measurement repeatability as the only source of uncertainty. The traceability chain to the SI units may or may not be established. With the newly proposed model for RM production, we hope that in the future, these materials can fully qualify as secondary CRMs.

13.5   Re-Envisioning the Production and Certification Model of the Seawater RMs for the Carbonate System

The last few years have illustrated that significant geopolitical and global public health crises can rapidly interrupt or halt regular distribution networks and supply chains. With that in mind, a new production and certification model for seawater RMs for the carbonate system is being discussed, taking as one of the main focuses providing sufficient resilience in the system to minimize the impact of any restricted movement between nations and regions.

The proposed production and certification model for seawater RMs for the carbonate system with improved resilience would consist of at least three globally distributed production and certification centers (regional hubs; Fig. 13.6). For each regional hub, production would be separated from certification, with the former likely occurring in institutions close to the sea (ensuring continuous access to clean low-nutrient seawater) and the latter likely occurring at NMIs (ensuring compliance with metrological standards). Production centers would perform stability and homogeneity assessments of the RM batches and would send a selected number of bottles within each batch to the certification centers where the batches would be value-assigned for the seawater carbonate system variables. The certification for each batch would also include the uncertainty budget. Procedures should conform to guidelines outlined in ISO 17034 (International Organization for Standardization, 2016) and the relevant ISO guides and standards to which it refers. Metrological traceability should all be well-documented, and homogeneity and stability testing should follow rigorous experimental and statistical design principles.

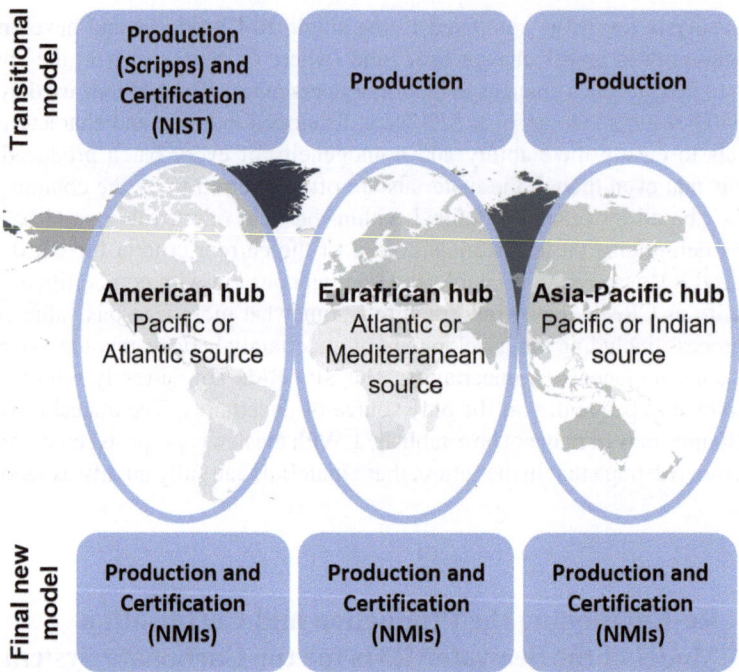

Fig. 13.6 Proposed final and transitional new model for production and certification of seawater RMs for the carbonate system

A model for the proposed program could follow the established NIST Traceable Reference Material (NTRM) Program for Gas Standards (Dorko et al., 2015). The program was designed to help meet the air quality monitoring requirements set forth by the U.S. EPA (U.S. Environmental Protection Agency) with the Clean Air Act. The demand for gas standards, similar to the seawater RMs for the carbonate system, far exceeded the ability of NIST to supply. Specialty gas companies, who produce the bulk supply of gas standards, work directly with NIST to provide SI traceable values on those gas standards and ensure their quality. NIST provides guidance and oversight of the production of NTRM gas standards using protocols defined by NIST, which include producer batch preparation, producer stability and homogeneity analyses, producer data validation, NIST batch value assignment, and evaluation of uncertainty according to the Guide to the Expression of Uncertainty in Measurement (GUM) (Williams, 2016). NIST staff members periodically conduct quality assessments of producer facilities, and NIST serves as the assessment laboratory for the mandatory EPA Protocol Gas Verification Program (PGVP), or EPA blind audit as it is commonly called, which was put in place to assure the accuracy of EPA Protocol gas standards produced and disseminated to end users by specialty gas companies utilizing NTRM standard.

In the oceanographic community, we envision that a similar structure is needed to aid in the sustained production of high-quality seawater RMs for the carbonate system and, in doing so, seawater RM programs will become integrated into the existing metrological infrastructure. NMIs, in partnership with production centers, would ideally participate in relevant KCs to ensure comparability of measurement capabilities and mutual recognition of the certificates provided with CRMs. Quality management systems for all participating bodies will undergo regular audits. NMIs and production centers would work together to ensure the user community has access to the best guidance for producing in-house or working RMs. These efforts will not only support the existing oceanographic research community but will also enhance efforts to expand support for monitoring CO_2 in coastal regions, which presents its challenges due to the proximity to terrestrial inputs.

In thinking of the new vision and knowing the challenges associated with setting up production centers, it is imperative to derive a transitional production model. Given the current impossibility of reproducing SIO's laboratory elsewhere in the near future, the transitional production and certification model is envisioned where the SIO's bottling facility is maintained to provide continuity in the seawater RM supply and will work in partnership with a metrology laboratory (most probably NIST in the U.S.) to begin certifying the RMs (Fig. 13.6). NIST is developing their measurement capabilities to certify seawater for TA and DIC and in doing this they will work closely with SIO to understand the measurement processes they have developed since the start of their RM program. Since the seawater carbonate system research community's measurements have been traceable to the SIO's laboratory for the past three decades, an important component of the transition will involve bilateral comparisons between SIO and NIST on every aspect of the measurement process (e.g., recovery of total CO_2 from a known mass of seawater and determination of total CO_2 from a purified CO_2 gas sample). These comparisons will document the concordance of the measurement approaches taken by both laboratories and will ensure sustained quality and continuity in data acquired by the seawater carbonate system research community. For example, if the measurement approach of NIST yields values that are not statistically in agreement with SIO, further investigation into the quality of the new measurement process and the SIO approach will be required. It is also important to note that the traceability of each variable will be assessed given the resources at the NMI (or NIST in the transition phase). In fact, the traceability of values for each variable extends beyond the simplified chain represented in Fig. 13.5, which excludes specific SI units and reference measurement procedures. For example, TA traceability could be established through coulometric titration, which is traceable to the SI units for current, time, and mass using Faraday's constant. Similar assessments will be conducted at the NMI to fully establish traceability of both TA and DIC measurements. With both parameters, the complete uncertainty budget will be derived.

During the implementation of the transitional model, regional hubs will concurrently be created in areas where there is access to seawater and a willingness to provide facilities and staff. These hubs will strengthen their capabilities to supply laboratories within their network with stable and homogeneous batches of seawater. With the help of either a centralized laboratory (such as NIST) or the implementation of an interlaboratory certification model, these hubs can supply batches of seawater with non-certified values for DIC and TA. In the absence of multiple qualified certification laboratories, a central NMI (most probably NIST in the U.S.) during this period would help to ensure consistency in the value assignment of these RMs on a global scale.

It is necessary to create best practices for RM production and certification for the regional hubs. Each producer must ensure that every batch is adequately tested for homogeneity and stability before certification. Once certification is performed, periodic stability assessments are needed to demonstrate the continued quality of the RMs based on the best assessment of seawater parameter stability. Additionally, guidance on the uncertainty quantification of each batch and, ideally, the pathway to SI traceability is critical to the ongoing success of these efforts. The certification testing process would ideally be refined through assessments involving the scientific community.

13.5.1 Advantages and Challenges of the New Model

The global distribution of production and certification for the seawater RMs for the carbonate system would result in a more resilient system with fewer bottlenecks and better statistics. Having multiple production and certification centers would improve international availability in addition to ensuring resilience. Besides, the new production scheme of seawater RMs for the carbonate system would ensure a wider composition of the RMs, with RMs produced with natural seawater from different ocean basins, therefore covering one of the identified community needs (see Sect. 13.2). Creating regional hubs would also help reduce the price of RMs by reducing logistics and distribution costs. Furthermore, the collaboration with NMIs would enable the use of the global metrological infrastructure in support of this new model of production and certification of seawater RMs for the carbonate system.

The partnership between oceanographic institutions, leading the production of RMs, and NMIs, leading the certification, would help to tie measurements into the global metrology system to ensure long-term traceability and extend the provision of RMs to other seawater carbonate system variables, such as pH and pCO_2, which currently lack seawater RMs. This would particularly benefit developing countries because pH is one of the most widely measured variables of the seawater carbonate system in areas where other equipment is not readily available. On the other hand, additional nonroutine variables that are not part of the seawater carbonate system, such as the $^{13}C/^{12}C$ isotope ratio, could also be certified in the seawater RMs. Access to technology to define isotope ratios is increasing in some laboratories since these ratios are tracers of water masses and biological processes.

The new scheme of production and certification of seawater RMs for the carbonate system comes with its challenges. The first one is the establishment of production centers. Those centers would require continuous access to clean low-nutrient seawater, adequate space, and equipment. This is especially important given the expected future demand for seawater RMs for the carbonate system. The current producer produces between 7 and 9 batches per year requiring approximately 720 L per batch. If a continuous supply of seawater is not possible, an integrative effort with international cruise programs, such as GO-SHIP (Sloyan et al., 2019), could provide a relatively continuous supply of open ocean seawater for RM production. In addition to seawater access, new production centers will need dedicated space to prepare RMs and store them once they are produced. For example, the current producer has over 370 m² of chemical laboratory space solely reserved for this purpose. However, an alternate model is one where the existing laboratory facilities are used. Additionally, specialized equipment is needed for batch sterilization and bottling (including the bottles) and performing acceptance testing to ensure stability and homogeneity assessments of the RMs. The new production centers would also need personnel exclusively dedicated to the production of the RMs. The current producer of RMs needs at least three full-time staff.

Beyond the establishment of regional production centers, another level of global organization will be needed to ensure the RMs produced by all centers are comparable. The robustness of the regional hubs will require periodic stability testing and a commitment or requirement for each production center to participate in interlaboratory exercises (proficiency testing). This quality assurance of the RMs produced by different hubs would reduce the user's concern of decreased data quality derived from using RMs produced by different manufacturers. Additionally, if the certification and traceability are established using NMIs or designated institutes, it will be critical to evaluate their measurement capabilities within KCs organized by the BIPM. Currently, one NMI (NIST) is developing its measurement capabilities for both DIC and TA, but an ideal model will involve multiple NMIs establishing these measurement procedures. The proposed regional model (Fig. 13.6) considers the current CMC claims documented by the BIPM such that within each region, at least two NMIs have technical capabilities that could be applied to certify seawater for both TA and DIC.

Finally, the biggest challenge to developing the proposed new model of production and certification of seawater RMs for the carbonate system is the investment needed. The required investment likely will come from either a governmental source or a commercial entity, and perhaps a combination of both. The current seawater RMs for the carbonate systems are partially subsidized by the National Science Foundation (NSF in the U.S.) (Dickson, 2010), allowing an affordable cost per bottle of ca. 65 USD (excluding shipping and customs). A similar subsidizing scheme would need to be in place for the proposed future production and certification model, so users are not cost-prohibited from acquiring RMs. Partnerships with NMIs may help to defer some of the analytical costs as there are NMIs, such as NIST in the U.S., PTB in Germany, and LNE in France, currently researching measurement services related to seawater carbonate system measurements. In the future, these measurement services could be provided within the global distribution network on a fee-per-service basis. As such, it will still be necessary to have partners willing to provide investments to both establish and likely supplement these distribution centers. Additionally, a subsidizing scheme to reduce RM prices for developing countries would significantly improve the accessibility of seawater RMs for the carbonate system to a wider number of laboratories, ensuring the production of high-quality measurements of the variables of the seawater carbonate system worldwide. One of the ways RM supply could be subsidized is by reducing or eliminating importation taxes, which can represent a price increase of around 20% of the bottle's original price. The Ocean Foundation (ToF; https://oceanfdn.org/) is interested in working on the elimination of duties of RM shipments through regional conventions (ToF pers. comm.). The reduced cost of seawater RMs for the carbonate system would improve our ability to meet the United Nations SDG Target 14.3 "Minimize and address the impacts of ocean acidification, including through enhanced scientific cooperation at all levels", and would contribute to a better understanding of how the ocean regulates our climate. The funding for a more resilient production and certification model for seawater RMs is fundamental to achieving climate-quality observations that allow the detection of long-term trends associated with ocean acidification (Newton et al., 2015).

13.5.2   Remaining Challenges

There are several ongoing innovations that are needed to improve the quality of the existing seawater carbonate system RMs. For example, finding alternative packaging to the glass bottles, which are currently used for the TA and DIC RMs, will lower costs and improve production capabilities, which are currently limited by the glass bottle supply chain. SIO encourages customers to return the bottles to receive a discount on subsequent orders. The glass bottles are not always returned and thus represent an ongoing bottleneck in the production process. Aside from supply chain nuances, the glass bottle contributes 40% of the RM weight and results in a substantial expense in shipping costs. Alternative lightweight, flexible, and/or single-use packaging will considerably reduce the costs of the RMs.

The preservation method is another area of research to be explored. Currently, mercury (II) chloride ($HgCl_2$) is used to effectively maintain the RM values for at least three years. However, this preservative is actively discouraged due to its toxicity (e.g., the Minimata Convention of Mercury), and it is a banned chemical that cannot be shipped into some countries. Therefore, finding alternative preservation for these RMs could increase their availability in nations that currently cannot receive such items. The potential preservative should be a biocide and not compromise the acid–base chemistry in seawater.

Alternatives to solve both packaging and preservative problems are being explored. For example, JAMSTEC (Japan Agency for Marine-Earth Science and Technology), in collaboration with KANSO TECHNOS CO., LTD, is developing non-toxic seawater RMs for TA and DIC using sterilized natural seawater by autoclaving and bottling in aluminum bottles similar to those used for sodas (Murata, 2010). Other laboratories have explored the possibility of pasteurization for seawater dissolved inorganic nutrient samples (Daniel et al., 2012); while another potential solution comes through the reformulation of the seawater RMs for the carbonate system to be in salt form, such as the method explored for dissolved inorganic nutrient standards (Pagliano et al., 2022).

Certifying values for pH and $p\text{CO}_2$ in the available seawater RMs for the carbonate system represents another challenge. For pH, technical issues are particularly problematic in seawater. First, seawater has a high ionic strength, which causes problems when using conventional pH calibration standards as the activity of hydrogen ion changes with ionic strength and, therefore, salinity. For this reason, spectrophotometric indicator dyes have often been recommended for seawater pH measurements because they can be calibrated over a range of temperatures and salinities using synthetic seawater buffers. The indicator dyes present their own challenges due to the presence of impurities, which cause pH-dependent offsets between the actual and measured pH (e.g., Carter et al., 2013, 2018; Fong & Dickson, 2019; Álvarez et al., 2020; Takeshita et al., 2021). Additionally, a massive effort is required to "certify" a batch of purified indicator dye against a range of artificial seawater buffers prepared at multiple pH values and salinities. When potentiometric electrode measurements are the only option available, electrodes can be calibrated using these in-house artificial seawater buffers. The challenge is that the buffer preparation is somewhat complex requiring seven different components, some of which require assays prior to use. With $p\text{CO}_2$, the biggest challenge is supplying users with a seawater matrix standard in a form easily integrated into existing measurement systems. As the efforts move forward such that these RMs are produced in multiple locations, we must also be aware that innovation is needed to improve the products supplied to the user community.

13.6 Conclusion

The seawater community will greatly benefit from the restructuring of the existing seawater RM model for the carbonate system. Over the past few years, it has become evident that the oceanographic community cannot depend on one single laboratory to provide global access to seawater RMs for the carbonate system. A new production and certification model, which involves the establishment of regional hubs in partnership with SIO and NMIs, is being discussed. Although this new model faces inherent challenges (e.g., finding locations close to seawater sources that have the facilities, personnel, and technical expertise to provide values on the materials), there is a great benefit to expanding the production to a global scale. Users will have access to multiple sources of seawater, which can provide matrix standards similar in composition to their areas of study. We also believe the new model may help to reduce the costs of production by eliminating some of the current shipping costs, which can represent a price increase of around 40% of the RM's original price. Partnerships with NMIs will ensure these RMs are connected to the global metrology system, are metrologically traceable, and that batches produced across the globe are delivered at a consistent quality. Only time and effort will determine if this new model provides to this important community, whose data are intrinsically linked to our assessment of climate change, long-term sustained access to these critical RMs.

13.7 Competing Interests

The authors have no conflicts of interest to declare that are relevant to the content of this chapter.

Acknowledgements This activity is supported by the International Ocean Carbon Coordination Project and the author team acknowledges funding to IOCCP provided by a grant to SCOR from the U.S. National Science Foundation (OCE-2140395).

References

Acquafredda, M., Cochran, C., Busch, D. S., Jewett, L., Edmonds, H., & Dickson, A. (2022). *Understanding the current use and future needs of CO_2 in seawater certified reference materials* (NOAA Technical Memorandum OAR-OAP-5; 21). U.S. Department of Commerce.

Álvarez, M., Fajar, N. M., Carter, B. R., Guallart, E. F., Pérez, F. F., Woosley, R. J., & Murata, A. (2020). Global ocean spectrophotometric pH assessment: Consistent inconsistencies. *Environmental Science & Technology, 54*(18), 10977–10988. https://doi.org/10.1021/acs.est.9b06932

Bakker, D. C. E., Pfeil, B., Landa, C. S., Metzl, N., O'Brien, K. M., Olsen, A., Smith, K., Cosca, C., Harasawa, S., Jones, S. D., Nakaoka, S., Nojiri, Y., Schuster, U., Steinhoff, T., Sweeney, C., Takahashi, T., Tilbrook, B., Wada, C., Wanninkhof, R., Alin, S. R., Balestrini, C. F., Barbero, L., Bates, N. R., Bianchi, A. A., Bonou, F., Boutin, J., Bozec, Y., Burger, E. F., Cai, W. -J., Castle, R. D., Chen, L., Chierici, M., Currie, K., Evans, W., Featherstone, C., Feely, R. A., Fransson, A., Goyet, C., Greenwood, N., Gregor, L., Hankin, S., Hardman-Mountford, N. J., Harlay, J., Hauck, J., Hoppema, M., Humphreys, M. P., Hunt, C. W., Huss, B., Ib´ anhez, J. S. P., Johannessen, T., Keeling, R., Kitidis, V., Körtzinger, A., Kozyr, A., Krasakopoulou, E., Kuwata, A., Landschützer, P., Lauvset, S. K., Lef' evre, N., Lo Monaco, C., Manke, A., Mathis, J. T., Merlivat, L., Millero, F. J., Monteiro, P. M. S., Munro, D. R., Murata, A., Newberger, T., Omar, A. M., Ono, T., Paterson, K., Pearce, D., Pierrot, D., Robbins, L. L., Saito, S., Salisbury, J., Schlitzer, R., Schneider, B., Schweitzer, R., Sieger, R., Skjelvan, I., Sullivan, K. F., Sutherland, S. C., Sutton, A. J., Tadokoro, K., Telszewski, M., Tuma, M., van Heuven, S. M. A. C., Vandemark, D., Ward, B., Watson, A. J., Xu, S. (2016). A multi-decade record of high-quality fCO2 version 3 of the surface ocean CO_2 data in atlas (SOCAT). *Earth System Science Data, 8,* 383–413. https://doi.org/10.5194/essd-8-383-2016

Bastkowski, F., Sander, B., Lozano, H., Puelles, M., Snedden, A., Deleebeeck, L., Asakai, T., Hwang, E., Jo, K., Ortiz-Aparicio, J. L., Montero-Ruiz, J., Roziková, M., Kozlowski, W., Quezada, H. T., Morales, L. V., Ahumada, D. A., Borges, P. P., Neves, R. S., Sobral, S. P., Uysal, E., Liv, L., Prokunin, S., Dobrovolskyi, V., Stoica, D., Wu, B., Ma, L., Máriássy, M., Hanková, Z., Sobina, A., & Shimolin, A. (2020). Key Comparison CCQM-K73.2018 Amount Content of

H^+ in Hydrochloric Acid (0.1 mol·kg$-$1). *Metrologia, 58*(1A), 08002. https://doi.org/10.1088/0026-1394/58/1A/08002

Bates, N., Astor, Y., Church, M., Currie, K., Dore, J., Gonaález-Dávila, M., Lorenzoni, L., Muller-Karger, F., Olafsson, J., & Santa-Casiano, M. (2014). A time-series view of changing ocean chemistry due to ocean uptake of anthropogenic CO_2 and ocean acidification. *Oceanography, 27*(1), 126–141. https://doi.org/10.5670/oceanog.2014.16

BIPM, IEC, IFCC, ILAC, IUPAC, IUPAP, ISO, and OIML (2012) International vocabulary of metrology – Basic and general concepts and associated terms (VIM), JCGM 200:2012 (3rd edition). http://www.bipm.org/vim

Bushinsky, S. M., Takeshita, Y., & Williams, N. L. (2019). Observing changes in ocean carbonate chemistry: Our autonomous future. *Current Climate Change Reports, 5*(3), 207–220. https://doi.org/10.1007/s40641-019-00129-8

Carter, B. R., Feely, R. A., Williams, N. L., Dickson, A. G., Fong, M. B., & Takeshita, Y. (2018). Updated methods for global locally interpolated estimation of alkalinity, pH, and nitrate. *Limnology and Oceanography: Methods, 16*(2), 119–131. https://doi.org/10.1002/lom3.10232

Carter, B. R., Radich, J. A., Doyle, H. L., & Dickson, A. G. (2013). An automated system for spectrophotometric seawater pH measurements. *Limnology and Oceanography: Methods, 11*(1), 16–27. https://doi.org/10.4319/lom.2013.11.16

Catherman, C. (2021). *The world's only source of critical seawater samples could dry up.* https://doi.org/10.1126/science.acx9252

Clayton, T. D., & Byrne, R. H. (1993). Spectrophotometric seawater pH measurements: Total hydrogen ion concentration scale calibration of m-cresol purple and at-sea results. *Deep Sea Research Part I: Oceanographic Research Papers, 40*(10), 2115–2129. https://doi.org/10.1016/0967-0637(93)90048-8

Daniel, A., Kérouel, R., & Aminot, A. (2012). Pasteurization: A reliable method for preservation of nutrient in seawater samples for inter-laboratory and field applications. *Marine Chemistry, 128–129*, 57–63. https://doi.org/10.1016/j.marchem.2011.10.002

Dickson, A. (2010). Standards for ocean measurements. *Oceanography, 23*(3), 34–47. https://doi.org/10.5670/oceanog.2010.22

Dickson, A. G., Sabine, C. L., & Christian, J. R. (2007). *Guide to best practices for ocean CO_2 measurements.* PICES.

Dobson, K. L., Newton, J. A., Widdicombe, S., Schoo, K. L., Acquafredda, M. P., Kitch, G., Bantelman, A., Lowder, K., Valauri-Orton, A., Soapi, K., Azetsu-Scott, K., & Isensee, K. (2022). Ocean acidification research for sustainability: Co-designing global action on local scales. *ICES Journal of Marine Science*, fsac158. https://doi.org/10.1093/icesjms/fsac158

Doney, S. C., Busch, D. S., Cooley, S. R., & Kroeker, K. J. (2020). The impacts of ocean acidification on marine ecosystems and reliant human communities. *Annual Review of Environment and Resources, 45*(1), 83–112. https://doi.org/10.1146/annurev-environ-012320-083019

Dorko, W. D., Kelley, M. E., & Guenther, F. R. (2015). The NIST traceable reference material program for gas standards. *NIST Special Publication 260–126 Rev 2013*, 41. https://doi.org/10.6028/NIST.SP.260-126

Flores, E., Viallon, J., Choteau, T., Moussay, P., Idrees, F., Wielgosz, R. I., Meyer, C., & Rzesanke, D. (2019). Report of the pilot study CCQM-P188 (in parallel with CCQM-K120.a and b). *Metrologia, 56*(1A), 08012. https://doi.org/10.1088/0026-1394/56/1A/08012

Fong, M. B., & Dickson, A. G. (2019). Insights from GO-SHIP hydrography data into the thermodynamic consistency of CO_2 system measurements in seawater. *Marine Chemistry, 211*, 52–63. https://doi.org/10.1016/j.marchem.2019.03.006

Friedlingstein, P., Jones, M. W., O'Sullivan, M., Andrew, R. M., Bakker, D. C. E., Hauck, J., Le Quéré, C., Peters, G. P., Peters, W., Pongratz, J., Sitch, S., Canadell, J. G., Ciais, P., Jackson, R. B., Alin, S. R., Anthoni, P., Bates, N. R., Becker, M., Bellouin, N., ... Zeng, J. (2022). Global carbon budget 2021. *Earth System Science Data, 14*(4), 1917–2005. https://doi.org/10.5194/essd-14-1917-2022

García-Ibáñez, M. I., Takeshita, Y., Guallart, E. F., Fajar, N. M., Pierrot, D., Pérez, F. F., Cai, W.-J., & Álvarez, M. (2022). Gaining insights into the seawater carbonate system using discrete fCO_2 measurements. *Marine Chemistry*, 104150. https://doi.org/10.1016/j.marchem.2022.104150

Gruber, N., Clement, D., Carter, B. R., Feely, R. A., van Heuven, S., Hoppema, M., Ishii, M., Key, R. M., Kozyr, A., Lauvset, S. K., Lo Monaco, C., Mathis, J. T., Murata, A., Olsen, A., Perez, F. F., Sabine, C. L., Tanhua, T., & Wanninkhof, R. (2019). The oceanic sink for anthropogenic CO_2 from 1994 to 2007. *Science, 363*(6432), 1193–1199. https://doi.org/10.1126/science.aau 5153

Hassoun, A. E. R., Bantelman, A., Canu, D., Comeau, S., Galdies, C., Gattuso, J.-P., Giani, M., Grelaud, M., Hendriks, I. E., Ibello, V., Idrissi, M., Krasakopoulou, E., Shaltout, N., Solidoro, C., Swarzenski, P. W., & Ziveri, P. (2022). Ocean acidification research in the Mediterranean Sea: Status, trends and next steps. *Frontiers in Marine Science, 9*. https://doi.org/10.3389/fmars. 2022.892670

International Organization for Standardization and International Electrotechnical Commission. (2007). *ISO/IEC Guide 99:2007 International vocabulary of metrology—Basic and general concepts and associated terms (VIM)*. https://www.iso.org/standard/45324.html

International Organization for Standardization. (2016). *General requirements for the competence of reference material producers (ISO Standard No. 17034:2016)*. International Organization for Standardization.

International Organization for Standardization. (2017). *General requirements for the competence of testing and calibration laboratories (ISO/IEC Standard No. 17025:2017)*. International Organization for Standardization.

Kroeker, K. J., Kordas, R. L., Crim, R., Hendriks, I. E., Ramajo, L., Singh, G. S., Duarte, C. M., & Gattuso, J.-P. (2013). Impacts of ocean acidification on marine organisms: Quantifying sensitivities and interaction with warming. *Global Change Biology, 19*(6), 1884–1896. https://doi.org/10.1111/gcb.12179

Lauvset, S. K., Lange, N., Tanhua, T., Bittig, H. C., Olsen, A., Kozyr, A., Alin, S. R., Álvarez, M., Azetsu-Scott, K., Barbero, L., Becker, S., Brown, P. J., Carter, B. R., da Cunha, L. C., Feely, R. A., Hoppema, M., Humphreys, M. P., Ishii, M., Jeansson, E., Jiang, L.-Q., Jones, S. D., Lo Monaco, C., Murata, A., Müller, J. D., Pérez, F. F., Pfeil, B., Schirnick, C., Steinfeldt, R., Suzuki, T., Tilbrook, B., Ulfsbo, A., Velo, A., Woosley, R. J., & Key, R. M. (2022). GLODAPv2.2022: The latest version of the global interior ocean biogeochemical data product. *Earth System Science Data Discussions*, 1–37. https://doi.org/10.5194/essd-2022-293

Magnusson, B., & Örnemark, U. (2014). *Eurachem Guide: The Fitness for Purpose of Analytical Methods—A Laboratory Guide to Method Validation and Related Topics, (2nd ed. 2014)*. www. eurachem.org.

Metrology for Climate Action 2022. Retrieved November 19, 2022, from https://www.bipmwmo22. org/

Milton, M. J. T., & Quinn, T. J. (2001). Primary methods for the measurement of amount of substance. *Metrologia, 38*(4), 289. https://doi.org/10.1088/0026-1394/38/4/1

Mostofa, K. M. G., Liu, C.-Q., Zhai, W., Minella, M., Vione, D., Gao, K., Minakata, D., Arakaki, T., Yoshioka, T., Hayakawa, K., Konohira, E., Tanoue, E., Akhand, A., Chanda, A., Wang, B., & Sakugawa, H. (2016). Reviews and syntheses: Ocean acidification and its potential impacts on marine ecosystems. *Biogeosciences, 13*(6), 1767–1786. https://doi.org/10.5194/bg-13-1767-2016

Murata, A. (2010). The development of non-toxic reference materials for oceanic CO_2 measurements: Current status and future plans. In *Comparability of Nutrients in the World's Ocean* (pp. 11–30). INSS international workshop, Paris, France, February 10–12, 2009 M. Aoyama ed., Mother Tank, Tsukuba, Japan.

National Academies of Sciences, Engineering, and Medicine. (2022). A Research Strategy for Ocean Carbon Dioxide Removal and Sequestration. *Washington, DC: The National Academies Press*. https://doi.org/10.17226/26278

Neill, C., Johnson, K. M., Lewis, E., & Wallace, D. W. (1997). Accurate headspace analysis of fCO_2 in discrete water samples using batch equilibration. *Limnology and Oceanography, 42*, 1774–1783.

Newton, J., Feely, R., Jewett, E., Williamson, P., & Mathis, J. (2015). *Global Ocean Acidification Observing Network: Requirements and Governance Plan.* https://archimer.ifremer.fr/doc/00651/76343/

OCADS - Information on Batches of CO_2 in Seawater Reference Material. Retrieved November 17, 2022, from https://www.ncei.noaa.gov/access/ocean-carbon-acidification-data-system/oceans/Dickson_CRM/batches.html

Olsen, A., Lange, N., Key, R. M., Tanhua, T., Bittig, H. C., Kozyr, A., Álvarez, M., Azetsu-Scott, K., Becker, S., Brown, P. J., Carter, B. R., Cotrim da Cunha, L., Feely, R. A., van Heuven, S., Hoppema, M., Ishii, M., Jeansson, E., Jutterström, S., Landa, C. S., Lauvset, S. K., Michaelis, P., Murata, A., Pérez, F. F., Pfeil, B., Schirnick, C., Steinfeldt, R., Suzuki, T., Tilbrook, B., Velo, A., Wanninkhof, R., & Woosley, R. J. (2020). An updated version of the global interior ocean biogeochemical data product, GLODAPv2.2020. *Earth System Science Data, 12*(4), 3653–3678. https://doi.org/10.5194/essd-12-3653-2020

Pagliano, E., Nadeau, K., Mihai, O., Pihillagawa Gedara, I., & Mester, Z. (2022). From sea salt to seawater: A novel approach for the production of water CRMs. *Analytical and Bioanalytical Chemistry, 414*(16), 4745–4756. https://doi.org/10.1007/s00216-022-04098-0

Pörtner, H.-O., Karl, D. M., Boyd, P. W., Cheung, W. W. L., Lluch-Cota, S. E., Nojiri, Y., Schmidt, D. N., & Zavialov, P. O. (2014). Ocean systems. In C. B. Field, V. R. Barros, D. J. Dokken, K. J. Mach, M. D. Mastrandrea, T. E. Bilir, M. Chatterjee, K. L. Ebi, Y. O. Estrada, R. C. Genova, B. Girma, E. S. Kissel, A. N. Levy, S. MacCracken, P. R. Mastrandrea, & L. L. White (Eds.), *Climate Change 2014: Impacts, Adaptation, and Vulnerability. Part A: Global and Sectoral Aspects. Contribution of Working Group II to the Fifth Assessment Report of the Intergovernmental Panel on Climate Change* (pp. 411–484). Cambridge University Press.

Pratt, K. W., Ortiz-Aparicio, J. L., Matehuala-Sanchez, F. J., Pawlina, M., Kozlowski, W., Borges, P. P., Junior, W. B. da S., Borinsky, M. B., Puelles, A. H.-M., Hatamleh, N., Acosta, O., Nunes, J., Lito, M. J. G., Camões, M. F., Filipe, E., Hwang, E., Lim, Y., Bing, W., Qian, W., Chao, W., Hioki, A., Asakai, T., Máriássy, M., Hanková, Z., Nagibin, S., Manska, O., & Gavrilkin, V. (2013). Final report on key comparison CCQM-K73: Amount content of H^+ in hydrochloric acid (0.1 mol kg^{-1}). *Metrologia, 50*(1A), 08001. https://doi.org/10.1088/0026-1394/50/1A/08001

Riebesell, U., Zondervan, I., Rost, B., Tortell, P. D., Zeebe, R. E., & Morel, F. M. M. (2000). Reduced calcification of marine plankton in response to increased atmospheric CO_2. *Nature, 407*(6802), 364. https://doi.org/10.1038/35030078

Sabine, C., Ducklow, H., & Hood, M. (2010). International carbon coordination: Roger Revelle's legacy in the intergovernmental oceanographic commission. *Oceanography, 23*(3), 48–61. https://doi.org/10.5670/oceanog.2010.23

Sloyan, B. M., Wanninkhof, R., Kramp, M., Johnson, G. C., Talley, L. D., Tanhua, T., McDonagh, E., Cusack, C., O'Rourke, E., McGovern, E., Katsumata, K., Diggs, S., Hummon, J., Ishii, M., Azetsu-Scott, K., Boss, E., Ansorge, I., Perez, F. F., Mercier, H., Williams, M. J. M., Anderson, L., Lee, J. H., Murata, A., Kouketsu, S., Jeansson, E., Hoppema, M., & Campos, E. (2019). The Global Ocean Ship-Based Hydrographic Investigations Program (GO-SHIP): A Platform for Integrated Multidisciplinary Ocean Science. *Frontiers in Marine Science, 6*. https://doi.org/10.3389/fmars.2019.00445

Spitzer, P., Bastkowski, F., Adel, B., Dimitrova, L., Gonzaga, F. B., Jakobsen, P. T., Fisicaro, P., Stoica, D., Asakai, T., Maksimov, I., Szilágyi, Z. N., Reyes, A., Monroy, M., Canaza, G. T., Kozlowski, W., Pawlina, M., Kutovoy, V., Vyskocil, L., Mathiasova, A., … Waters, J. (2013). Final report on CCQM-K91: Key comparison on pH of an unknown phthalate buffer. *Metrologia, 50*(1A), 08016. https://doi.org/10.1088/0026-1394/50/1A/08016

Takahashi, T., Williams, R., & Bos, D. (1982). Carbonate chemistry. GEOSECS Pacific. *Expedition, 3*, 1973–1974.

Takeshita, Y., Warren, J. K., Liu, X., Spaulding, R. S., Byrne, R. H., Carter, B. R., DeGrandpre, M. D., Murata, A., & Watanabe, S. (2021). Consistency and stability of purified meta-cresol purple for spectrophotometric pH measurements in seawater. *Marine Chemistry, 236*, 104018. https://doi.org/10.1016/j.marchem.2021.104018

United Nations (2015). *Goal 14—Conserve and sustainably use the oceans, seas and marine resources for sustainable development*. https://sdgs.un.org/goals/goal14

Wanninkhof, R., & Thoning, K. (1993). Measurement of fugacity of CO_2 in surface water using continuous and discrete sampling methods. *Marine Chemistry, 44*(2), 189–204. https://doi.org/10.1016/0304-4203(93)90202-Y

Williams, J. H. (2016). Guide to the expression of uncertainty in measurement (the GUM). *Quantifying Measurement: The Tyranny of Numbers*. https://doi.org/10.1088/978-1-6817-4433-9ch6

Woolliams, E., Pascale, C., Fisicaro, P., & Fox, N. (2019). The European metrology network for climate and ocean observation. In *19th International Congress of Metrology (CIM2019)*, 05001. https://doi.org/10.1051/metrology/201905001

World Meteorological Organization (2021). *State of the Global Climate 2021*. WMO-No. 1290.

Chapter 14
Production and Usage of Reference Materials for Total Dissolved Inorganic Carbon and Total Alkalinity

Akihiko Murata, Nagisa Fujiki, and Hitoshi Mitsuda

Abstract Certified Reference Materials (CRMs) are essential for high-quality measurements. However, CRMs are sometimes unavailable. In such cases, the comparability of measurements can be maintained to some extent if homogeneous and stable reference materials (RMs) are available. In this chapter, we present examples of the preparation and use of RMs for total dissolved inorganic carbon (DIC) and total alkalinity (TA) in seawater. One example is a working RM produced in-house by JAMSTEC, whose use has been limited to R/V *Mirai* cruises. The other is the KANSO TECHNOS CO., LTD (KANSO) RM, which is commercially available in Japan. Both RMs have been used to check the performance of instruments for DIC and TA analyses and to maintain the comparability of the measurements together with the globally used RM from Scripps Institution of Oceanography. We also discuss issues related to maintaining RM production and the possibility of wider distribution of RMs.

Keywords CRM · In-house RM · Total dissolved inorganic carbon · Total alkalinity

A. Murata (✉)
Physical and Chemical Oceanography Research Group, Global Ocean Observation Research Center, Japan Agency for Marine-Earth Science and Technology, Yokosuka, Japan
e-mail: murataa@jamstec.go.jp

N. Fujiki
Marine Chemical Analysis Section, Office of Marine and Earth Environmental Analysis, Department of Marine and Earth Sciences, Marine Works Japan LTD., Yokosuka, Japan
e-mail: fujiki.nagisa@mwj.co.jp

H. Mitsuda
Laboratory for Instrumentation and Analysis, KANSO TECHNOS CO., LTD., Osaka, Japan
e-mail: mituda_hitosi@kanso.co.jp

© The Author(s) 2025
M. Aoyama et al. (eds.), *Chemical Reference Materials for Oceanography*, Springer Oceanography, https://doi.org/10.1007/978-981-96-2520-8_14

14.1   Introduction

Since it was recognized that the increase of CO_2 in the atmosphere is the main cause of global warming, CO_2 has been vigorously measured worldwide. For global-scale estimates of the carbon budget, the data collected at different times and by different groups must be highly comparable. To achieve such comparability, standards or reference materials (RMs) that are sufficiently reliable and widely used in research and among monitoring groups are essential. In the oceanographic community, this issue was discussed as early as the 1960s in international communities such as the Intergovernmental Oceanographic Commission (IOC) and the Scientific Committee on Oceanic Research (SCOR) (Dickson, 2010). However, it was not until the late 1980s, when large international ocean science programs such as the Joint Global Ocean Flux Study (JGOFS) and the World Ocean Circulation Experiment (WOCE) were initiated, that RMs for oceanic CO_2 became available to the oceanographic community (Dickson, 2001). Since then, and even after those science programs ended, RMs for total dissolved inorganic carbon (DIC) and total alkalinity (TA) in seawater have been provided by Prof. Andrew G. Dickson at Scripps Institution of Oceanography (SIO). As of 2 May 2024, the latest such RM is Batch 217. The development and maintenance of the RM for DIC and TA are detailed in Dickson (2010).

In Japan, the RMs produced at SIO began to be used in the early 1990s, for example, on a Southern Ocean cruise aboard the icebreaker *Shirase* (Ishii et al., 1998) and the WOCE/JGOFS-P2 cruise aboard the R/V *Kaiyo-Maru* (Ono et al., 1998). On these cruises, the RMs were used only for DIC measurements, because TA was not measured when the batches of the RM were originally produced. The TA values were later added to old SIO RM batches based on archived samples of the batches. It was probably during an R/V *Mirai* cruise in the late 1990s that RMs were first used in Japan for TA in conjunction with DIC measurements (Murata et al., 2002). On these cruises, in-house RMs were generally measured together with SIO RMs to supplement the SIO RM measurements. Although the production methods of the in-house RMs differ from laboratory to laboratory, they are traditionally prepared using stable water, e.g., deep water > 2000 m, to ensure comparability of measurements at least during a cruise.

In this chapter, we introduce RMs that have been used during our research cruises on the R/V *Mirai*. One is the RM produced by JAMSTEC, originally prepared to facilitate high-quality data for carbonate system properties in samples collected from the R/V *Mirai*. The other RM is commercially produced and distributed by KANSO TECHNOS CO., LTD. (hereafter abbreviated as KANSO). The use of the KANSO RM is currently limited to Japan, where it complements the SIO RM, which we will refer to in this chapter as the Dickson RM.

The RMs are defined here as sufficiently homogeneous and stable with respect to one or more specified properties, in this case, DIC, and TA, whose values are not necessarily certified. In the following chapters, in-house RM refers to RM produced by JAMSTEC.

14.2   In-House RM by JAMSTEC

14.2.1   Materials and Procedures

14.2.1.1   Sampling of Seawater

Seawater used for in-house RM production should have low concentrations of dissolved inorganic nutrients to minimize biological activity that could alter the DIC concentration. For this reason, surface seawater in the subtropical gyre of the North Pacific, where dissolved inorganic nutrient concentrations are low, is collected from a research vessel. Seawater at ~5 m depth is taken from the taps of a surface seawater distributing system on the R/V *Mirai* into 20-L plastic containers (Cubitainer®, FUJI-MORI KOGYO CO., LTD., Japan). The containers are rinsed with a small amount of seawater before sampling. Then seawater is poured into the containers. The DIC concentration at the time of collection is not necessarily measured, because the seawater is re-equilibrated with atmospheric CO_2 concentrations in a laboratory on land. After sampling, the containers are stored in a cool (~4 °C), dark place on board and on land until RM production.

14.2.1.2   Sterilization of Seawater

A schematic flow diagram and a photo of the device used for sterilization are shown in Figs. 14.1 and 14.2, respectively. The design of the device is based on Dickson (2001, 2010).

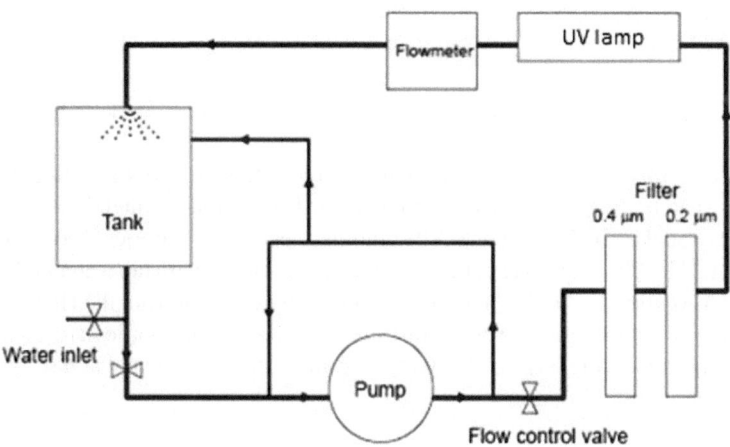

Fig. 14.1 Flow diagram of the device used for sterilization

Fig. 14.2 Photo of the apparatus used for sterilization. The right-hand device is the water flow sterilizer (UVC-WSGL-10L, KURIKOU, Japan). The yellow tank is a 200 L polyethylene tank

The tank can hold ~200 L of seawater. Sterilization is done by ultraviolet radiation and the addition of mercuric chloride (4 g per 100 L of seawater). Filters (pore size 0.4 μm and 0.2 μm) remove particles that are biologically produced in seawater during long-term storage in plastic containers. A shower is set to promote the equilibration of the CO_2 in the seawater with that in the laboratory air. The device runs continuously at a flow rate of 10 L min^{-1} for 8 h. Upon completion of the operation, the seawater is left overnight with the surface of the seawater covered by a polyethene film to stabilize the seawater.

14.2.1.3   Preparation of Bottles for RM

Borosilicate glass bottles (500 mL) with ground glass stoppers are used for RM bottling. The bottles are washed with tap water and acetone sequentially. Next, the bottles are soaked in 5% Decon® 90 solution (Decon Laboratories, Ltd., England) for about 24 h. Afterward, the bottles are thoroughly rinsed with tap water and soaked in an HCl solution (1 mol L^{-1}) for about 24 h. After removal from the HCl solution, the bottles are thoroughly rinsed with Milli-Q® water. The bottles are then dried thoroughly in a drying oven at 45 °C. Lastly, the bottles are heated at 480 °C for at least 3 h in a muffle furnace.

Fig. 14.3 Greased glass
stopper

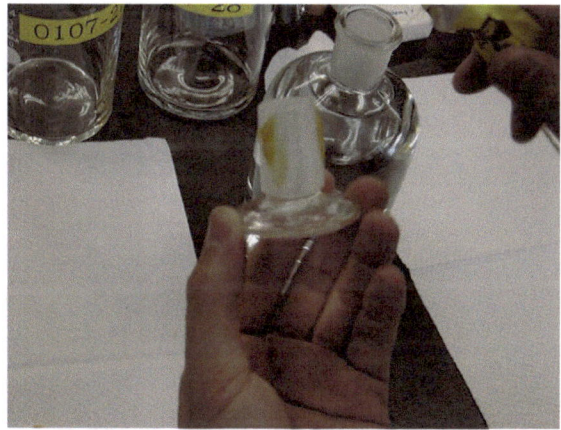

14.2.1.4 Seawater Bottling

The sterilized seawater is transferred from the tank to the glass bottles by means
of a peristaltic pump, while the seawater is filtered through filters (Whatman,
POLYCAP75TC). The filter is replaced every 50–100 L, depending on the filtration area. When introducing the sterilized seawater into the bottles, the pump speed
must be adjusted to prevent bubbles and to gently fill each bottle to ~1 cm below the
bottle neck through a drawing tube placed at the bottom of the bottle. The drawing
tube is then gently removed from the bottle to avoid wetting the ground part of bottle
neck. The bottle is then closed with a ground glass stopper lubricated with Apiezon®
M Grease (M&I Materials, Ltd.) (Fig. 14.3). Finally, the stopper is secured with a
rubber band and hose clip (Fig. 14.4).

14.2.1.5 Assigning Values

The DIC and TA values are assigned by comparison with Dickson RM. For example,
in-house RM (batch Q37) was compared with Dickson RM (batch 183) (see
Sect. 14.3) to be assigned as 2008.64 μmol kg^{-1} and 2254.0 μmol kg^{-1} for DIC and
TA, respectively.

14.3 Quality of In-House RM

In JAMSTEC, we use our in-house RM to complement the Dickson RM. Here, we
compare the DIC and TA measurements of the in-house RM with those of the Dickson
RM to assess the quality of the in-house RM.

Fig. 14.4 Batch Q43 RM
serial no. 042

14.3.1 DIC

Figure 14.5b shows a control chart (X-bar chart) of the JAMSTEC in-house RM bottled on 19 July 2019. The $\overline{X}$ is the mean value, Upper control limit (UCL) $= \overline{X} + 3 \times$ standard deviation (s), upper warning limit (UWL) $= \overline{X} + 2$ s, lower warning limit (LWL) $= \overline{X} - 2$ s and lower control limit (LCL) $= \overline{X} - 3$ s. The DIC measurements of Dickson RM (Fig. 14.5a) and the in-house RM were repeated from December 1994 until the end of January 2020, whenever the coulometer cathode and anode cell solutions were changed (see Appendix for details of the DIC measurement). The mean and standard deviation of Dickson RM measurements were 2034.24 $\pm$ 1.43 μmol kg^{-1}, while those of the in-house RM were 2008.64 $\pm$ 1.63 μmol kg^{-1}. Thus, the DIC measurements of the in-house RM are comparable to those of the Dickson RM, indicating that the in-house RM is as homogeneous and stable with respect to DIC and as useful for verifying the performance of an analytical instrument as is the Dickson RM.

14.3.2 TA

TA was measured based on single-point spectrophotometry (see Appendix for the details of TA measurement). The Dickson RM was analyzed after every 70–80 seawater samples, while the in-house RM was analyzed after every 35–40 seawater

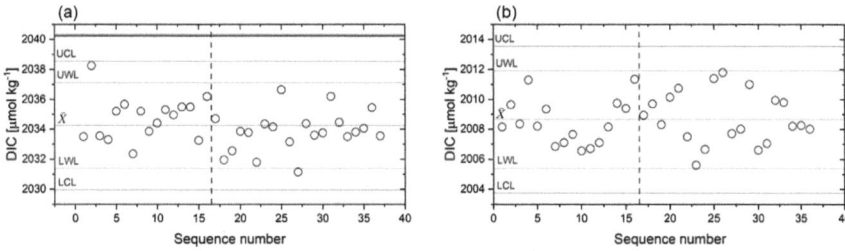

Fig. 14.5 X-bar charts of DIC measurements of (**a**) Dickson RM (batch 183) and (**b**) in-house RM (batch Q37) during cruise MR19-04. The broken vertical line in each panel indicates a time gap of about one week between the first and second legs of the cruise. The red line of panel (**a**) indicates the value of 2040.25 ± 0.43 μmol kg^{-1} given by SIO

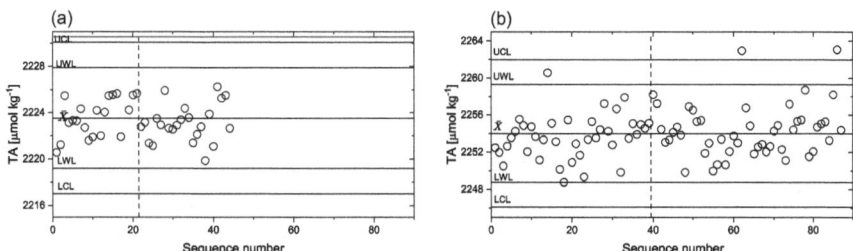

Fig. 14.6 X-bar charts of TA measurements of (**a**) Dickson RM (batch 183) and (**b**) in-house RM (batch Q37) during cruise MR19-04. The broken vertical line in each panel indicates the time gap of about one week between the first and second legs of MR19-04. The red line of panel (**a**) indicates the certified value of 2230.52 ± 0.28 μmol kg^{-1}

samples. The lower number of Dickson RM measurements compared to the in-house RM is due to the use of remnants of the RM for DIC analyses. The mean and standard deviation of Dickson RM measurements of TA were 2223.5 ± 2.2 μmol kg^{-1}, while those of the in-house RM were 2254.0 ± 2.6 μmol kg^{-1}. The dispersion of the TA measurements of the in-house RM was comparable to that of the Dickson RM (Fig. 14.6); thus, as for DIC, the in-house RM works well for checking TA measurements as well.

14.4 Commercially-Available RM by KANSO

KANSO has been supplying RM for DIC mainly within Japan since March 2000. As of 20 May 2024, an RM bottle costs ¥5,000 plus 10% sales tax, with users paying the return shipping cost. The seawater for RM production is collected from a depth of 397 m in Suruga Bay, Japan. The collected seawater is sequentially filtered through 5 and 0.2 μm media composition cartridge filters (ADVANTEC Toyo Kaisha, Ltd., Japan) and transferred to a polyethylene tank (600 L, Suiko Co., Ltd., Japan). Then, mercuric chloride is added to the tank to a final concentration of 37 mg kg^{-1}. The seawater is recirculated from the tank, through a 0.2 μm pore size filter and a UV irradiation apparatus, and back to the tank. The outlet for the return of the seawater to the tank is at the top of the tank so that the seawater falls from the top of the tank headspace to the seawater surface. The headspace of the tank is supplied with room air through a 0.2 μm pore size filter to equilibrate the CO_2 concentration of the seawater in the tank with that in the room air. After this circulation has continued for at least 3 days, it is stopped, and the seawater is allowed to stand for 30–60 min to allow bubbles in the tank to dissipate.

Once the bubbles in the tank have dissipated, the seawater is dispensed into clean bottles (500 mL, CORNING® PYREX). Seawater is transferred to each bottle from the tank through a peristaltic pump and 0.1 μm pore size filter (Whatman® Polycap75TC). At this point, approximately 6 mL of headspace should remain in each bottle. After filling the bottle with seawater, it is immediately closed with a stopper coated with Apiezon® L Grease (M&I Materials, Ltd.). To prevent the stopper from lifting, it is pressed down from above with a rubber band that is attached to the neck of the bottle with a hose clip (Fig. 14.7).

Each batch of RM is accompanied by a certificate of measurement that describes the average of 10 measurements of DIC in μmol kg^{-1}, the standard deviation in μmol kg^{-1}, the coefficient of variation in percentage, and the maximum and minimum of the measurements in μmol kg^{-1}. DIC is measured by coulometry. For reference, the values of salinity and TA, each derived from an average of 10 measurements, are also provided. The values of DIC and TA are assigned by comparison with Dickson

Fig. 14.7 KANSO RM, batch AX bottle no. 063

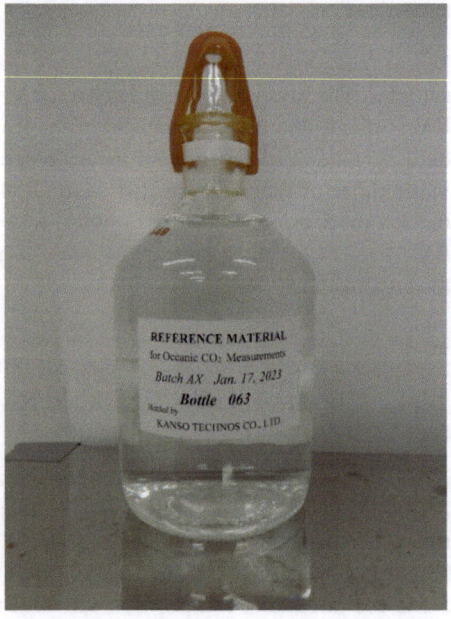

RM. The salinity value is assigned on the basis of IAPSO (International Association for the Physical Sciences of the Oceans) standard seawater.

14.5   Usage of RM

14.5.1   Monitoring of the DIC-Measuring System

The in-house RM is used to check the condition of the DIC-measuring system for each set of coulometer cell solutions. For each set, we usually analyze 70–80 seawater samples, CO_2 gas, blank, RM, and Dickson RM. The condition of the DIC-measuring system is monitored by repeatedly measuring the in-house RMs. For each set of coulometer cell solutions (Fig. 14.8), the operating conditions of the DIC-measuring system remained fairly stable for approximately 30 h (70–80 seawater samples).

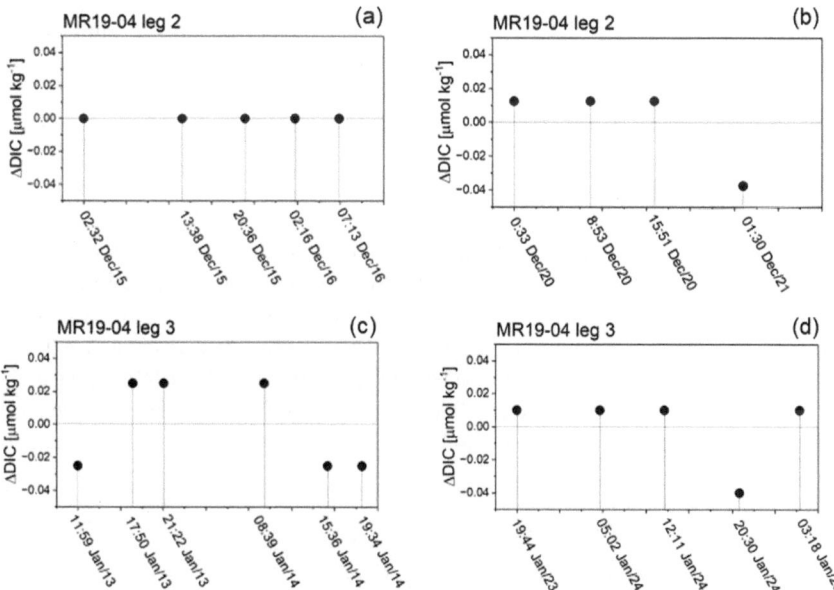

Fig. 14.8 Monitoring of the DIC-measuring system by measuring in-house RM (batch Q37) during cruise MR19-04. DIC was measured with 4 individual sets (**a–d**) of the coulometer cell solutions. ΔDIC indicates the difference from the average of DIC measurements in each coulometer cell solution

14.5.2   Measurements During COVID-19

We usually measure and report DIC and TA values aboard based on Dickson RMs. However, for cruise MR21-04 in 2021, we were unable to obtain the Dickson RM in time due to disruptions caused by the COVID-19 pandemic. Therefore, we measured DIC and TA based on KANSO RM (batch AV). The mean and standard deviation of DIC and TA for measurements of KANSO RM batch AV were 2042.92 ± 1.56 µmol kg^{-1} (n = 22) and 2311.2 ± 1.7 µmol kg^{-1} (n = 31), respectively. The reference values of DIC and TA for KANSO RM batch AV are 2043.6 ± 1.0 µmol kg^{-1} and 2311.6 ± 1.2 µmol kg^{-1}, respectively. The stability of the repeated analyses of the KANSO RM demonstrates that our analyses of both DIC and TA remained consistent during cruise MR21-04 (Fig. 14.9). From these results, we reported DIC and TA values re-calibrated by the reference values of the KANSO RM (dividing sample measurements by the factors = reference values of RM/RM measurements). According to the data synthesis of the Global Ocean Data Analysis Project (GLODAP) (Lauvset et al., 2024), adjustments of the reported DIC and TA values are ± 0 µmol kg^{-1} and $+ 3$ µmol kg^{-1}, respectively; therefore, KANSO RM could be an alternative to Dickson RM to maintain data consistency, although we have to discuss how to assign DIC and TA values.

14.6   Discussion

In this chapter, we introduced two Japanese-produced RM used on cruises on the R/V *Mirai*. We do not typically use the RMs as a measurement standard for calibration: the purpose of using the RMs has been to complement measurements based on Dickson RM, and the benefit of doing so has been to reduce the costs associated with the Dickson RM, such as purchasing costs ($65.00 per bottle of Dickson RM) and freight. We started producing RMs around 2000 and since then we have over 20 years of experience. As shown in Sects. 14.3 and 14.4, because the RMs are as homogeneous

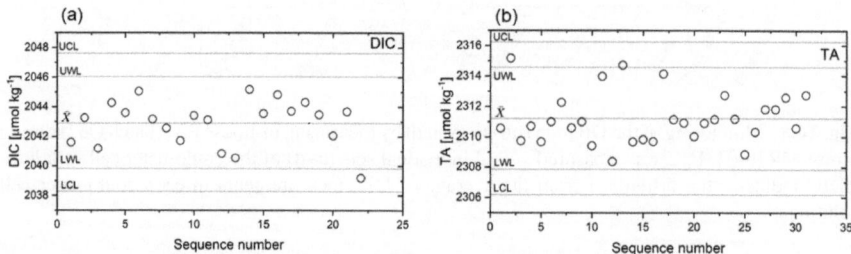

Fig. 14.9 The X-bar charts of (**a**) DIC and (**b**) TA measurements of KANSO RM (batch AV) in MR21-04

and stable as the Dickson RM, the Japanese RMs could be used as an alternative to the Dickson RM during the COVID-19 pandemic (see Subsection 14.5.2).

The shortage of Dickson RM during the COVID-19 pandemic due to shutdown of SIO made us aware of the risks of relying on one laboratory to produce and distribute RMs (Acquafredda et al., 2022; Catherman, 2021). To overcome this constraint, it is advisable to establish regional hubs to provide Certified Reference Materials (CRMs) in each ocean basin (e.g., Pacific, Atlantic, etc., see Chapter 13 in this book). Such hubs would also represent ocean basins with different seawater characteristics (Jeansson et al., 2023). In the following section, we discuss some constraints on providing the RMs presented in this chapter, at least for regional oceans, e.g., northwestern Pacific.

14.6.1 Assigning Values

In JAMSTEC, we produce in-house RMs for seawater DIC and TA and use them on research cruises. We assign the values of DIC and TA by comparison with Dickson RM Since we assume that the in-house RM will be used together with Dickson RM on ships, we do not measure the DIC and TA of the in-house RM on land at the time of RM production. However, for distribution of the RMs outside JAMSTEC, it would be better to assign the values of DIC and TA, although according to the definition of metrology: *"material, sufficiently homogeneous and stable with respect to one or more specified properties which has been established to be fit for its intended use in measurement or in examination"*, the RM does not necessarily need to be evaluated. Actually, we can assign the values based on Dickson RM on land. The values of DIC and TA of KANSO RM are assigned by KANSO based on Dickson RM (Sect. 14.4).

14.6.2 Batch Size

The in-house JAMSTEC and KANSO RM are prepared in 200 L and 600 L tanks, respectively, allowing to produce 400 and 1200 bottles (500 mL) for each batch of the in-house JAMSTEC and KANSO RM, respectively. It would not be difficult to use a larger tank to produce a larger amount of in-house RM, but this would be more expensive. Because the production of in-house RM depends on research funds, the increased expenses could cause a shortage of research funds. This raises the question of whether we can adequately meet the needs of users.

14.6.3 Funding

Research groups that produce in-house RM frequently face challenges in securing funds for this activity. As introduced in Sect. 14.2, the production of RM is a labor-intensive process. Consequently, technician labor represents a significant portion of the cost. In JAMSTEC, the funds necessary for producing the in-house RMs are shared by several research groups that host research cruises, e.g., Global Ocean Ship-based Hydrographic Investigations Program (GO-SHIP) cruises, time-series cruises, and Arctic cruises. The scarcity of funds is a persistent challenge, making it difficult to sustain these efforts.

14.6.4 Shipment

The RMs presented in this chapter are packaged in borosilicate glass bottles with a nominal volume of 500 mL. Each bottle weighs approximately 300 g when empty and 900 g when full. Weight is a significant factor contributing to transport fees (Carter et al., 2024; Huang et al., 2012). Additionally, the RMs contain mercuric chloride to halt biological activity. Mercuric chloride is classified as a hazardous chemical by the United Nations (UN number 1624), making it difficult for RMs to clear customs in many countries (Acquafredda et al., 2022). In Japan, mercuric chloride is classified as a poisonous substance by the Poisonous and Deleterious Substances Control Act, leading to transportation restrictions event within Japan. Due to these legal constraints, KANSO does not provide their RM outside Japan. To provide RMs regionally, this legal constraint must be overcome.

14.7 Future Perspectives

The RMs for seawater DIC and TA presented in this chapter are homogeneous and stable, with values assigned based on the Dickson RM. Therefore, they can be used as secondary DIC and TA standards for calibration, provided that each laboratory has the necessary proficiency to assign values accurately. To validate this proficiency, regular interlaboratory comparisons should be performed.

The shipping constraint associated with solutions containing mercuric chloride can be solved by replacing it with a disinfection method that is used for CRMs for seawater dissolved inorganic nutrients (Ota et al., 2010; Chapter 2 in this book). In fact, the production of RM for seawater DIC and TA has been carried out using this alternate method (Murata, 2010; Chapter 12 in this book). By utilizing lighter aluminum bottles for RM without mercuric chloride, such RMs can be shipped more economically. Consequently, in the near future, we anticipate being able to provide our RM worldwide.

We have discussed the feasibility of providing the RMs presented in this chapter, mainly in terms of logistics. Although there are still scientific issues to be solved in the field of RMs (Carter et al., 2024; Mos et al., 2021), we hope that our report will help facilitate a stable supply of RMs.

Competing Interest The authors have no conflicts of interest to declare that are relevant to the content of this chapter.

Appendix

Measurement of DIC

The DIC-measuring system (Nihon ANS, Inc., Japan) installed on the R/V *Mirai* follows the design outlined by Dickson et al. (2007). The system uses a coulometer (Nihon ANS, Inc., Japan) as a detector, and its seawater dispensing unit has an auto-sampler (6 ports), which dispenses seawater from a 250 mL borosilicate glass bottle (DURAN® ORIGINAL GL 45, Germany) into a pipette with a volume of about 15 mL under control of a personal computer (PC). The pipette is kept at 20 °C ($s = \pm 0.01$ °C) by a water jacket through which water from a water bath set at 20 °C is circulated. CO_2 dissolved in a seawater sample is extracted in a stripping chamber of the CO_2 extraction system by adding about 2 mL of phosphoric acid (~9% v/v). The stripping chamber is approximately 25 cm long and has a fine frit at the bottom. The acid is added to the stripping chamber from the bottom of the chamber by pressurizing an acid-containing bottle with nitrogen gas (99.9999%) for the amount of time needed to push out the 2 mL of acid. Then the seawater sample in the pipette is introduced to the stripping chamber by the same method that was used to add the acid. The seawater reacts with phosphoric acid and is stripped of CO_2 by bubbling nitrogen gas through the fine frit at the bottom of the stripping chamber. The CO_2 stripped in the chamber is then carried by nitrogen gas (flow rate ~140 mL min^{-1}) to the coulometer through a dehydrating module, which consists of two electric dehumidifiers (kept at ~2 °C) and a chemical desiccant ($Mg(ClO_4)_2$).

The DIC-measuring system is first calibrated with six concentrations of sodium carbonate solution ranging from 0–2500 μmol dm^{-3} at 500 μmol dm^{-3} intervals. Then, the system is re-calibrated with Dickson RMs by multiplying a factor of the certified value divided by the measured value.

Measurement of TA

The TA measuring-system installed on the R/V *Mirai* is based on spectrophotometry with a custom-made system (Nihon ANS, Inc., Japan) following Yao and Byrne (1998). The system comprises a water dispensing unit, an HCl titration unit (Hamilton 250 μL syringe, USA), and a spectrophotometer detection unit (TM-UV/ VIS C10082CAH, Hamamatsu Photonics, Japan) with an optical source (Ocean Photonics, Japan). The system is automatically controlled by a PC. The water-dispensing unit consists of a water-jacketed pipette and a water-jacketed glass titration cell. About 42 mL of seawater is transferred from a sample bottle (100 mL borosilicate glass bottle, DURAN® ORIGINAL GL 45, Germany) into the water-jacketed (25 °C, $s = \pm 0.01$ °C) pipette by pressurizing the sample bottle (nitrogen gas), and then it is introduced into the water-jacketed (25 °C) glass titration cell. This introduced seawater is used to rinse the titration cell. The seawater used for rinsing is dumped, and then Milli-Q water is introduced into the titration cell twice for two additional rinses. Then, another 42 mL of seawater is transferred by the pipette into the titration cell. Absorbances of the seawater blank are measured at three wavelengths (730 nm, 616 nm, and 444 nm). After this measurement, an acid titrant, which is a mixture of ~0.05 M HCl in 0.65 M NaCl and bromocresol green (BCG), is added to the titration cell. The volume of acid titrant solution is changed according to the expected values of TA from 1.980 mL to 2.190 mL. The seawater and acid titrant are mixed for 5 min with a stirring tip and bubbling by nitrogen gas in the titration cell. Absorbances at the three wavelengths are then measured. From these absorbances, the absorbance ratio (R) is calculated as $R = (A_{616} - A_{730})/(A_{444} - A_{730})$, where A_i is the absorbance of the seawater mixed with acid titrant at wavelength i nm after subtracting the absorbance of the seawater blank. The pH or total hydrogen ion concentration $[H^+]_T$ is then calculated from A_i and the salinity following Yao and Byrne (1998). TA is finally obtained as $TA = (-[H^+]_T V_{SA} + M_A V_A)/V_S$, where V_S, V_A, and V_{SA} are the initial seawater volume, the added acid titrant volume, and the combined seawater plus acid titrant volume, respectively, and M_A is the molarity of the acid titrant added to the seawater sample. In the same way as for DIC, the system is calibrated with Dickson RM.

References

Acquafredda, M., Cochran, C., Busch, D. S., Jewett, L., Edmonds, H., & Dickson, A. G. (2022). Understanding the current use and future needs of CO_2 in seawater certified reference materials. NOAA Technical Memorandum OAR-OAP-5. https://doi.org/10.25923/anc4-gj33

Carter, B. R., Sharp, J. D., Dickson, A. G., Álvarez, M., Fong, M. B., García-Ibáñez, M. I., Woosley, R. J., Takeshita, Y., Barbero, L., Byrne, R. H., Cai, W.-J., Chierici, M., Clegg, S. L., Easley, R. A., Fassbender, A. J., Fleger, K. L., Li, X., Martín-Mayor, M., Schockman, K. M., & Wang, Z. A. (2024). Uncertainty sources for measurable ocean carbonate chemistry variables. *Limnology and Oceanography, 69*, 1–21. https://doi.org/10.1002/lno.12477

Catherman, C. (2021). Ocean scientists confront a critical bottleneck. *Science, 374*, 17.

Dickson, A. G. (2001). Reference materials for oceanic CO_2 measurements. In: R. A. Feely, C. L. Sabine, T. Takahashi & R. Wanninkh (Eds.), Uptake and storage of carbon dioxide in the ocean: The global CO_2 survey. *Oceanography, 14*, 21–22.

Dickson, A. G. (2007). Determination of total dissolved inorganic carbon in sea water. In: A. G. Dickson, C. L., Sabine & J. R., Christian (Eds.), *Guide to best practices for ocean CO_2 measurements* (pp. 191). PICES Special Publication

Dickson, A. G. (2010). Standards for ocean measurements. *Oceanography, 23*(3), 34–47. https://tos.org/oceanography/issue/volume-23-issue-03

Huang, W.-J., Wang, Y., & Cai, W.-J. (2012). Assessment of sample storage techniques for total alkalinity and dissolved inorganic carbon in seawater. *Limnology and Oceanography, 10*, 711–717. https://doi.org/10.4319/lom.2012.10.711

Ishii, M., Inoue, H. Y., Matsueda, H., & Tanoue, E. (1998). Close coupling between seasonal biological production and dynamics of dissolved inorganic carbon in the Indian Ocean sector and the western Pacific Ocean sector of the Antarctic Ocean. *Deep-Sea Research I, 45*, 1187–1209.

Jeansson, E., McDonagh, E., Álvarez, M., Cantoni, C., Carse, F., Velo, A., Cusack, C., Firing, Y., Karstensen, J., Schroeder, K., Steinhoff, T., Weber, R., & Woodward, M. (2023). D2.1. Initial scope of shared facilities: New services and access opportunities. EuroGO-SHIP.

Lauvset, S. K., Lange, N., Tanhua, T., Bittig, H. C., Olsen, A., Kozyr, A., Álvarez, M., Azetsu-Scott, K., Brown, P. J., Carter, B. R., Cotrim da Cunha, L., Hoppema, M., Humphreys, M. P., Ishii, M., Jeansson, E., Murata, A., Müller, J. D., Pérez, F. F., Schirnick, C., Steinfeldt, R., Suzuki, T., Ulfsbo, A., Velo, A., Woosley, R. J. & Key, R. M. (2024). The annual update GLODAPv2.2023: the global interior ocean biogeochemical data product. *Earth System Science Data, 16*, 2047–2072. https://doi.org/10.5194/essd-16-2047-2024

Mos, B., Holloway, C., Kelaher, B. P., Santos, I. R., & Dworjanyn, S. A. (2021). Alkalinity of diverse water samples can be altered by mercury preservation and borosilicate vial storage. *Scientific Reports, 11*, 9961. https://doi.org/10.1038/s41598-021-89110-w

Murata, A. (2010). The development of non-toxic reference materials for oceanic CO_2 measurements: Current status and future plans. In M. Aoyama, A. G. Dickson, D. J. Hydes, A. Murata, J. R. Oh, P. Roose, & E. M. S. Woodward (Eds.), *Comparability of nutrients in the world's ocean* (pp. 31–41). Mother Tank.

Murata, A., Kumamoto, Y., Saito, C., Kawakami, H., Asanuma, I., Kusakabe, M., & Inoue, H. Y. (2002). Impact of a spring phytoplankton bloom on the CO_2-system in the mixed layer of the Northwestern North Pacific. *Deep-Sea Research, II, 49*(24–25), 5531–5555

Ono, A., Watanabe, S., Okuda, K., & Fukasawa, M. (1998). Distribution of total carbonate and related properties in the North Pacific along 30°N. *Journal of Geophysical Research, 103*(C13), 30873–30883.

Ota, H., Mitsuda, H., Kimura, M., & Kitao, T. (2010). Reference materials for nutrients in seawater: Their development and present homogeneity and stability. In M. Aoyama, A. G. Dickson, D. J. Hydes, A. Murata, J. R. Oh, P. Roose, & E. M. S. Woodward (Eds.), *Comparability of nutrients in the world's ocean* (pp. 11–30). Mother Tank.

Yao, W., & Byrne, R. B. (1998). Simplified seawater alkalinity analysis: Use of linear array spectrometers. *Deep-Sea Research, 45*, 1383–1392.

Chapter 15
Development of Dissolved Organic Matter Reference Material

Dennis A. Hansell and Takeshi Yoshimura

Abstract Dissolved organic matter (DOM) is an important component for the cycling of bioactive elements such as carbon, nitrogen, and phosphorus. However, confidence in the measurement of dissolved organic carbon (DOC), nitrogen (DON), and phosphorus (DOP) in marine systems is limited compared to inorganic nutrients due to complex analytical processes, the absence of certified standards, and the associated uncertainty of analytical results. To improve the latter issue, reference materials have been developed for DOC and total dissolved nitrogen and distributed by the Hansell laboratory at the University of Miami, but not yet for DON and DOP. Here we begin with a description and history of the DOC reference material (RM) program, then consider progress and needs in the development of RMs necessary for organic nutrients. Although currently available RMs for inorganic nutrients were expected to function as RMs for DOM, our previous work negates this option. We have sought an appropriate bottle container material for a DOM-RM, successfully for DOP but still to be developed for simultaneous achievement for DOC, DON, and DOP. We found a potential in both perfluoroalkoxy and glass bottles to achieve DOM-RM that remains to be examined. On the other hand, producing a single-purpose DOP-RM may make a significant contribution to elevating the comparability of DOP data, which are of growing interest among marine research communities.

Keywords Dissolved organic matter · Carbon · Nitrogen · Phosphorus · Reference material · CRM · Bottle material · Elution

D. A. Hansell
Department of Ocean Sciences, University of Miami, Miami, FL, USA
e-mail: dhansell@miami.edu

T. Yoshimura (✉)
Faculty of Fisheries Sciences, Hokkaido University, Sapporo, Japan
e-mail: yoshimura-t@fish.hokudai.ac.jp

© The Author(s) 2025
M. Aoyama et al. (eds.), *Chemical Reference Materials for Oceanography*, Springer Oceanography, https://doi.org/10.1007/978-981-96-2520-8_15

15.1   The Necessity of Reference Materials for DOM

Dissolved organic matter (DOM) is involved in the cycling of bioactive elements such as carbon, nitrogen, and phosphorus. A portion of the organic matter produced during photosynthesis by phytoplankton, the dominant oceanic photoautotroph, is released into seawater as DOM through extracellular release by phytoplankton and via grazing processes by zooplankton (Carlson & Hansell, 2015; Nagata, 2000). DOM is mainly made with a carbon backbone for its structure, but it commonly includes nitrogen and phosphorus at lower abundances. The inventory of carbon in marine DOM (i.e., DOC) is comparable to that of atmospheric carbon dioxide (Hansell, 2005), making it one of Earth's major reservoirs of reactive carbon. Dissolved organic nitrogen (DON) and phosphorus (DOP) are present at concentrations exceeding inorganic nutrients in surface oligotrophic waters (Karl et al., 2001). These organic nutrients are typically found at higher concentrations in upwelling systems where the nutrients are introduced to the euphotic zone, and at lower concentrations at the surface of subtropical gyres after autotrophic use of the nutrients (Letscher et al., 2013). DOM can be mineralized by biotic or abiotic processes to produce carbon dioxide, ammonium, and phosphate. Thus, DOM plays a major role both as standing stocks and in regeneration of bioactive elements in the ocean.

The number of measurements of marine DOM is limited compared to inorganic nutrients, largely because of the great analytical efforts required. DOC can be measured directly by high temperature combustion of seawater after removal of dissolved inorganic carbon by acidification and sparging (Halewood et al., 2022). On the other hand, determining concentrations of DON and DOP requires measurements of total dissolved nitrogen and phosphorus (TDN and TDP) as well as nutrients (nitrate, nitrite, ammonium, and phosphate), with the dissolved organic nutrient concentrations (i.e., DON and DOP, respectively) calculated as the difference in the concentrations between total (i.e., organic plus inorganic) and inorganic phases. Thus, DOM measurements include time consuming and analytically challenging protocols, with commensurate challenges in measurement reliability and large analytical errors (i.e., uncertainty).

To improve the reliability and reduce uncertainty in DOM measurements, reference materials (RMs) have been developed. RMs for DOM have primarily focused on DOC since carbon is a key element in climate change (Hansell, 2005). Due to the importance of understanding the amount of carbon present as DOC in the ocean, an RM for DOC was briefly developed in the mid-1990s (Sharp et al., 2002a). Acidified seawater was sealed in 10 mL (then 20 mL) glass ampoules and distributed to users. In 1998, the Hansell laboratory at the University of Miami, with support from the U.S. National Science Foundation, initiated a consensus reference material (CRM) program that continues to the time of this writing (~25 years). The DOC-CRM is primarily intended for use in high temperature combustion methods using a total organic carbon analyzer, typically coupled with a total nitrogen analyzer (Halewood et al., 2022). The latest RMs are supplied in screw-capped glass vials, which can be placed directly into the total organic carbon analyzer's autosampler. However, these

DOC/TDN RMs are not evaluated for nitrate, nitrite, and ammonium contents, and thus do not strictly serve as DON-RMs. In the DON determinations, large analytical variabilities are produced by poorly controlled analyses of nutrients, as well as in TDN (Sharp et al., 2002b). Therefore, establishment of DON-RM should be considered. Moreover, similar studies on DOP have not yet been conducted, therefore a DOP-RM must be developed. Hereafter, we describe the successful history of the DOC-CRM program and trials for producing a new DOM-RM targeting DON and DOP.

15.2 Development of the DOC Reference Material Program

There is an important distinction to be noted between the terms *certified* and *consensus* reference materials. The marine CO_2 community uses *certified* references materials. The CO_2 analysis employed at the time of certification is traceable to NIST standards, thus making the analysis *certifiable* (NIST is the National Institute of Standards and Technology at the U.S. Department of Commerce). There is not a similarly certified standard for marine DOC, so the analysis is not certifiable. Instead, *consensus* determinations are made. That is, it is the consensus of several expert laboratories that determines the acceptable range of concentrations for the reference water distributed. Hence, when the marine CO_2 community uses the term CRM, they mean *certified* reference material (Dickson, 2001). When the DOM community employs *CRM*, they mean *consensus* reference material (Hansell, 2005).

The challenge of analytical comparability between the dozens of laboratories making marine DOC measurements with non-uniform methodologies and variable expertise is a primary motivation for creating CRMs. Given the many international laboratories making the measurements at many places and times through the global ocean, it is imperative that the resulting data be made useful and reliable to all users, both now and in the future. The differences in the results from these labs are narrowed when each laboratory uses common reference water against which to judge their results. Without such, differences in DOC concentrations between visits to a particular oceanic region cannot be assigned to natural variability any more than they can be clearly assigned to analytical variability. For example, DOC had been reported by Thomas et al. (1995) to have a mean of 46 ± 7 µmol L^{-1} in the deep water of the equatorial Atlantic, a value that is high given measurements done in that system during a 2013 occupation of the A16N line (Hansell, 2023). Without common reference material we cannot know if the reported differences are analytical or natural, leaving the two sets of data incomparable. As another example, DOC was originally measured in the northern Arabian Sea in the early 1960s when Menzel (1964) visited the region; Kumar et al. (1990) and Hansell and Peltzer (1998) made subsequent visits. The results of DOC analyses by these three groups from proximal sites are shown in Fig. 15.1. While it is unlikely that DOC varied naturally by >80 µmol L^{-1} at depths >3000 m, it cannot be determined with any certainty which part of the variability was analytical (though it is likely to be all of it). If these three

Fig. 15.1 Depth profiles of organic carbon in the northern Arabian Sea (near 20°N, 65°E) reported by three analysts not linked by a common reference material. Data from Menzel (1964; diamonds), Kumar et al., (1990; squares), and Hansell and Peltzer (1998; circles)

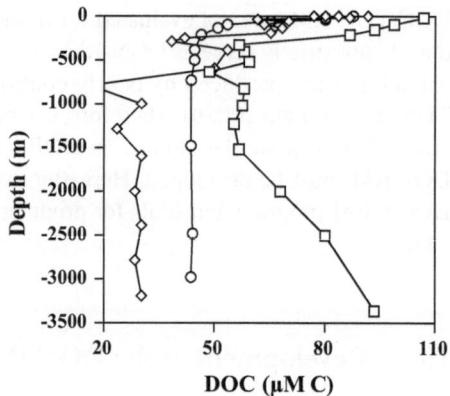

groups had a common reference material, we would today have three useful sets of data spanning more than 30 years. But without the reference material, the data cannot be compared reliably. Such had been the status for most DOC measurements made globally. Between-laboratory variability cannot be reduced without the dedicated use of a common reference material, and long-term trends in concentrations cannot be resolved without a long-term investment in those reference materials.

The Hansell CRM program has distributed seawater and low carbon water (LCW) to the international community of marine DOC analysts for ~25 years. Early in the program, deep seawater (DSW) was collected from 2600 m in the Sargasso Sea; since then, water has been collected on cruises out of Miami in the Florida Strait. Marine and LCW, the latter for determining instrument blanks, are added to vials, acidified, and then distributed to a group of expert DOC analysts for consensus analysis. Their results are used to assign a small range (normally 2 μmol C L^{-1}) of acceptable concentrations for the reference waters. Low carbon water is generated in a Milli-Q® A10 system, which reduces DOC in deionized water to very low levels by utilizing high intensity, multi-wavelength (185 and 284 nm) UV light. Specifications call for the unit to generate water with a DOC content of < 0.4 μmol C L^{-1} (<5 ppb).

The community of users of the carbon CRMs asked for a wider range of concentrations to work with, so the provider began distributing water collected from 3 depths: "deep" (DSR; 700 m in Florida Strait, capturing a strong signal of Antarctic Intermediate Water, with a few hundred years since ventilation), mid-depth (MSR; 150 m, capturing North Atlantic Mode Water, with about a decade since ventilation) and surface (SSR; 5 m, with a strong South Atlantic component) (Schmidt, 1996). Also increased was the number of volumes to order, with both 20- and 40-mL vials made available. These acidified reference waters were assessed to be stable for at least two years, with storage in the dark between 4 and 20 °C.

The provider also determined that there was a need for complete sets of standards and CRMs, to benefit new (or infrequent) DOC analysts. To this end, they created and offered "DOC-TN Determination Kits". Each kit comes with all that is necessary (other than hardware) to process samples for the determination of DOC

Table 15.1 Cumulative number of boxes of CRMs ordered by nation from 1998–2021. Forty-five nations have used ~ 2600 boxes (374K vials) of CRMs

Nation	Boxes	Nation	Boxes	Nation	Boxes
Australia	20	India	16	Poland	4
Austria	5	Ireland	4	Portugal	2
Belgium	7	Indonesia	2	Saudi Arabia	10
Bermuda	2	Israel	14	Singapore	14
Brazil	5	Italy	70	Slovenia	5
Canada	131	Japan	111	South Korea	98
Chile	8	Malaysia	10	Spain	141
China	313	Mexico	14	Sweden	15
Ethiopia	1	Monaco	4	Switzerland	1
Denmark	22	Netherlands	36	Taiwan	37
Finland	6	New Caledonia	2	Trinidad/Tobago	2
France	104	New Zealand	25	United Kingdom	103
Germany	204	Norway	10	United States	981
Granada	3	Oman	3	Uruguay	1
Greece	15	Philippines	7	Vietnam	8

and TN (where TN is total nitrogen). Included are 4 calibration standards for carbon and nitrogen and 4 reference materials (DSR, MSR, SSR, and LCW). Provided too is an Excel template describing the recommended analytical sequence of standards, CRMs, and samples for each analytical run. The kit offers a complete written description with diagrams/images of the 'standard operating procedures' used in the Craig Carlson (University of California Santa Barbara) and Dennis Hansell laboratories for collecting, processing, and analyzing DOM samples (Halewood et al., 2022). The intent is to bring new analytical laboratories up to speed quickly using common protocols.

From the beginning of operations to the end of 2021, the provider shipped ~ 2600 boxes (374,400 vials at 144 vials/box); as for recent demand, ~180 boxes were shipped to 55 laboratories in 2021. Forty-five nations have participated in the program, with the USA, Japan, China, Canada, Germany, Spain, and France being the largest users (Table 15.1).

15.3   Consideration of a Nutrient Reference Material as a DOM Reference Material

As shown in this book, the inorganic nutrient RMs, produced and distributed by KANSO, Japan (Ota et al., 2010), are quite homogeneous in a batch and stable over a long period for inorganic nutrients including nitrate, nitrite, and phosphate as well

Table 15.2 Selected analytical results of dissolved organic carbon, nitrogen, and phosphorus in some lots of KANSO nutrient RMs

	Lot	Mean (μmol L^{-1})	Standard deviation (μmol L^{-1})	Coefficient of variance (%)
DOC	BF	77.8	21.6	28
	BG	89.4	13.3	15
DON	BF	6.6	2.5	38
	BG	7.4	0.45	6
DOP	AU	0.07	0.05	70
	BD	0.10	0.03	26

as silicate. They were anticipated to be used as a DOP-RM as well. If they could be also used as DOC- and DON-RM, the usefulness of the RMs would expand greatly. However, KANSO nutrient RMs had not been examined for DOC, DON, and DOP.

To evaluate the potential use of KANSO nutrient RMs as DOM-RMs, concentrations of DOC, DON, and DOP were measured for several lots of nutrient RMs available as of 2009 (Yoshimura & Sharp, 2010). The concentrations of all analytes were higher than values expected from the ambient seawaters used for the batch and varied widely among multiple bottles within the same lot (Table 15.2). Therefore, it was concluded that the nutrient RM is not suitable as a DOM-RM. We need additional considerations to remove DOM contamination in the current KANSO nutrient RMs.

15.4 Trials for Producing a New Reference Material for DOM

We conducted experiments to select appropriate storage bottle containers (Yoshimura, 2013). Various plastic and glass bottle containers (Table 15.3) were either ordered pre-cleaned or cleaned by us with alkaline detergent and 1 N HCl. The containers were filled with subtropical surface water, depleted in inorganic nutrients, that had been sterilized through a 0.22 μm pore size capsule filter. They were stored at 25 °C in the dark, and then DOC and DOP concentrations were determined after time had elapsed. DOC was contaminated in most materials within at least 400 days and DOP was contaminated in HDPE, LDPE, and PAN bottles within at least 250 days. In these experiments, only glass and PFA bottles showed no significant change in DOC and DOP concentrations.

Based on these results, we attempted to create a DOM-RM using the same manufacturing process as the nutrient RM but with different bottle containers. Eighty prototype bottles (DOM-RM#1) were made using subtropical surface and coastal deep waters, each stored in three types of bottles (PC, PFA, and PP). The nutrient

Table 15.3 Plastic materials of sample bottles for which stabilities of dissolved organic carbon and phosphorus concentrations were examined in Yoshimura (2013)

Bottle material	Abbreviation
Fluorinated High-Density Polyethylene	FLPE
High-Density Polyethylene	HDPE
Low-Density Polyethylene	LDPE
Polyacrylonitrile	PAN
Polycarbonate	PC
Polyethylene Terephthalate	PET
Perfluoroalkoxy	PFA
Polypropylene	PP
Polyvinyl Chloride	PVC
Polymethylpentene	TPX

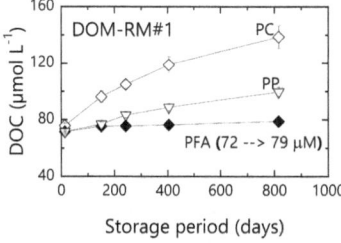 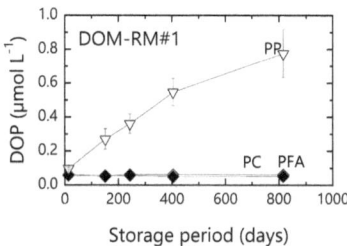

Fig. 15.2 Time courses of dissolved organic carbon (DOC) and phosphorus (DOP) concentrations (mean ± 1 s.d., n = 3) during the storage of subtropical surface waters in DOM-RM#1

and DOM concentrations were measured over time. Inorganic nutrient concentrations did not change after 3 months, and no microbial contamination was observed. On the other hand, contamination of DOC occurred in all bottles and of DOP in PP bottles (Fig. 15.2).

Next, about a hundred prototype bottles (DOM-RM#2) were filled using a subtropical surface water, then stored in five types of bottles (PET, two different PFA, PP, and glass). The DOC and DOP concentrations were measured over time. Contamination occurred in all bottles for DOC and only in PP bottles for DOP (Fig. 15.3). Note that DOC contaminations in PFA and glass bottles are significant, but at a relatively small extent with 7 μmol L^{-1} and 4 μmol L^{-1}, respectively. DOC contamination in 100 mL volume glass bottles of DOM-RM#2 contrasts with the stable DOC concentrations in the CRMs distributed by the Hansell lab, which use relatively small glass vials (20–40 mL). The composition of the cap may be the key factor in determining whether a contamination occurs in glass bottles.

Given the results of DOM-RM#1 and #2, we speculate that the PP bottles, which are used in the KANSO nutrient RMs, are a source of DOC and DOP contamination. The high and variable DOM concentrations in the current nutrient RMs, as shown in Table 15.2, originate by elution of DOM from the bottle material. Our results indicate

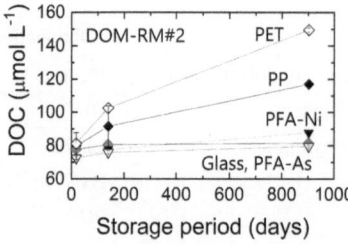 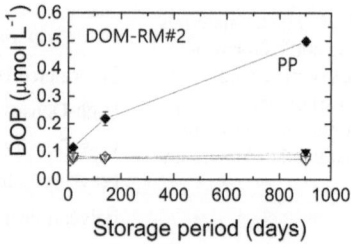

Fig. 15.3 Time courses of dissolved organic carbon (DOC) and phosphorus (DOP) concentrations (mean $\pm$ 1 s.d., n = 3) during the storage of subtropical surface waters in DOM-RM#2

that many other bottle materials besides PP also cause serious DOM contamination. Although we did not measure DON in these experiments, similar results may be the case for DON in these bottle materials.

Since we found DOP is not contaminated by PFA as examined in DOM-RM#1 and #2 (Figs. 15.2 and 15.3), we conducted a small-scale laboratory intercomparison experiment for DOP analysis using seawater samples stored in PFA bottles by a subset of the participants in the 2012 inter-laboratory comparison study of a reference material for nutrients in seawater (Murata et al., 2020). Samples for DOP analysis were produced by KANSO in January 2011 using a subtropical surface water (sample #6) and a coastal deep water (sample #7), stored in cleaned PFA bottles. Nine laboratories reported their DOP analysis results for these samples. The datasets are not large enough to conduct significant statistical analyses, thus we present a simple data analysis here.

We found unacceptably large differences in the reported DOP concentrations among the laboratories (Fig. 15.4), ranging from 0.03 to 0.43 μmol kg^{-1} with mean and 1 SD of 0.15 ± 0.13 μmol kg^{-1} for sample #6 and from 0.01 to 0.42 μmol kg^{-1} with 0.19 ± 0.17 μmol kg^{-1} for sample #7. In this study, reported phosphate concentrations showed good agreements with 0.02 ± 0.02 μmol kg^{-1} for sample #6 and 2.18 ± 0.04 μmol kg^{-1} for sample #7. Therefore, the large differences in DOP concentration among the laboratories are derived from the variations in TDP measurements. In general, DOP in the ocean shows highest concentration in surface layers and gradually decreases in deeper layers (Karl & Björkman, 2015). Our results showed the opposite trend with lower DOP concentration in sample #6 (surface water) than sample #7 (deep water). Improvements in TDP analyses will be required to reduce the differences in DOP concentration among the laboratories. This study is the first step of inter-laboratory comparison for DOP analysis. More laboratories should be recruited to participate in the next step. In future experiments, appropriate DOP-RMs will be analyzed in many worldwide laboratories and significant analyses must be conducted not only for concentration values but for analytical methods to demonstrate the cause of the differences in TDP values among the laboratories. Further work is needed to achieve comparability of DOP measurement among worldwide laboratories.

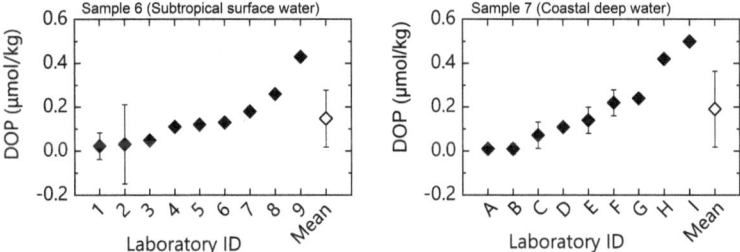

Fig. 15.4 Inter-laboratory comparisons of dissolved organic phosphorus (DOP) analytical results from nine laboratories conducted in the 2012 study of a reference material for nutrients in seawater. Closed plots represent mean and one standard deviation in each laboratory if they reported multiple analytical results; the open symbol shows the mean and one standard deviation of the nine laboratories

15.5   Future RM Development

Based on the results of previous studies, no container material has been found to be free of both DOC and DOP, and probably DON, contaminations. However, with respect to DOC, contamination by PFA and glass bottles is relatively small; more aggressive cleaning may lessen the contamination. For example, the method used to clean bottle containers for trace metal measurements (Cutter et al., 2017) could be attempted. However, since the contamination may not be from the container material but from the gas phase through the screwed cap, additional careful experiments are required to establish a DOM-RM container suitable for all of DOC, DON, and DOP analyses.

On the other hand, it is shown here that DOP concentrations can be safely stored in bottles of many materials. If DOP is the only target analyte, new RMs can be produced now. Recently, the role of DOP in low latitude phosphorus limited regions has been considered (Karl & Björkman, 2015). Under such circumstances, DOP analytical data and their reliability with comparability are growing in importance. Historically, DOC and TDN analytical qualities have been advanced through inter-laboratory comparison experiments (Sharp et al., 2002a, 2002b). The lack of inter-laboratory comparison experiments, due to the absence of DOP-RMs, is a major issue. Our tentative inter-laboratory comparison experiment showed that DOP results show unacceptably large variability among laboratories. An appropriate DOP-RM is required to be developed based on the current knowledge demonstrated here and should be utilized by the marine research community as soon as possible.

15.6   Competing Interests

The authors have no conflicts of interest to declare that are relevant to the content of this chapter.

Acknowledgements DAH thanks the U.S. National Science Foundation for its long and consistent support of the Consensus Reference Material program for Dissolved Organic Carbon operated at the University of Miami (Award OCE-2218815). TY acknowledges CRIEPI (#060215 and #090313) and JSPS KAKENHI (JP22681004) for supporting the DOP-RM research.

References

Carlson, C. A., & Hansell, D. A. (2015). DOC sources, sinks and fluxes. In D. A. Hansell & C. A. Carlson (Eds.), *Biogeochemistry of Marine Dissolved Organic Matter* (pp. 65–126). Academic Press.

Cutter, G., Casciotti, K., Croot, P., Geibert, W., Heimbürger, L., Lohan, M., Planquette, H., van der Fliert, T. (2017). *Sampling and Sample-Handling Protocols for GEOTRACES Cruises.* Version 3.0. (p. 139).

Dickson, A. G. (2001). Reference materials for oceanic CO_2 measurements. *Oceanography, 14,* 21–22.

Halewood, E., Opalk, K., Custals, L., Carey, M., Hansell, D.A., & Carlson, C. A. (2022). GO-SHIP repeat hydrography: Determination of dissolved organic carbon (DOC) and total dissolved nitrogen (TDN) in seawater using high temperature combustion analysis. Methods. *Frontiers in Marine Science-Ocean Observation 9.* https://doi.org/10.3389/fmars.2022.1061646

Hansell, D. A. (2005). Dissolved organic carbon reference material program. *Eos, Transactions American Geophysical Union, 86,* 318.

Hansell, D. A. (2023). DOC data from Cruise 33RO20130803, exchange version. Retrieved April 8, 2023, from CCHDO https://cchdo.ucsd.edu/cruise/33RO20130803.

Hansell, D. A., & Peltzer, E. T. (1998). Spatial and temporal variations of total organic carbon in the Arabian Sea. *Deep-Sea Research II, 45,* 2171–2193.

Karl, D. M., & Björkman, K. M. (2015). Dynamics of DOP. In D. A. Hansell & C. A. Carlson (Eds.), *Biogeochemistry of Marine Dissolved Organic Matter* (pp. 233–334). Academic Press.

Karl, D. M., Björkman, K. M., Dore, J. E., Fujieki, L., Hebel, D. V., Houlihan, T., Letelier, R. M., & Tupas, L. M. (2001). Ecological nitrogen-to-phosphorus stoichiometry at station ALOHA. *Deep Sea Research Part II, 48,* 1529–1566.

Kumar, M., Rajendran, A., Somasundar, K., Haake, B., Jenisch, A., Shuo, Z., Ittekkot, V., & Desai, B. N. (1990). Dynamics of dissolved organic carbon in the northwestern Indian Ocean. *Marine Chemistry, 31,* 299–316.

Letscher, R. T., Hansell, D. A., Carlson, C. A., Lumpkin, R., & Knapp, A. N. (2013). Dissolved organic nitrogen in the global surface ocean: Distribution and fate. *Global Biogeochemical Cycles, 27,* 141–153.

Menzel, D. W. (1964). The distribution of dissolved organic carbon in the Western Indian Ocean. *Deep-Sea Research, 11,* 757–765.

Murata, A., Aoyama, M., Cheong, C., Miura, T., Fujii, T., Mituda, H., Kitao, T., Sasano, D., Nakano, T., Nagai, N., Kodama, T., Kasai, H., Kiyomoto, Y., Setou, T., Ono, T., Yokogawa, S., Arii, Y., Sone, T., Ishikawa, Y., & Wakita, M. (2020). Current situation and future perspective for environmental standards of seawater: Commencing with certified reference materials (CRMs) for nutrients. *Oceanography in Japan, 29,* 153–187. (in Japanese with English abstract).

Nagata, T. (2000). Production mechanisms of dissolved organic matter. In: D. L., Kirchman (Eds.), *Microbial ecology of the oceans* (pp. 121–152). New York, USA: Wiley-Liss.

Ota, H., Kitao, T., Mitsuda, H., Kimura, M., Kitao, T. (2010). Reference materials for nutrients in seawater: Their development and present homogeneity and stability. In: M. Aoyama (Eds.), *Comparability of nutrients in the World's ocean* (pp. 11–30). Tsukuba, Japan: Mother Tank.

Schmidt, W. J., Jr. (1996). On the World ocean circulation: Volume I, Some Global Features/ North Atlantic Circulation. *Woods Hole Oceanographic Institution Technical Report* (p 141), WHOI-96–03.

Sharp, J. H., Carlson, C. A., Peltzer, E. T., Castle-Ward, D. M., Savidge, K. B., & Rinker, K. R. (2002a). Final dissolved organic carbon broad community intercalibration and preliminary use of DOC reference materials. *Marine Chemistry, 77*, 239–253.

Sharp, J. H., Rinker, K. R., Savidge, K. B., Abell, J., Yves Benaim, J., Bronk, D., Burdige, D. J., Cauwet, G., Chen, W., Doval, M. D., Hansell, D., Hopkinson, C., Kattner, G., Kaumeyer, N., McGlathery, K. J., Merriam, J., Morley, N., Nagel, K., Ogawa, H., & Wong, C. S. (2002b). A preliminary methods comparison for measurement of dissolved organic nitrogen in seawater. *Marine Chemistry, 78*, 171–184.

Thomas, C., Cauwet, G., & Minster, J. F. (1995). Dissolved organic carbon in the equatorial Atlantic ocean. *Marine Chemistry, 49*, 155–169.

Yoshimura, T. (2013). Appropriate bottles for storing seawater samples for dissolved organic phosphorus (DOP) analysis: A step towards the development of DOP reference materials. *Limnology and Oceanography: Methods, 11*, 239–246.

Yoshimura, T., & Sharp, J. H. (2010). Additional calibration of nutrient reference materials for dissolved organic carbon, nitrogen, and phosphorus. In: M. Aoyama (Eds.), *Comparability of nutrients in the World's ocean* (pp. 91–100). Tsukuba, Japan: Mother Tank.

Chapter 16
Certified Reference Material of Trace Elements in Seawater, NMIJ CRM 7204-a

Yanbei Zhu

Abstract A certified reference material (CRM) for the determination of trace elements in seawater samples was developed by the National Metrology Institute of Japan (NMIJ) and coded as NMIJ CRM 7204-a, in which the concentrations of trace elements were elevated to facilitate the quality control in analysis for environment conservation purpose. It is a complement to CRMs aimed at quality control in natural seawater samples, e.g. NASS and CASS series by National Research Council of Canada, as well as GEOTRACES standards and reference materials. NMIJ CRM 7204-a was characterized for the analysis of 10 regulated trace elements (Cr, Mn, Fe, Ni, Cu, Zn, As, Se, Cd, and Pb) and provided in 500 mL polyethylene bottle. Homogeneity test and stability study on the elements were carried out by inductively coupled plasma tandem quadrupole mass spectrometry. A property value was calculated as the weighted mean of the results obtained by multiple methods, whose weights were obtained as the reciprocal of their standard uncertainties. Combined uncertainty of a property value was the root mean square of the standard uncertainties of homogeneity, stability, analysis reproducibility, method-to-method variance, and calibrating standard. The concentration of Cd in NMIJ CRM 7204-a is approximately 3 µg/kg, while those of other trace elements are around 10 µg/kg. The concentrations of Na, Mg, K, and Ca were provided as information values, along with the density values at 15, 20, and 25° Celsius.

Keywords Certified reference material · Trace elements · Seawater · Homogeneity · Stability · Property value · Uncertainty · Traceability

Y. Zhu (✉)
National Metrology Institute of Japan (NMIJ), National Institute of Advanced Industrial Science and Technology (AIST), Tsukuba, Ibaraki, Japan
e-mail: yb-zhu@aist.go.jp

© The Author(s) 2025   289
M. Aoyama et al. (eds.), *Chemical Reference Materials for Oceanography*, Springer
Oceanography, https://doi.org/10.1007/978-981-96-2520-8_16

16.1 Introduction

Determination of trace elements (or trace metals) in natural seawater is essential to understand processes and fluxes of trace elements (Sohrin et al., 2008), as well as to invest their biogeochemical cycles (Morel & Price, 2003) in the ocean. Meanwhile, analysis of trace elements in seawater samples is difficult work due to the extremely low concentrations of analytes (usually lower than 10 µg/L) with quite high concentrations of salt contents around 3.5% (Sturgen et al., 1980).

Even with the most powerful instrument for elemental analysis, e.g. inductively coupled plasma mass spectrometer (ICP-MS), trace elements in seawater are often analyzed with pretreatment processes to separate analytes from the salt contents (Samanta et al., 2021). At the same time, the accuracy of analyzing trace elements usually depends on multiple factors such as sample matrix, instrumental operating conditions, and pretreatment process. Therefore, certified reference materials (CRMs) for trace elements in seawater are required for quality control of the analytical protocol.

According to ISO 17034:2016 developed by the International Organization for Standardization (ISO), a CRM is *"a reference material characterized by a metrologically valid procedure for one or more specified properties, accompanied by a reference material certificate that provides the value of the specified property, its associated uncertainty, and a statement of metrological traceability"* (ISO, 2016). National Research Council of Canada has continuously issued seawater CRMs for the analysis of trace elements, namely NASS and CASS for open and coastal seawater samples. Certified and reference values of trace elements in CASS-6 and NASS-7 are summarized in Table 16.1.

In addition to such CRMs issued by metrological institutes, toward aims to improve the understanding of biogeochemical cycles and large-scale distribution of trace elements and their isotopes in the marine environment, the international program GEOTRACES developed standards to ensure the results are accurate, precise, and intercomparable. Consensus values of trace elements in two representative GEOTRACES standards, namely GSP and GSC, are summarized in Table 16.2. Consensus values of trace elements are also available for another two sets of reference samples, i.e. SAFe and North Atlantic reference seawater samples (GEOTRACES, 2019).

It is notable that the concentrations of most trace elements in Tables 16.1 and 16.2 are under 1.0 µg/L, which is extremely low, especially considering the salt contents in seawater samples. The determination of trace elements at such low concentrations is necessary for the research on their behavior and cycles in natural ocean environment.

On the other hand, the concentrations of multiple trace elements, e.g. As, Cd, and Pb, are regulated as criteria for water quality (including seawater) to protect aquatic life as well as the water environment of human being (AMSAT, 2008; Cole et al., 1999; MEGJ, 2022; USEPA, 2022a, 2022b). In these regulations, the maximum allowable concentrations are usually at lower µg/L levels, which are much higher

Table 16.1 Certified and reference values of trace elements in CASS-6 and NASS-7 issued by NRCC (unit, µg/L)

Element	CASS-6[a]	NASS-7[b]
	Mean ± Expanded uncertainty	Mean ± Expanded uncertainty
B	4090 ± 100	3750 ± 120
V	0.50 ± 0.12	1.30 ± 008
Cr	0.100 ± 0.016	0.107 ± 0.016
Mn	2.22 ± 0.12	0.75 ± 0.006
Fe	1.56 ± 0.12	0.351 ± 0.026
Co	0.0672 ± 0.0052	0.0146 ± 0.0014
Ni	0.418 ± 0.040	0.248 ± 0.018
Cu	0.530 ± 0.032	0.199 ± 0.014
Zn	1.27 ± 0.18	0.42 ± 0.08
As	1.04 ± 0.10	1.26 ± 0.06
Mo	9.15 ± 0.52	9.29 ± 0.40
Cd	0.0217 ± 0.0018	0.0161 ± 0.0016
Pb	0.0106 ± 0.0040	0.0026 ± 0.0008
U	2.92 ± 0.42	2.87 ± 0.16

Source Modified from Brophy et al. (2022) and Nadeau et al. (2016) [a] Nearshore seawater [b] Open seawater

Table 16.2 Consensus values of trace elements in GEOTRACES standards (unit, µg/L)

Element	GSP[a]		GSC[b]	
	Mean ± Standard deviation	n	Mean ± Standard deviation	n
Mn	0.0427 ± 0.0019	9	0.1198 ± 0.0041	8
Fe	0.0087 ± 0.0025	11	0.0857 ± 0.0064	137
Ni	0.1523 ± 0.0059	11	0.2578 ± 0.0120	7
Cu	0.0365 ± 0.0034	9	0.0698 ± 0.0095	11
Zn	0.0020 ± 0.0034	10	0.0937 ± 0.0067	12
Cd	0.0002 ± 0.0002	4	0.0409 ± 0.0025	9
Pb	0.0128 ± 0.0010	11	0.0081 ± 0.0008	9

Source Calculated from GEOTRACES (2019), initial unit given in nmol/L [a] 2009 GEOTRACES Pacific surface seawater [b] 2009 GEOTRACES coastal surface seawater

than the concentrations in natural seawater. A one-order of magnitude lower concentration indicates much challenging analytical requirements, such as a better blank control and/or a higher sensitivity, especially for seawater with its high salt contents. Therefore, considering the concentrations of trace elements in NASS-7, CASS-6, GSP, and GSC (Tables 16.1 and 16.2), they are mismatching with the concentrations for confirmation of practical criteria for water quality.

Recently, National Metrology Institute of Japan (NMIJ) developed a CRM for elevated levels of trace elements in seawater, providing a fit-for-purpose quality control in analysis of seawater to fulfill the requirements of environmental protection (Zhu et al., 2022). This CRM is coded as NMIJ CRM 7204-a, in which the concentrations of trace elements were set to near or slightly lower than the concentrations regulated for water environment, permitting the direct determination by ICP-MS after approximately 10 times dilution.

This chapter presents technical details about the development of NMIJ CRM 7204-a.

16.2 Chemicals and Materials

Single-element standard solutions (1000 mg/L, for atomic analysis) were purchased from Kanto Chemical Co., Inc. The concentrations, chemical species, and matrix of these standard solutions are summarized in Table 16.3. Metrological traceability to the International System of Units was guaranteed via Japan Calibration Service System (JCSS).

Calibrating solutions and samples for analysis were all prepared based on mass fraction by using an electronical balance, which is yearly calibrated by JCSS. The following enriched isotopes were used in isotope dilution analysis: ^{111}Cd (96.44%), ^{53}Cr (96.98%), ^{65}Cu (99.70%), ^{57}Fe (92.88%), ^{61}Ni (99.44%), ^{67}Zn (94.60%), and ^{206}Pb (98.66%), which were purchased from the Oak Ridge National Laboratory (ORNL) except for ^{206}Pb from the National Institute of Standards and Technology

Table 16.3 Single element standard solutions used in the present work

Element	Concentration (mg/L)	Chemical species	Matrix
Na	999	Na^+	Water
Mg	999	Mg^{2+}	0.1 mol/L HNO_3
K	1001	K^+	Water
Ca	1001	Ca^{2+}	0.1 mol/L HNO_3
Cr	1002	$Cr_2O_7^{2-}$	0.1 mol/L HNO_3
Mn	996	Mn^{2+}	0.1 mol/L HNO_3
Fe	1002	Fe^{2+}	0.2 mol/L HNO_3
Ni	999	Ni^{2+}	0.1 mol/L HNO_3
Cu	997	Cu^{2+}	0.1 mol/L HNO_3
Zn	995	Zn^{2+}	0.1 mol/L HNO_3
As	1002	AsO_3^{3-}	0.05% NaCl
Se	1000	SeO_3^{2-}	0.1 mol/L HNO_3
Cd	1002	Cd^{2+}	0.1 mol/L HNO_3
Pb	998	Pb^{2+}	0.1 mol/L HNO_3

(NIST). The values in the bracket following each isotope address the degrees of enrichment.

Ultrapure® grade HNO_3 (60%), acetic acid (>99.0%), and ammonia solution (28% as NH_3) were also purchased from Kanto Chemical Co. Inc. and used for preparing samples and calibrating standard solutions. Cica-reagent grade of HNO_3 (60%) and acetone (>99.5%) used for washing bottles (polyethylene, *i.e.* PE 500 mL) to get rid of possible contaminants were also purchased from Kanto Chemical Co. Inc. Membrane filters (pore sizes of 0.45 μm and 0.20 μm, respectively, Whatman GD/X) for filtration of seawater were purchased from GE Healthcare Japan. Single-use plastic tube (VIO-15BN, VIO-50BN) and all-plastic syringe (24 mL) were purchased from AS ONE Corp. and used for preparing samples. Chelating resin columns (NOBIAS Chelate-PA1, M-size) were purchased from Hitachi High-Tech Fielding Corp and used for solid phase extraction.

A seawater CRM (CASS-5) disseminated by the NRCC was analyzed to confirm the validity of the analytical method. Natural seawater (ca. 200 L) was collected near the shore of Nikkawa Beach (Kamisu, Ibaraki, Japan) and used as the raw material for the present candidate reference material.

16.3 Preparation of Candidate Reference Material

The raw material for the candidate reference material was first filtrated in turn by using membrane filters with pore sizes 0.45 μm and 0.2 μm, respectively. Then it was acidified to ca. 0.1 mol/L HNO_3 to ensure the stability for preservation. A preliminary test was carried out to check the initial concentrations of trace elements in this filtrated seawater. Then, a mix elemental standard was added to the filtrated seawater to elevate the concentrations of regulated trace elements to ca. 10 μg/kg (exc. Cd, to ca. 3 μg/kg). The candidate reference material was stirred for homogenization and bottled in 500-mL PE bottles. It is notable that prior to bottling of the candidate reference material, the PE bottles were cleaned in advance in turns of acetone washing, drying, 1.0 mol/L HNO_3 soaking for one week, and pure water washing. Each of the candidate reference material was 500 mL of seawater (in 0.1 mol/L HNO_3) in a PE bottle with a serial number and sealed in a clear plastic bag. The candidate materials were stored at room temperature (*i.e.* 5 °C to 35 °C) and shielded from light. The procedure for preparation of candidate reference material is illustrated in Fig. 16.1.

16.4 Analytical Strategy and Instruments

Due to the high salt contents in seawater, coprecipitation (Freslon et al., 2011; Komjarova & Blust, 2006; Nicolaysen et al., 2003; Raso et al., 2013) and solid phase extraction (SPE) (AlSuhaimi et al., 2019; Itoh et al., 2018; Komjarova & Blust, 2006; Otero-Romani et al., 2005; Rao et al., 2006; Sakamoto et al., 2006;

Fig. 16.1 Preparation of
candidate reference material
(Tank volume, 200 L; Bottle
volume, 500 mL; F1 and F2,
filters with pore sizes of
0.45 μm and 0.2 μm,
respectively.)

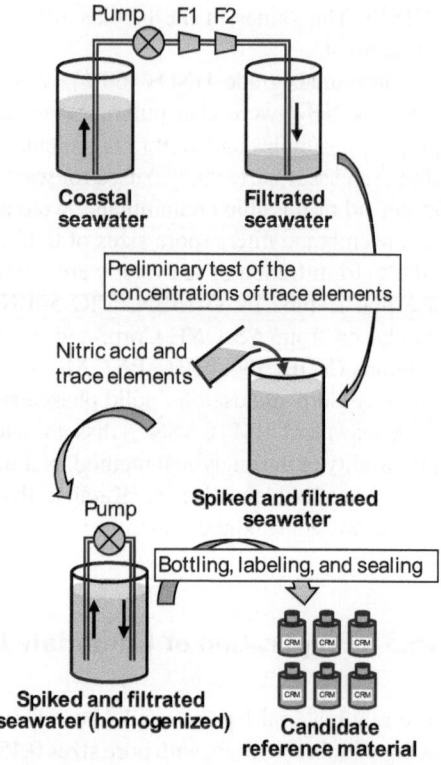

Wells & Bruland, 1998) had been usually used to separate trace elements prior to their analysis. Coprecipitation and SPE were used as pretreatments in the present work as well as simple dilution.

Isotope dilution (ID-) ICP-MS, as one of the primary methods for quantitative analysis, has been widely used in development of CRMs (Ariga et al., 2018; Zhu et al., 2011, 2013). Meanwhile, standard addition with internal standard correction was effective in canceling non-spectral interferences in the measurement of trace elements in high matrix samples by ICP-MS (Zhu et al., 2018). In the present work, concentrations of trace elements in seawater were obtained by multiple analytical methods, combining pretreatment operations (dilution, coprecipitation, and SPE) with different analytical instruments (ICP-MS and graphite furnace atomic absorption spectrometry, i.e. GF-AAS) and different calibrating methods (standard addition and ID). It is notable that at least two independent methods were applied to determine the concentration of each trace element in the candidate CRM.

Two analytical instruments, i.e. an ICP-MS instrument (Agilent 8800s, Agilent Technologies) and a GF-AAS instrument (AAnalyst 800, PerkinElmer), were majorly used for measuring trace elements in the present work. The operating conditions for both instruments were daily optimized, for which the typical values are summarized in Tables 16.4 and 16.5 along with the elements measured. The ICP-MS measurement can be operated in on-mass mode and mass-shift mode with various reaction gases including H_2 and O_2. It is capable of separating spectral interferences to permit the direct analysis of trace elements in seawater. A Millipore purification system (Milli-Q Advantage A10, Nihon Millipore Kogyo) was used to provide pure water throughout the present work. All analyses for certification of the present CRM were carried out by the analysts in NMIJ.

Table 16.4 Operating conditions for ICP-MS

Instrument	Agilent 8800 s
Plasma conditions:	
Incident Rf power	1.55 kW
Reflected power	< 1.0 W
Outer gas flow rate	Ar 15 dm^3/min
Intermediate gas flow rate	Ar 0.90 dm^3/min
Carrier gas flow rate	Ar 0.85 dm^3/min
Make-up gas flow rate	Ar 0.45 dm^3/min
Sampling conditions:	
Sampling cone	Pt
Skimmer cone	Pt
Nebulizer	MicroMist 100 mm^3/min (natural aspirate)
Spray chamber	Scott type (2 °C)
Sampling depth	8.0 mm from work coil
Reaction gas [a]	H_2 7.0 cm^3/min (on-mass) O_2 0.3 cm^3/min (mass-shift)
Data acquisition:	
Scanning mode	Peak hopping
Data points	3 points / peak
Dwell time	0.33 ms / point
Integration	10 times
Measured isotopes:	^{75}As, ^{111}Cd, ^{52}Cr, ^{63}Cu, ^{56}Fe, ^{55}Mn, ^{60}Ni, ^{208}Pb, ^{82}Se, ^{88}Sr, ^{66}Zn, ^{23}Na, ^{24}Mg, ^{39}K, ^{40}Ca
Measured isotope pair:	^{111}Cd/^{110}Cd, ^{53}Cr/^{52}Cr, ^{65}Cu/^{63}Cu, ^{57}Fe/^{56}Fe, ^{61}Ni/^{60}Ni, ^{206}Pb/^{204}Pb, ^{206}Pb/^{207}Pb, ^{206}Pb/^{208}Pb, ^{77}Se/^{78}Se, ^{67}Zn/^{66}Zn

a, Cd, Pb, Na, Mg, K, Ca were measured at on-mass mode with H_2 as the reaction gas, the other elements were measure at mass-shift mode with O_2 as the reaction gas.

Table 16.5 Operating
conditions for GF-AAS

Instrument	AAnalyst800
Wavelength [a]	193.7/279.5 nm
Lamp [a]	EDL/HCL
Lamp current [a]	350/10 mA
Slit width [a]	0.7/0.2 mm
Signal type	AA-BG
Measurement	Peak area
Background correction	Zeeman
Furnace	THGA
Data acquisition:	
Read time	3 s
Delay time	0 s
BOC time	2 s
Replicates	3 times
Autosampler conditions:	
Sample volume	10 mm^3
Diluent volume [a]	10/0 mm^3
Matrix modifier volume [a]	10/0 mm^3
Injection temperature	20 °C

[a] The values before and after the "/" symbol are respectively for
As and Mn

16.5 Analytical Procedures and Uncertainty Estimation

Dilution by HNO_3, coprecipitation, and SPE were respectively applied to determine trace elements in seawater of the present work. For dilution by HNO_3, an aliquot of the sample was diluted 1/10 with 0.3 mol/L HNO_3 solution. Procedures for coprecipitation and SPE are respectively shown in Fig. 16.2a, b.

As can be seen from Fig. 16.2a, an aliquot of La solution was added into 5 mL of seawater sample, which was diluted with pure water. The sample was laid for overnight for precipitation after adding 2 mL of ammonia solution. The precipitate was separated by decantation after centrifuged at 3500 rpm for 10 min. Thus obtained precipitate was a mixture of $Mg(OH)_2$ and $La(OH)_3$, which was dissolved with 2 mL of 2 mol/L HNO_3 and diluted to 15 mL with pure water. It is notable that a solution of enriched isotopes was added prior to the addition of La standard solution in case of isotope dilution (ID-) ICP-MS analysis, while samples for standard addition were made prior to the analysis by ICP-MS. To investigate the composition of the precipitate, an aliquot of the precipitate was dissolved in 0.3 mol/L HNO_3 and subjected to the determination of Mg and La by ICP-MS. The results showed that approximately 95% and 99%, respectively, of Mg and La were precipitated

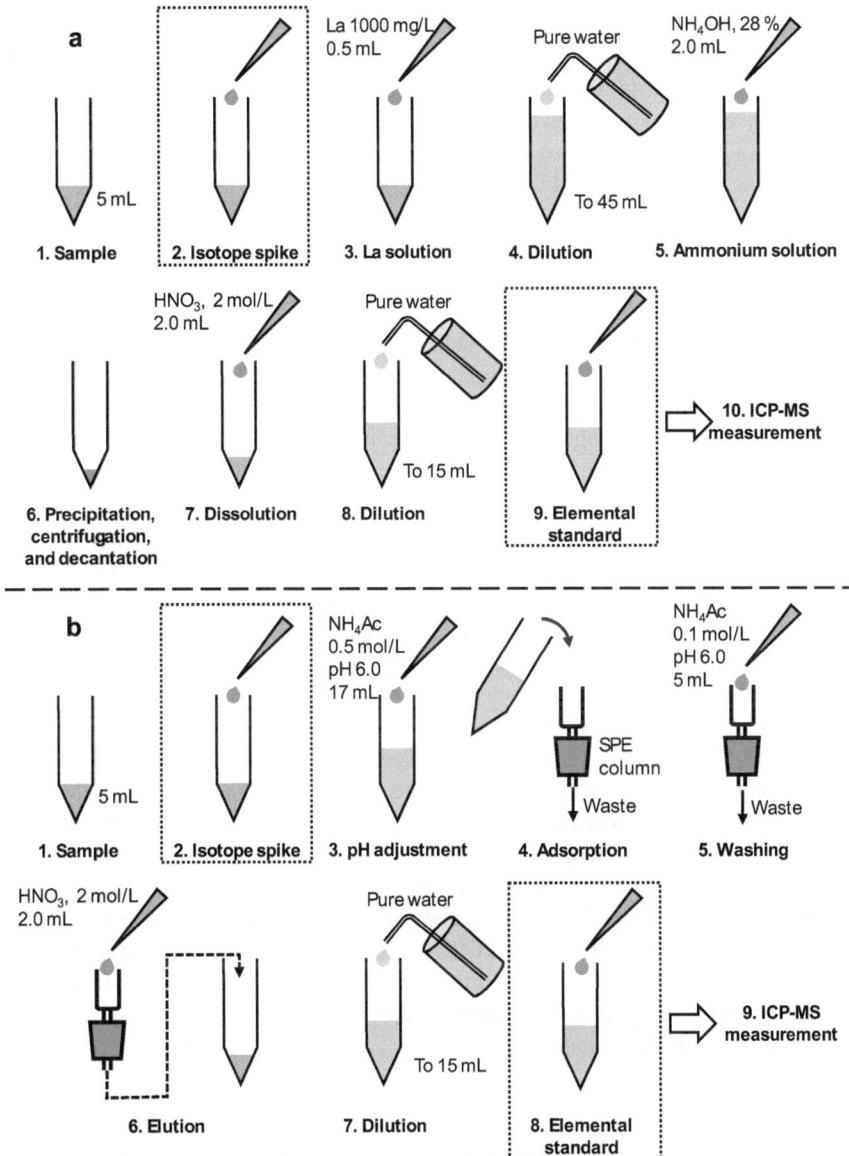

Fig. 16.2 Operations for precipitation (**a**) and SPE (**b**) (Operations surrounded by dotted squares were conducted only for ID-ICP-MS or standard addition ICP-MS.)

from the seawater sample. The chemical formula of the precipitate can be shown as $Mg_{1.22}La_{0.18}(OH)_3$.

As can be seen from Fig. 16.2b, ammonium acetate (0.5 mol/L, pH 6.0, 17 mL of) was mixed with 5 mL seawater sample to adjust the pH condition prior to being passed through an SPE column at a flow rate of ca. 10 mL/min. After that, 5 mL of ammonium acetate (0.1 mol/L, pH 6.0) was passed through the column to elute the salt contents trapped by the column. Finally, trace elements were eluted with 2 mol/L HNO_3 (2 mL), which was diluted to 15 mL by adding pure water. In a similar way to that for coprecipitation, addition of enriched isotopes and preparation of spiked sample for standard addition were respectively carried for ID-ICP-MS and ICP-MS.

It is notable that the procedures for coprecipitation and SPE resulted in 1/3 dilution of the initial seawater sample. Because the concentrations of trace elements of interest in the present candidate reference material were elevated, the concentrations of trace elements in the solutions obtained after coprecipitation and SPE were sufficient high to be measured by an ICP-MS instrument. The present procedures for coprecipitation and SPE were effective for the removal of salt contents, not aimed to enrich elements of interest.

Recovery test and blank test were carried for both coprecipitation and SPE, in parallel with the samples. By adding a mix standard, the concentrations of trace elements in the obtained test sample for recovery test had elemental concentrations doubled those in the initial sample. A solution of 0.1 mol/L HNO_3 was used as the sample for blank test.

Measurements of trace elements were carried out by ICP-MS (dilution, coprecipitation, and SPE) and GF-AAS (dilution), respectively. Standard addition ICP-MS was applied to all elements of interest in the present work, while ID-ICP-MS was applied to elements having multiple stable isotopes. GF-AAS was applied to the analysis of two mono-isotopic elements, i.e. As (standard addition) and Mn (linear regression), which cannot be determined by ID-ICP-MS.

Uncertainty of the result was estimated based on the ISO/IEC Guide 98–3 (ISO/IEC 2008) and the EURACHEM/CITAC Guide CG4 (Ellison & Williams, 2012). A combined uncertainty was obtained as the root mean squares of the standard uncertainties of quantitative analysis, method-to-method variance, homogeneity, long term stability, and calibration standard.

16.6 Concentrations of Dominating Metals in the Seawater Sample

Salt contents in seawater are major obstacles for determining trace elements due to both spectral interferences and non-spectral interferences. Dominating metals in seawater are Na, Mg, K, and Ca. The concentrations of these elements were determined by ICP-MS after proper dilution and provided basic information for the present seawater sample.

Table 16.6 Recovery values of spiked trace elements in a seawater CRM, CASS-5

Element	Spiked concentration (μg/kg)	Recovery
		Mean ± SD (n = 3)
Cr	10	100 ± 3
Mn	10	101 ± 1
Fe	10	101 ± 1
Ni	10	99.8 ± 0.8
Cu	10	102 ± 2
Zn	10	99.2 ± 1.6
As	10	102 ± 1
Se	10	100 ± 2
Cd	3.0	99.7 ± 1.6
Pb	10	101 ± 1

The relative expanded uncertainties for the present results (see 16.9 for detail) of Na, Mg, K, and Ca were 1.3%, 1.8%, 1.7%, and 1.6% respectively. These results are in good relationship with those reported by Pichler et al., 1999, indicating that the present candidate material is proper as a sample matrix for seawater analysis.

16.7   Confirmation of the Validity of Analytical Method

Measurement with ICP-MS after proper dilution (10 times with 0.3 mol/L HNO_3) was the dominant method used in the present work. A spiked sample of seawater CRM, *i.e.* CASS-5, was analyzed to check the validity of this method, for which the results are summarized in Table 16.6. The results showed recovery values close to 100% with standard deviation less than 3%, which were sufficient for the analysis in the present work. The concentrations spiked for recovery test were close to the certified values aimed in the present work, feasible for the analysis for environmental conservation purpose.

16.8   Analytical Results of Trace Elements with Various Methods

Multiple methods were applied to the quantitation of each element in the present work. At least two methods were used to determine the concentration of each element if ID-ICP-MS is applicable, while minimum methods of three were used for each element (Mn and As in the present work) if ID-ICP-MS is not applicable. The results for each element obtained by multiple methods are summarized in Table 16.7, along with the standard uncertainties of measurements.

As can be seen from Table 16.7, the results for each element obtained by various methods agreed with one another in the range of the standard uncertainties. The weighted arithmetic mean of results from different methods was calculated as the property value for each element. The weight (w_i) was obtained based Eq. (16.1)

$$w_i = \frac{\frac{1}{u_{x_i}}}{\sum \frac{1}{u_{x_i}}} \tag{16.1}$$

where, u_{x_i} is the standard uncertainty (u value in Table 16.7) of a mean value x_i.

As summarized in Table 16.7, the property values for most elements are around 10 μg/kg, except for Cd (3.07 μg/kg) and Zn (12.6 μg/kg). The concentration of Cd was intentionally prepared to meet its lower criteria for water quality. The concentration of Zn can be partially attributed to the initial content in the seawater material.

Table 16.7 Analytical results by different methods

Element	Concentration (Mean ± u; Unit: μg/kg)				Property value (μg/kg)
	ICP-MS	**SPE ICP-MS**	**GF-AAS**	**Coprecipitation ICP-MS**	
Mn	9.25 ± 0.09	9.27 ± 0.27	9.38 ± 0.30	9.19 ± 0.18	9.26
As	10.9 ± 0.2		11.4 ± 0.4	10.7 ± 0.1	10.9
	ICP-MS	**SPE ID-ICP-MS**	**Coprecipitation ID-ICP-MS**		
Cr	9.46 ± 0.29		9.39 ± 0.08		9.41
Fe	9.65 ± 0.12	9.76 ± 0.14	9.59 ± 0.15		9.67
Ni	9.73 ± 0.26	9.93 ± 0.26	9.78 ± 1.04		9.82
Cu	9.51 ± 0.58	9.61 ± 0.13			9.59
Zn	13.1 ± 0.7	12.3 ± 0.6			12.6
Se	9.88 ± 0.43		9.60 ± 0.26		9.70
Cd	3.03 ± 0.08	3.09 ± 0.05	3.08 ± 0.18		3.07
Pb	9.36 ± 0.16	9.40 ± 0.15	9.44 ± 0.14		9.40

16.9 Uncertainties for Property Values

The following parameters were considered to calculate a combined uncertainty (u_c) value for each property value: u_{anal}, uncertainty of analysis; u_{meth}, method-to-method variance; u_{std}, uncertainty of calibration standard; u_{homo}, uncertainty of homogeneity; and u_{lts}, uncertainty of stability.

The weight defined by Eq. (16.1) was also applied to calculate a value of u_{anal} as the weighted root mean square (RMS) of the standard uncertainties obtained with different methods in Table 16.7. The relative value of between-method-variance obtained by an analysis of the variance (ANOVA) of individual results from different methods was considered as u_{meth} for each element. Each u_{std} was cited from the certificate of elemental solution.

In response to the requirement of ISO 17034 (2016), homogeneity and stability were evaluated for the present candidate reference material following the ISO Guide 35 (2017). Technical details about calculations for homogeneity and stability tests can be found in reports about other CRMs (Ariga et al., 2018; Zhu et al., 2011, 2013). Results for homogeneity and stability tests were obtained by measuring with ICP-MS after proper dilution (10 times with 0.3 mol/L HNO_3).

Based on an ANOVA of the results of homogeneity test, the relative values of s_{bb} (between-bottle variance) and u_{bb} (between-bottle variance incorporating the influence of analytical variation) were obtained and summarized in Table 16.8. A larger value (shown with bold fonts in Table 16.8) of s_{bb} and u_{bb} was assigned to u_{homo}.

The results for the stability test were subjected to a linear regression analysis against the time stored, which results showed that the slopes for linear regression were

Table 16.8 Budgets of combined standard uncertainties

Element	u_{anal} (%)	u_{meth} (%)	u_{std} (%)	u_{homo}		u_{lts}[a]				u_c (%)
				s_{bb} (%)	u_{bb} (%)	b_1[b]	$s(b_1)$[b]	$T \times s(b_1)$ (µg/kg)	Relative $T \times s(b_1)$ (%)	
Cr	0.95	0.30	0.30	**0.92**	0.29	0.109	0.25	0.21	**2.20**	2.64
Mn	0.91		0.20	**0.44**	0.23	− 0.071	0.21	0.18	**1.91**	2.21
Fe	0.81	0.80	0.25	**1.18**	0.68	0.003	0.30	0.25	**2.57**	3.09
Ni	2.04		0.20	0.44	**0.63**	0.254	0.44	0.36	**3.71**	4.30
Cu	1.58		0.20	**1.24**	0.75	0.192	0.37	0.31	**3.24**	3.83
Zn	3.44	4.30	0.25	**1.99**	0.65	0.287	0.47	0.39	**3.09**	6.64
As	1.15	3.10	0.20	**1.76**	0.29	−0.167	0.25	0.21	**1.91**	4.23
Se	2.37	2.00	0.60	**1.22**	0.54	−0.249	0.40	0.33	**3.42**	4.84
Cd	1.48	0.80	0.20	-	**1.16**	0.015	0.09	0.08	**2.46**	3.23
Pb	0.92	0.10	0.20	**0.32**	0.21	− 0.149	0.25	0.21	**2.25**	2.49

insignificant at a level of confidence of 95%. These results showed that the present candidate material is sufficiently stable for the elements of interest. Furthermore, the slope of the regression (b_1) and the standard deviation of the slope ($s(b_1)$) were obtained and summarized in Table 14.8. Because $|b_1|{<}s(b_1)$ was true for each element, $T \times s(b_1)$ was calculated and assigned to u_{lts}. The value of T in the present work was 832 days, which is the period from the end of the stability test to the expiration date of the certification. The relative u_{lts} values in Table 16.8 were used to calculate the relative combined standard uncertainties of property values.

The square root of the sum of squares of u_{anal}, u_{meth}, u_{std}, u_{homo}, and u_{lts} was obtained as the relative combined standard uncertainty (u_c (%)) of each element as summarized in Table 16.8, covering a range from 2.21 for Mn to 6.64 for Zn.

16.10   Certified Values and Additional Technical Information.

The present certified reference material was coded as NMIJ CRM 7204-a, for which the certified values are summarized in Table 16.9. A certified value is given as (property value ± expanded uncertainty). Property values are cited from Table 16.7. An expanded uncertainty value was calculated as the product of the property value multiplied by the u_c value in Table 16.8 and a coverage factor ($k = 2$) corresponding to a level of confidence of approximately 95%.

Dominant metals, *i.e.* Na, Mg, K, and Ca, are usually concerned for the determination of trace elements in seawater due to spectral and non-spectral interferences. The concentrations of these metals in the present CRM were determined by ICP-MS and provided as technical information, which is summarized in Table 16.9.

The certified values assigned for the present CRM were given as mass fractions with a gravimetric unit. Because volumetric units are also often used in many laboratories, the density of the present CRM at 15, 20, and 25 °C were measured and provided as technical information, which are summarized in Table 16.9, too. The users can obtain the elemental concentrations with volumetric units based on these density values.

Established by NMIJ (Japan), Korea Research Institute of Standards and Science (KRISS, Korea), and National Institute of Metrology (NIM, China), Asian Collaboration on Reference Materials (ACRM) is an organization for cooperating in metrology. As one of the actions of ACRM, co-analysis of the candidate materials of the present CRM was carried out by scientists from NMIJ, KRISS, and NIM. The results for Cd by KRISS and NIM, Ni by KRISS, and As by NIM respectively agreed with the certified values of the present CRM in the range of expanded uncertainties, contributing to confirmation of the validity of the present CRM.

Table 16.9 Certified values and additional technical information for NMIJ CRM 7204-a

Certified values (unit, µg/kg)	
Element	Property value ± Expanded uncertainty ($k = 2$)
Cr	9.4 ± 0.5
Mn	9.3 ± 0.4
Fe	9.7 ± 0.6
Ni	9.8 ± 0.8
Cu	9.6 ± 0.7
Zn	12.6 ± 1.7
As	10.9 ± 0.9
Se	9.7 ± 0.9
Cd	3.1 ± 0.2
Pb	9.4 ± 0.5

Technical information: concentration of dominant metals		
Element	Concentration	Unit
Na	10.2	g/kg
Mg	1.22	g/kg
K	340	mg/kg
Ca	380	mg/kg

Technical information: density	
Temperature (°C)	Density (g/cm^3)
15	1.028
20	1.026
25	1.025

16.11   Conclusion

A seawater CRM, NMIJ CRM 7204-a, was certified for measuring the concentrations of Cr, Mn, Fe, Ni, Cu, Zn, As, Se, Cd, and Pb. To facilitate the determination for environmental monitoring purpose, the concentrations of these elements were elevated to approximately 10 µg/kg, except for Cd which was elevated to approximately 3 µg/kg. The property value for each element was obtained as the weighted arithmetic mean of results from multiple methods. The contributions of analysis reproducibility, method-to-method variance, calibrating standard, homogeneity, and stability were considered to calculate a combined uncertainty of a property value. The relative values of combined uncertainty were in the range from 2 to 7%, sufficient for quality control in environmental analysis of seawater samples. The following technical information is provided for the present CRM: concentrations of Na, Mg, K, and Ca, the density at 15, 20, and 25 degrees Celsius. A co-analysis of the candidate material by NMIJ, KRISS, and NIM contributed to the confirmation of the validity of the present CRM.

16.12 Competing Interests

The authors have no conflicts of interest to declare that are relevant to the content of this chapter.

References

AlSuhaimi, A. O., AlRadaddi, S. M., Ali, A. K. A. S., Shraim, A. M., & AlRadaddi, T. S. (2019). Silica-based chelating resin bearing dual 8-Hydroxyquinoline moieties and its applications for solid phase extraction of trace metals from seawater prior to their analysis by ICP-MS. *Arabian Journal of Chemistry, 12*, 360–369.

AMSAT. (2008). Asean Marine Water Quality: Management guidelines and monitoring manual. Australian Marine Science and Technology Ltd.

Ariga, T., Zhu, Y. B., Ito, M., Takatsuka, T., Terauchi, S., Kurokawa, A., & Inagaki, K. (2018). Quantification of elemental area densities in multiple metal layers (Au/Ni/Cu) on a Cr-coated quartz glass substrate for certification of NMIJ CRM 5208-a. *Analytical and Bioanalytical Chemistry, 410*, 2849–2857.

Brophy, C., Nadeau, K., Yang, L., Grinberg, P., Pihillagawa, G. I., Meija, J., Pagliano, E., McRae, G., & Mester, Z. (2022). *CASS-6: Nearshore seawater certified reference material for trace metals and other constituents.* National Research Council Canada.

Cole, S., Codling, I. D., Parr, W., & Zabel, T. (1999). Guidelines for managing water quality impacts within UK European marine sites. English Nature.

Ellison, S. L. R., & Williams, A. (2012). EURACHEM/CITAC Guide CG4: Quantifying uncertainty in analytical measurement. EURACHEM/CITAC working group.

Freslon, N., Bayon, G., Birot, D., Bollinger, C., & Barrat, J. A. (2011). Determination of rare earth elements and other trace elements (Y, Mn Co, Cr) in seawater using Tm addition and Mg(OH)2 co-precipitation. *Talanta, 85*, 582–587.

GEOTRACES. (2019). Summary of consensus values for samples collected on the 2009 GEOTRACES intercomparison cruise. Retrieved November 18, 2022, from https://www.geo traces.org/wp-content/uploads/2020/03/2019_Consensus_Values_2009_samples.pdf.

Komjarova, I., & Blust, R. (2006). Comparison of liquid-liquid extraction, solid-phase extraction and co-precipitation preconcentration methods for the determination of cadmium, copper, nickel, lead and zinc in seawater. *Analytica Chimica Acta, 576*, 221–228.

ISO 17034 (2016). General requirements for the competence of reference material producers. Geneva: International Organization for Standardization

ISO/IEC Guide 98–3 (2008). Uncertainty of measurement — Part 3: Guide to the expression of uncertainty in measurement (GUM: 1995). Geneva: International Organization for Standardization

Itoh, A, Ono, M., Suzuki, K., Yasuda, T., Nakano, K., Kaneshima, K., & Inaba, K. (2018). Simultaneous determination of Cr, As, Se, and other trace metal elements in seawater by ICP-MS with hybrid simultaneous preconcentration combining iron Hydroxide Coprecipitation and solid phase extraction using chelating resin. *International Journal of Analytical Chemistry, 2018*, 9457095

MEGJ. (2022). Environmental quality standards for water pollution. Retrieved November 18, 2022, from https://www.env.go.jp/en/water/wq/wp.pdf.

Morel, F. M. M., & Price, N. M. (2003). The biogeochemical cycles of trace metals in the oceans. *Science, 300*, 944–947.

Nadeau, K., Brophy, C., Yang, L., Grinberg, P., Gedara, I. P., Meija, J., Pagliano, E., McRae, G., & Mester, Z. (2016). *NASS-7: Seawater certified reference material for trace metals and other constituents.* National Research Council Canada.

Nicolaysen, P. M., Steinnes, E., & Sjobakk, T. E. (2003). Pre-concentration of selected trace elements from seawater by co-precipitation on $Mg(OH)_2$. *Journal De Physique IV, 107*, 945–948.

Otero-Romani, J., Moreda-Pineiro, A., Bermejo-Barrera, A., & Bermejo-Barrera, P. (2005). Evaluation of commercial C18 cartridges for trace elements solid phase extraction from seawater followed by inductively coupled plasma-optical emission spectrometry determination. *Analytica Chimica Acta, 536*, 213–218.

Rao, G. P. C., Veni, S. S., Pratap, K., Rao, Y. K., & Seshaiah, K. (2006). Solid phase extraction of trace metals in seawater using Morpholine Dithiocarbamate-loaded amberlite XAD-4 and determination by ICP-AES. *Analytical Letters, 39*, 1009–1021.

Raso, M., Censi, P., & Saiano, F. (2013). Simultaneous determinations of Zirconium, Hafnium, Yttrium and lanthanides in seawater according to a co-precipitation technique onto Iron-hydroxide. *Talanta, 116*, 1085–1090.

Sakamoto, H., Yamamoto, K., Shirasaki, T., & Inoue, Y. (2006). Pretreatment method for determination of trace elements in seawater using solid phase extraction column packed with Polyamino-Polycarboxylic acid type chelating resin. *Bunseki Kagaku, 55*, 133–139. (in Japanese with English abstract).

Samanta, S., Cloete, R., Loock, J., Rossouw, R., & Roychoudhury, A. N. (2021). Determination of trace metal (Mn, Fe, Ni, Cu, Zn Co, Cd and Pb) concentrations in seawater using single quadrupole ICP-MS: A comparison between offline and online preconcentration setups. *Minerals, 11*, 1289.

Sohrin, Y., Urushihara, S., Nakatsuka, S., Kono, T., Higo, E., Minami, T., Norisuye, K., & Umetani, S. (2008). Multielemental determination of GEOTRACES key trace metals in seawater by ICPMS after preconcentration using an Ethylenediaminetriacetic acid chelating resin. *Analytical Chemistry, 80*, 6267–6273.

Sturgeon, R. E., Berman, S. S., Desaulniers, J. A. H., Mykytiuk, A. P., Mclaren, J. W., & Russell, D. S. (1980). Comparison of methods for the determination of trace-elements in seawater. *Analytical Chemistry, 52*, 1585–1588.

Pichler, T., Veizer, J., & Hall, G. E. M. (1999). The chemical composition of shallow-water hydrothermal fluids in Tutum Bay, Ambitle Island, Papua New Guinea and their effect on ambient seawater. *Marine Chemistry, 64*, 229–252.

USEPA (2022a). National recommended water quality criteria - aquatic life criteria table. Retrieved November 18, 2022, from https://www.epa.gov/wqc/national-recommended-water-quality-criteria-aquatic-life-criteria-table.

USEPA (2022b). National recommended water quality criteria - human health criteria table. Retrieved November 18, 2022, from https://www.epa.gov/wqc/national-recommended-water-quality-criteria-human-health-criteria-table.

Wells, M. L., & Bruland, K. W. (1998). An improved method for rapid preconcentration and determination of bioactive trace metals in seawater using solid phase extraction and high resolution inductively coupled plasma mass spectrometry. *Marine Chemistry, 63*, 145–153.

Zhu, Y., Narukawa, T., Miyashita, S., Ariga, T., Kudo, I., Koguchi, M., Nonose, N., Baharom, N. B., Lee, K.-S., Yim, Y.-H., Wang, Q., & Chao, J.-B. (2022). Development and co-validation of a certified reference material (NMIJ CRM 7204-A) for the analysis of trace elements in seawater sample. *Bulletin of Chemical Society of Japan, 95*, 208–215.

Zhu, Y. B., Narukawa, T., Inagaki, K., Kuroiwa, T., & Chiba, K. (2011). Development of a certified reference material (NMIJ CRM 7505-a) for the determination of trace elements in tea leaves. *Analytical Sciences, 27*, 1149–1155.

Zhu, Y. B., Narukawa, T., Numata, M., Kitamaki, Y., Matsuo, M., Hioki, A., Kato, K., & Chiba, K. (2013). Characterization of a certified reference material (NMIJ CRM 8301-a) for determination of Cu in bio-ethanol. *Fuel, 103*, 736–741.
Zhu, Y. B., Nakano, K., Wang, Z. Y., Shikamori, Y., Chiba, K., Kuroiwa, T., Hioki, A., & Inagaki, K. (2018). Applications and uncertainty estimation of single level standard addition method ICP-MS for elemental analysis in various matrix. *Analytical Sciences, 34*, 701–710.

Appendix

Appendix: List of Where Readers Can Obtain CRM/RM for Parameters Listed within This Book

Identifier	Matrix	Analyzed for	Source	Information
Nutrients RMs	Seawater	Nitrate, Nitrite, Phosphate, Silicate	KANSO Technos Co., Ltd. https://www.kanso.co.jp/eng/	RMNS
	Seawater	Nitrate, Nitrite, Phosphate, Silicate	National Research Council Canada https://doi.org/10.4224/crm.2014.moos-3	MOOS-3
	Pure water, Silica dioxide, Sodium carbonate	Silicate	KANSO Technos Co., Ltd. https://www.kanso.co.jp/eng/	
	Seawater	Nitrate, Nitrite, Phosphate, Silicate	National Metrology Institute of Japan https://unit.aist.go.jp/nmij/english/refmate/	NMIJ CRM 7601-a, 7602-a, 7603-a
	Pure water, Silica dioxide, Sodium carbonate	Silicate	National Metrology Institute of Japan https://unit.aist.go.jp/nmij/english/refmate/	NMIJ CRM 3645-a (Available only for national metrology institutes and designated institutes which are participating in the CIPM MRA)

(continued)

(continued)

Identifier	Matrix	Analyzed for	Source	Information
	Sea salt	Phosphate, Silicate, Nitrate	National Research Council Canada https://doi.org/10.4224/crm.2022.salt-1	SALT-1
	Pure water	Ammonia, Nitrate, Nitrite, Phosphate, Silicate	Ocean Scientific International Ltd. https://osil.com	
CO_2 RM	Seawater	Total dissolved inorganic carbon, Total alkalinity	Scripps Institution of Oceanography co2crms@ucsd.edu	
	Seawater	Total dissolved inorganic carbon, Total alkalinity	KANSO Technos Co., Ltd. https://www.kanso.co.jp/eng/	Available in Japan
Standard Seawater (SSW)	Seawater	Practical salinity	Ocean Scientific International Ltd. https://osil.com	IAPSO standard seawater
Density SSW	Seawater	Density, Absolute salinity	KANSO Technos Co., Ltd. https://www.kanso.co.jp/eng/	MSSW (Not yet certified, but could be used for multi-parameters)
DOM RM	Seawater	Dissolved organic carbon, Total dissolved nitrogen	Rosenstiel School of Marine, Atmospheric, and Earth Science, University of Miami https://hansell-lab.earth.miami.edu/consensus-reference-material/index.html	Consensus reference material
Trace elements CRM	Seawater	Cr, Mn, Fe, Ni, Cu, Zn, As, Se, Cd, Pb	National Metrology Institute of Japan https://unit.aist.go.jp/nmij/english/refmate/	NMIJ CRM 7204-a